Springer Tracts in Civil Engineering

Springer Tracts in Civil Engineering (STCE) publishes the latest developments in Civil Engineering - quickly, informally and in top quality. The series scope includes monographs, professional books, graduate textbooks and edited volumes, as well as outstanding PhD theses. Its goal is to cover all the main branches of civil engineering, both theoretical and applied, including:

- Construction and Structural Mechanics
- Building Materials
- Concrete, Steel and Timber Structures
- Geotechnical Engineering
- Earthquake Engineering
- Coastal Engineering; Ocean and Offshore Engineering
- Hydraulics, Hydrology and Water Resources Engineering
- Environmental Engineering and Sustainability
- Structural Health and Monitoring
- Surveying and Geographical Information Systems
- Heating, Ventilation and Air Conditioning (HVAC)
- Transportation and Traffic
- Risk Analysis
- Safety and Security

Indexed by Scopus

To submit a proposal or request further information, please contact:

Pierpaolo Riva at Pierpaolo.Riva@springer.com (Europe and Americas)
Wayne Hu at wayne.hu@springer.com (China)

Angela Mejorin · Nebojsa Jakica · Mikkel K. Kragh
Editors

Performance-Based Façade Design

Discovery, Development, and Delivery

Springer

Editors
Angela Mejorin
Performance-Based Façade Design
Venice, Italy

Mikkel K. Kragh
Arup
Copenhagen, Denmark

Nebojsa Jakica
Department of Civil and Architectural Engineering
University of Southern Denmark
Odense, Denmark

ISSN 2366-259X ISSN 2366-2603 (electronic)
Springer Tracts in Civil Engineering
ISBN 978-3-032-28834-9 ISBN 978-3-032-28835-6 (eBook)
https://doi.org/10.1007/978-3-032-28835-6

This Springer imprint is published by the registered company Springer Nature Switzerland AG
The registered company address is: Gewerbestrasse 11, 6330 Cham, Switzerland

Preface

The design of building envelopes is entering a new phase of transformation. Increasing environmental demands, evolving performance expectations, and the growing complexity of façade systems are challenging conventional approaches and requiring new ways of thinking. In this context, performance-based façade design has emerged as a critical methodology—one that enables designers and engineers to move beyond prescriptive rules and instead define, evaluate, and validate performance through measurable criteria and project-specific strategies.

At the same time, this transition reflects a deeper shift within the discipline. The façade can no longer be understood as a discrete building component, but rather as a multi-scalar system embedded within the lifecycle of the building. Its performance is intrinsically linked to energy behavior, carbon footprint, fabrication logic, maintenance strategies, and long-term adaptability. Decisions taken in early design stages now carry disproportionate influence, requiring new forms of integration across traditionally separated domains.

This evolution, however, is not without its challenges. Performance metrics are often conflicting, difficult to evaluate simultaneously, and dependent on assumptions that introduce uncertainty. There is a growing risk that performance discourse becomes overly reductionist—privileging what can be measured while overlooking qualitative dimensions such as spatial experience, architectural intent, and cultural value. In this context, performance must be understood as a multi-dimensional and negotiated construct, rather than a purely optimized outcome.

This book, *Performance-Based Façade Design: Discovery, Development, and Delivery*, presents a curated collection of contributions that reflect this ongoing transformation. The volume is structured around three interconnected domains. Discovery addresses the foundational principles and experimental investigations that underpin façade performance. Development explores emerging tools, systems, and methodologies, including computational approaches, digital technologies, and innovative materials. Delivery demonstrates how these advancements are translated into built projects, often under demanding and unconventional conditions.

The origins of this publication are closely tied to the Performance-Based Façade Design (PBFD) initiative, founded in 2023 with the ambition of creating a dedicated platform for advancing façade engineering through rigorous technical discourse and international collaboration. From the outset, PBFD has positioned itself as a venue for deep, technical discussions centered on the performance of building envelopes—bringing together experts from industry, academia, and standardization bodies to address the challenges facing the built environment today.

The PBFD initiative convenes once a year through an in-person gathering held each September in Venice, bringing together the international façade community for three days of focused exchange and networking opportunities. The discussion is articulated around seminars, interactive workshops, and curated technical visits, creating a dynamic environment where ideas are not only presented but critically discussed and observed in practice. Within this compact timeframe, the initiative effectively brings together discovery, development, and practice—fostering a continuous dialogue that connects theoretical advancements, methodological innovation, and real-world application.

A defining objective of PBFD is to contribute meaningfully to the industry by fostering a dialogue that bridges research, practice, and standardization. Through its activities, the initiative seeks not only to disseminate knowledge, but also to support the translation of technical advancements into methodologies, guidelines, and ultimately into standards that shape practice at a global scale. In doing so, PBFD aims to have a tangible impact on both the profession and society, particularly in relation to resilience, sustainability, and the long-term performance of the built environment.

This international perspective is reflected throughout the book. The contributions included in this volume originate from authors across Europe, North America, Asia, and the Middle East, each bringing insights rooted in specific climatic, regulatory, and cultural contexts. One of the core missions of PBFD is precisely to provide a platform where regional and local advancements can be shared and understood within a broader global framework. In doing so, the initiative enables the cross-fertilization of ideas and promotes a more comprehensive understanding of façade performance worldwide.

The Discovery part establishes the scientific and methodological foundations of performance-based façade design. It includes contributions on advanced structural modeling approaches, such as the extension of multilayer beam theory for façade elements, as well as experimental frameworks for assessing curtain wall performance in super-high-rise buildings. These are complemented by investigations into extreme loading conditions, including wind-borne debris impact, where performance-based testing methodologies are developed to address scenarios not fully covered by existing standards.

The Development part reflects the rapid evolution of tools, systems, and design methodologies shaping contemporary façade engineering. It highlights how Industry 4.0 technologies—including computational design, digital simulation, and machine-learning-based optimization—are redefining how façade systems are conceived and evaluated. Contributions address advanced numerical modeling, predictive maintenance through cyber-physical systems, and performance-driven approaches to façade access and regulatory compliance. Material and system innovation is further explored through next-generation timber façades and experimental low-carbon envelope solutions, illustrating how new concepts are being tested and refined beyond conventional boundaries.

The Delivery part brings these advancements into the built environment through a series of case studies that demonstrate the application of performance-based design in practice. These projects span a wide range of conditions, from material-driven façade solutions and climate-responsive designs in extreme environments to complex high-rise systems and highly engineered glass structures. The selected case studies highlight how performance requirements are addressed through integrated design processes, often involving bespoke testing, advanced analysis, and close coordination between disciplines.

Across all three parts, a common thread emerges: the increasing need to bridge the gap between theory and practice. As façade systems become more sophisticated and as performance requirements continue to evolve, the integration of discovery, development, and delivery is no longer optional—it is essential. Performance-based design provides the framework through which this integration can occur, enabling informed decision-making while navigating uncertainty and complexity.

Ultimately, this volume reflects a broader transition within the discipline. Innovation is no longer a linear process, but a continuous exchange between research, design, and construction, increasingly supported by digital tools, data-driven methodologies, and emerging artificial intelligence applications. Within this context, the façade is evolving into a dynamic interface—mediating between environment, structure, technology, and human experience—while remaining grounded in the need for robustness, durability, and long-term performance.

We would like to express our gratitude to all contributing authors for their valuable work, their time, and their commitment to advancing the field. Their contributions form the foundation of this volume, which represents the first technical publication of the PBFD initiative.

We hope that this book will serve as a reference for practitioners and researchers, and that it will contribute to shaping the next generation of façade engineering—one that is grounded in performance, informed by innovation, and connected at a global scale.

Venice, Italy — Angela Mejorin
Odense, Denmark — Nebojsa Jakica
Copenhagen, Denmark — Mikkel K. Kragh

Contents

About the Editors

Dr. Angela Mejorin is the founder and curator of the Performance-Based Façade Design (PBFD) initiative, the international platform dedicated to advancing technical excellence in building envelope design through multidisciplinary collaboration. Alongside PBFD, she operates as an independent façade consultant, advising on façade strategies, performance-based design approaches, wind-related risk assessment, testing protocols, and compliance with international standards.

Angela holds dual Ph.D. degrees from Western University (Canada) and the Università Iuav di Venezia, where her research focused on wind-borne debris impact resistance and performance-based testing methodologies for façade systems. Her work has contributed to international standards development, including ISO 16316, where her design framework is referenced.

She is actively involved in ASTM International, ISO, and CEN committees, contributing to the evolution of façade performance standards. Through both her consulting practice and PBFD, she connects research, industry, and standardization to translate technical innovation into real-world impact.

Dr. Nebojsa Jakica is an associate professor at the University of Southern Denmark (SDU). He holds a Ph.D. in Technology and Design for the Environment and Buildings from Politecnico di Milano, where he designed, developed, and tested building-integrated photovoltaic façade systems.

His career bridges academia and professional practice. He has worked as an architect, computational designer, environmental design specialist, and façade/lightweight envelope consultant at leading architectural practices, including HENN in Berlin, Vittorio Grassi Architetto and Partners in Milan, and the TEXTILES Hub at Politecnico di Milano, contributing to award-winning projects and international competitions.

Nebojsa currently leads FACETS Lab, an interdisciplinary research laboratory specialising in digital technologies for façade design. The lab developed FacadeGenius, the world's first AI-powered, no-code computational façade design app, which enables façade stakeholders to collaborate seamlessly in real time across devices within a multiplayer cloud-based environment, connecting design and engineering.

Dr. Mikkel K. Kragh is an associate director at Arup, where he leads the Building Design and Advisory Practice and the Technical Specialist Services Team in Denmark. With extensive experience spanning consultancy, academia, and specialist façade engineering, he plays a key role in advancing high-performance building envelope solutions.

Prior to returning to Arup, Mikkel served as Head of the Department of Civil and Architectural Engineering at Aarhus University, where he led the department following its establishment in 2021 as part of a major institutional restructuring. He previously held the role of Head of Civil and Architectural Engineering at the University of Southern Denmark and was appointed Full Professor in 2017. Mikkel has worked internationally as a façade specialist on numerous high-profile projects with organizations including Arup, Dow Corning, and the Permasteelisa Group, leading collaborative R&D initiatives focused on advanced façade performance and building physics.

Mikkel has led numerous collaborative R&D initiatives in façade engineering and building physics. He holds an M.Sc. and Ph.D. from the Technical University of Denmark and is a Chartered Engineer, MASHRAE, MCIBSE, FSFE, and member of ATV, the Danish Academy of the Technical Sciences. In 2014, he was awarded the CIBSE Silver Medal for outstanding service to the Chartered Institution of Building Services Engineers.

The Life of a Façade: From Performance-Based Design to Lifecycle Intelligence

Nebojsa Jakica, Mikkel K. Kragh, and Angela Mejorin

Abstract Façade design is entering a decisive phase of transformation. Environmental imperatives, carbon constraints, digitalisation, and the growing complexity of building systems are reshaping not only how façades are designed, but how they are understood. The traditional view of the façade as a discrete exterior component, assessed primarily through static performance criteria and delivered as a finished product, is no longer adequate. In its place emerges a different disciplinary logic: the façade as a lifecycle system. From this standpoint, the façade is not merely the outer layer of a building, but a temporal and multi-scalar system mediating energy, carbon, fabrication, maintenance, adaptation, and human experience. Its significance does not reside in a single moment of completion; it unfolds and evolves across time.

Keywords Performance-based façade design · Lifecycle intelligence · Serviceability · Durability · Circular economy · Transient performance · Cyber-physical façades · Material-data coupling · Artificial intelligence · Robotic maintenance · Traceability · Design for disassembly

Façade design is entering a decisive phase of transformation. Environmental imperatives, carbon constraints, digitalisation, and the growing complexity of building systems are reshaping not only how façades are designed, but how they are understood. The traditional view of the façade as a discrete exterior component, assessed primarily through static performance criteria and delivered as a finished product, is no longer adequate. In its place emerges a different disciplinary logic: the façade as a lifecycle system. From this standpoint, the façade is not merely the outer layer of a

N. Jakica (✉)
University of Southern Denmark, Odense, Denmark
e-mail: nja@iti.sdu.dk

M. K. Kragh
Arup, Copenhagen, Denmark

A. Mejorin
Performance-Based Façade Design, Venice, Italy

A. Mejorin et al. (eds.), *Performance-Based Façade Design*, Springer Tracts in Civil Engineering, https://doi.org/10.1007/978-3-032-28835-6_1

building, but a temporal and multi-scalar system mediating energy, carbon, fabrication, maintenance, adaptation, and human experience. Its significance does not reside in a single moment of completion; it unfolds and evolves across time.

The title of this volume, Performance-Based Façade Design: Discovery, Development, and Delivery, clearly maps this transition. Discovery names the foundational questions, experimental methods, and critical investigations through which façade performance becomes knowable. Development denotes the expanding field of tools, materials, systems, and computational methods through which that knowledge is converted into design capability. Delivery identifies the ultimate test of these ideas in practice, where performance must be translated into buildable, maintainable, and robust architectural outcomes of the desired aesthetics. Together, these terms describe more than the structure of a book; they define a disciplinary continuum. In contemporary façade design, knowledge can no longer remain confined to research, and innovation can no longer remain detached from delivery. The central challenge is to connect inquiry, technological development, and built applications within a shared lifecycle perspective. This has been PBFD's ambition from the outset: to serve as a platform linking research, practice, and standardisation.

This shift matters because the field is still too often framed through point-in-time thinking. Performance is demonstrated at handover, justified against initial capital cost, and tied to assumed design horizons of twenty to fifty years. Yet façade design does not reach the end of its cycle when buildings are completed, nor even at the end of the first life cycle. Design consequences continue through operation, weathering, maintenance, repair, upgrading, disassembly, reuse, and material recovery. The conventional notion of lifespan is therefore increasingly inadequate. A nominal fifty-year design life should no longer be treated as a fixed endpoint, but as a provisional threshold within a much longer and more uncertain material trajectory. What has conventionally been called end-of-life is better understood as transition—a kind of half-life—not disappearance but change of state with a long tail of material value.

This requires a broader understanding of performance. Performance has rightly become the central organising principle of façade design, but it can no longer be reduced to a narrow set of criteria verified at completion. The façade affects operational energy, embodied carbon, daylight, thermal comfort, durability, constructability, maintenance intensity, and long-term adaptability. These criteria frequently conflict, and they do not operate on the same timescales. A solution that performs exceptionally at handover may later prove brittle, burdensome, or wasteful if it cannot be serviced, repaired, or reconfigured. Conversely, a façade designed for inspection, access, replacement, and adaptation may sustain value for far longer, even if its initial state appears less tightly optimised. The task, then, is not simply to maximise performance, but to redefine it as a lifecycle condition: negotiated, temporal, and inseparable from material and product continuity and future use.

A second conceptual shift follows. A façade is not a single product with a single lifespan; it is an assembly of products with unequal temporalities. Frames, sealants, gaskets, coatings, glazing units, fasteners, sensors, actuators, shading devices, and subassemblies age differently, fail differently, and demand intervention on different schedules. The façade should thus not be treated as a singular object moving along a uniform path of decline, but as an ecosystem of components with distinct service lives. Some elements are inherently long-life, others medium-life, and others intentionally short-life. Some may remain in place for decades, while others require periodic renewal. Durability, in this context, is no longer the capacity to remain unchanged; it is the capacity to remain valuable through change. A durable façade is not one that is never touched, but one designed to survive through selective maintenance, replacement, upgrading, and reassembly. Serviceability thus becomes a design principle rather than an afterthought. High quality, careful detailing, and uncompromising aesthetics are essential aspects of a façade, which is appreciated, looked after, and consequently retains its value and worth over time.

Accordingly, the goal of façade design is not to preserve peak performance for a fixed period, but to sustain—or even increase—value through changing performance curves over time. Much contemporary practice still assumes that a façade begins at its highest technical and economic value and then declines steadily towards replacement. That model is too crude for the realities of long-term stewardship. Façades do not simply deteriorate; they evolve. Performance may fall in one domain while remaining adequate or even beneficial in another. Components can lose value in their original application while gaining value in an alternative application. Materials can be repaired, reclassified, reused, downgraded, upgraded, and redirected. Designing for the future thus means designing beyond first use—towards second, third, and more distant lives. A component may reach the end of one performance life without reaching the end of its economic life.

This is where digitalisation and Industry 4.0 open a new horizon. Computational design, simulation, digital fabrication, sensing, and cyber-physical systems have already transformed the means of façade design. Parametric tools have shifted practice from object-making to the exploration of solution spaces; simulation has enabled earlier performance feedback; robotics and digital fabrication have linked geometry to production constraints; sensors and IoT systems have begun to connect the built façade to its operational life. The deeper opportunity, however, lies not merely in more advanced tools, but in integrating digital and physical intelligence across the entire life of the façade. Increasingly, façades must be understood as coupled systems of material and data—more broadly, of atoms and bits. Materials age physically while data accumulates digitally. Sensors translate physical change into signals; digital models turn those signals into diagnosis, prediction, maintenance planning, and future design knowledge. Circularity, in this view, is not only a matter of material loops, but of data continuity.

At this point, the façade may also be understood through the concept of autopoiesis, reframed for a cyber-physical continuum. The façade does not persist simply as a fixed object, but through the continuous reproduction of its identity across interactions between matter, data, environment, and human intervention. Its continuity is maintained through recursive processes of sensing, interpretation, maintenance, replacement, adaptation, and reclassification. In this sense, the façade is not merely a physical assembly enhanced by digital tools; it is a cyber-physical system that continually reconstitutes itself across time.

This coupling defines what may be called lifecycle intelligence. Lifecycle intelligence is not simply real-time control or smart operation. It is a collective and continuous cyber-physical capacity: the ability to track where materials come from, monitor how components perform and degrade, predict when thresholds are being approached, plan interventions, and identify what elements can become next. It means designing façades not only to perform, but to remain legible, traceable, serviceable, and re-assignable across multiple lives. Its importance lies in continuity, because continuity allows knowledge to compound. Knowledge is no longer lost at the end of each project cycle, but retained, shared, and extended across projects, interventions, and material transitions. The longer materials remain visible in data, the more effectively they can be maintained, reused, or redirected into new value chains and traded. Without matter there is no façade; without data there is no memory, no traceability, and no intelligent stewardship. The future façade is therefore not simply a smarter enclosure, but a lifecycle-aware cyber-physical system in which material loops and information loops continuously inform one another.

Artificial intelligence intensifies this possibility. AI should not be framed merely as a faster way to generate forms or approximate simulations, but as a means of coordinating extreme complexity across time. Façade lifecycle problems are distributed, multi-scalar, and dynamic. They involve degradation, replacement timing, maintenance planning, carbon accounting, cost analysis, logistics, reuse matching, and shifting performance thresholds across many components. These are conditions suited not only to predictive models, but also to swarm agentic workflows capable of coordinating distributed reasoning across actors, assemblies, and timescales. Their significance lies not simply in local optimisation, but in enabling a form of collective intelligence that can operate across fragmented responsibilities. What has traditionally been dispersed across design teams, contractors, operators, and supply chains can begin to function as a shared and compounding intelligence. Under such conditions, transient performance becomes operational—not as a single design calculation, but as a continuously updated field of coordinated decisions. Rather than optimising a façade once, intelligence can help manage it continuously as an evolving system. The role of AI, then, is not to replace architectural judgement, but to augment lifecycle stewardship at the scale and speed contemporary complexity demands.

The next stage of lifecycle intelligence may involve physical execution as much as digital coordination. Robotics is no longer significant only as a fabrication tool; it is emerging as a practical workforce for the longer life of the façade. The tasks that underpin serviceability and circularity—inspection, cleaning, repair, component exchange, selective disassembly, sorting, and material recovery—are among the most repetitive, hazardous, and physically demanding in the lifecycle of the built envelope. For that very reason, they are difficult to organise economically at scale through conventional labour alone. Robotic systems, and eventually coordinated robotic fleets or swarms, may assume these burdens with greater consistency, safety, and efficiency. Their importance lies less in replacing labour than in making intelligent lifecycle management economically viable and institutionally attractive. In such a scenario, multi-agent intelligence in software is coupled with multi-agent action in matter. AI no longer merely interprets the life of the façade; robotic agents begin to act within it. Robotics may thus become one of the key mechanisms through which lifecycle intelligence moves from analysis to action.

Yet the economic implications of material-data coupling remain under-explored in the façade industry. Even when a material cannot be directly reused, the data generated through its specification, installation, monitoring, degradation history, maintenance record, and failure modes may retain substantial value for future projects. Lifecycle intelligence creates not one circular economy, but two interrelated economies: a material economy and a data economy. This also begins to dissolve the conventional distinction between designing new and refurbishing existing building stock. A site may be physically void yet still dense with inherited data, while a new project may already be composed of recirculated resources and knowledge derived from prior buildings. New construction and refurbishment are therefore not different paradigms, but different entry points into the same coupled continuum of material recirculation and data reuse.

Similarly, façade design does not begin from an empty page, nor does it culminate in a final design. It begins from inherited conditions: existing datasets, accumulated technical knowledge, and locally available stocks of materials and products that may already be suitable for reuse. Compared with entirely new products, such resources can offer multiple and overlapping advantages. Economically, recovered and locally recirculated materials may become attractive where procurement costs, processing pathways, and supply-chain dependencies are reduced. Culturally, they may align more closely with the genius loci, reinforcing continuity between new interventions and local material histories. Politically, they can lessen dependence on volatile global trade networks and strengthen regional autonomy and resilience in construction supply. Technologically, they are not necessarily inferior to new products; when coupled with sensing, traceability, and performance data, they may in fact be better characterised than conventionally specified materials with uncertain long-term behaviour. Ethically and environmentally, their value is equally clear: they extend material life, reduce waste, and lower demand for virgin extraction.

Such stewardship also requires taking seriously the circular economy as just that: an economy, not only as ethics of circularity. Too often, service intervals are treated primarily as costs and end-of-life as disposal. Yet in lifecycle systems, material and data follow different value trajectories. Material performance in its original application may decline over time, while the data accumulated about that material—its composition, history, contamination level, residual capacity, and reuse potential—can increase in value. At the point of service, these trajectories intersect. Data can be used to recover, redirect, or even increase the value of materials by moving them into new performance domains where their remaining properties still matter. A gasket that has reached the end of its usefulness as a sealing element may still serve as feedstock for an acoustics product; stone removed from one architectural application may gain new value as aggregate in composite systems; glass cullet may reappear in secondary cladding or surface products. Service intervals thus become value-transfer points: moments at which degrading matter and accumulating data are converted into renewed utility, prolonged life, and new economic opportunity. Carbon reduction is a crucial co-benefit, but circularity will scale massively only when supported by economic incentive as well as environmental ethics, regulation, and terms and conditions offered by the financial sector.

The remaining challenge may thus be less technical or economic than cultural, aesthetic, and aspirational. Before circular products become standard, they must become desirable. Once reuse and recirculation are understood not as a compromise, but as expressions of quality, innovation, and identity, the question becomes one of design culture, visual language, and branding. The weathering of Corten steel, for example, can transform what might once have been classified as deterioration into an aesthetically valued condition. Lifecycle intelligence will not become transformative if it remains primarily a moral argument against waste, carbon, and climate change. It must render alternative models progressively obsolete by offering a demonstrably superior mode of design, operation, value creation, and profitability. Industrial transitions rarely succeed because incumbent systems are ethically criticised; they succeed because emerging systems deliver clear advantages in efficiency, precision, adaptability, safety, and economic return. The same principle applies here. If lifecycle-oriented façade systems can outperform linear models by making construction more efficient, maintenance more predictable, reuse pathways more valuable, material flows more legible, and service interventions cheaper and safer, then the transition ceases to depend on persuasion alone. It becomes a more competitive and more compelling model. Under such conditions, materials and assemblies currently perceived as residual or visually inferior may be redefined as part of a new beauty standard: one grounded in a renewed Bauhaus-like synthesis of technology, ethics, aesthetics and form.

At a deeper level, this argument also challenges the fiction of zero-based design. Dominant paradigms of recent decades—zero-energy, zero-waste, zero-carbon—have been important as ambitions, but they can also imply a misleading image of architecture as something that begins from a clean slate and concludes in neutralised absence. In reality, buildings do not begin from zero, and they do not end in zero. Every project starts from inherited conditions: inherited matter, inherited carbon, inherited components, inherited waste streams, inherited drawings, scans, records, sensor histories, and other forms of accumulated data. Design is therefore not the creation of a wholly new object from nothing, but the temporary reorganisation of what already exists into a new building state. What we call a building is less an isolated product than a provisional condensation of matter, data, value, and performance. It is not born from nothing, and it does not die absolutely; it is reconfigured from prior states and later dissolves into others.

This perspective also expands the meaning of nature-based design. Nature-based should not be limited to materials that are merely organic in origin. A more substantive interpretation includes systems capable of re-entering broader ecological, technical, and economic cycles through reuse, upcycling, downcycling, and revaluation. In that sense, a building does not simply end; it dissolves back into larger metabolisms. A genuinely net-zero and nature-based approach would therefore focus not only on reducing impacts at the point of delivery, but on enabling materials and data to remain active beyond first use. End-of-life would be replaced by productive dissolution and redistribution. Architecture and façades would no longer be understood as a sequence of isolated beginnings and endings, but as a series of transient states within a continuous material and informational continuum.

This chapter adopts precisely this position. It argues that façade design must move from point-in-time evaluation to transient performance, from fixed lifespan to managed lifecycle, from singular objects to assemblies of unequal component lives, and from static envelopes to coupled systems of matter and data. It argues that durability lies not in resistance to change, but in the capacity to remain valuable through change. It suggests that circularity becomes meaningful when linked to serviceability, traceability, economic value recovery, and worth. And it proposes that the new frontier of the discipline will depend on integrating digital methods, AI, sustainability, circularity, and human-centred judgement within a lifecycle-oriented framework.

In this context, the 2026 PBFD conference theme, Serviceability and Durability, gives particular focus to the broader argument developed here, while extending the trajectory established by earlier editions. If Discovery asks how façade performance is understood, Development asks how it can be modelled, simulated, designed, and tested, and Delivery asks how it can be realised in practice, then Serviceability and Durability asks how that performance can be sustained, renewed, and extended across time. It moves the conversation beyond the moment of completion towards the longer life of the façade as a system of maintenance, adaptation, replacement, and reuse. The theme does not sit alongside the logic of Discovery, Development, and Delivery; it deepens that logic by making lifecycle stewardship its explicit consequence.

The future of PBFD will therefore not be defined by performance or technology alone. It will be defined by the ability to connect intelligence with stewardship, performance with time, material circularity with data, and architectural meaning with economic value. Façades will no longer be conceived as finished objects designed from zero for fixed lifespans, but as temporary and adaptable condensations of inherited matter and data, managed through transient performance, service, and renewal, and gradually transformed into new material, informational, and economic forms. In this view, the façade becomes more than a boundary. It becomes a dynamic interface through which buildings, occupants, industry, and society negotiate continuity through transformation.

Discovery—Foundations of Performance-Based Design

Rethinking EN 14024 by Extending Allen's Multilayer Beam Theory

Carlos Pascual and Jacopo Montali

Abstract The mechanical behavior of thermally broken metal profiles is standardised by EN 14024. Appendix C of the standard provides equations to model simply supported composite multilayer beams subjected to uniform loads, assuming their cross-sections are composed of two metallic parts separated by a single thermal break (3-layer assemblies). These equations, which are derived from Feldmeier's (1987) adaptation of Stamm and Witte's (1974) sandwich theory, simplify the actual mechanical response of multilayer assemblies. In practice, façade systems exhibit more complex supporting conditions and load configurations that demand a broader theoretical framework. This work revisits EN 14024 through Allen's (1969) multilayer beam theory, demonstrating that its formulation matches the results of the standard under equivalent static conditions. Building upon Allen's foundation, the extension proposed by Pascual, Montali, and Overend (2017) is applied to composite multilayer beams featuring two thermal barriers (5-layer assemblies), also enabling accurate prediction of stresses and deflections. This study bridges the gap between classical multilayer beam theories and modern façade applications, encouraging the development of new standards to support the design of optimised thermally broken profiles.

Keywords Façade structural analysis · Analytical model · Composite action · Multilayer beam · Shear transfer · Thermally broken metal profiles

C. Pascual
Department of Architecture, University of Cambridge, 1-5 Scroope Terrace, Cambridge CB2 1PX, UK
e-mail: cp570@cam.ac.uk

J. Montali (✉)
Algorixon Srl, Parma, Italy
e-mail: jacopo@algorixon.com

A. Mejorin et al. (eds.), *Performance-Based Façade Design*, Springer Tracts in Civil Engineering, https://doi.org/10.1007/978-3-032-28835-6_2

1 Introduction

1.1 Overview on Thermally Broken Profiles for Façade Applications

Thermally broken metallic profiles used in building façade applications are multilayer assemblies in which polymer-based thermal barriers, also referred to as thermal breaks, are sandwiched between outer metal profiles, see Fig. 1. Such composite profiles are very often 3-layer assemblies, i.e. one thermal barrier (Fig. 2a, b), but can also adopt 5-layer configurations, i.e. two thermal barriers (Fig. 2c). The metal components are generally made of aluminium, which eases manufacturability by extruding highly custom geometries while providing stiff and lightweight façade systems. Alternatively, cold-rolled steel profiles can be employed whenever higher levels of slenderness, stiffness and strength are required. Thermal break profiles are commonly made of glass fibre-reinforced polymers e.g., extruded polyamide featuring a thermal conductivity of circa two to three orders of magnitude lower than steel and aluminium respectively, thus providing thermal insulation and reducing the heat transfer through the cross-section of the composite profile.

Generally, in glazed facades, a grid of vertical (mullions) and horizontal (transoms) thermally broken composite profiles form the skeleton of the building envelope and insulated glazing units (IGUs) or infill panels complete the spaces within the grid. Regardless of the level of prefabrication used (unitized or stick), every composite profile of the grid is supported in a statically determined configuration i.e., mullions and transoms behave mechanically as simply supported, with optional cantilevers beyond their supports. These cantilevers may provide a structural benefit by mitigating deflections as discussed later. Structurally, the IGUs and infills behave

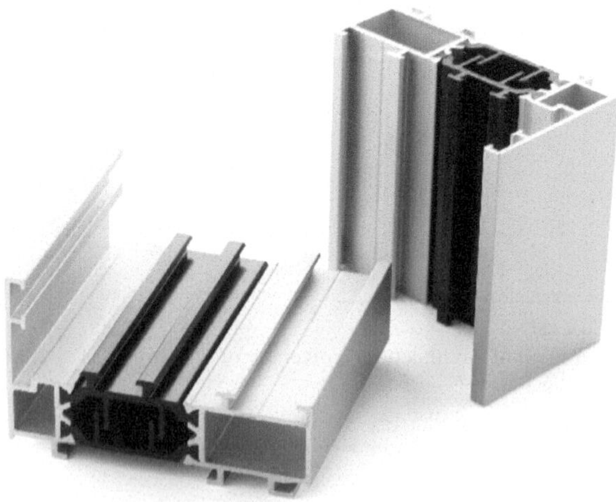

Fig. 1. 3-layer thermally broken profiles made from a polymer-based thermal barrier (black) sandwiched in between two external aluminium (grey) profiles. Credits: Algorixon srl

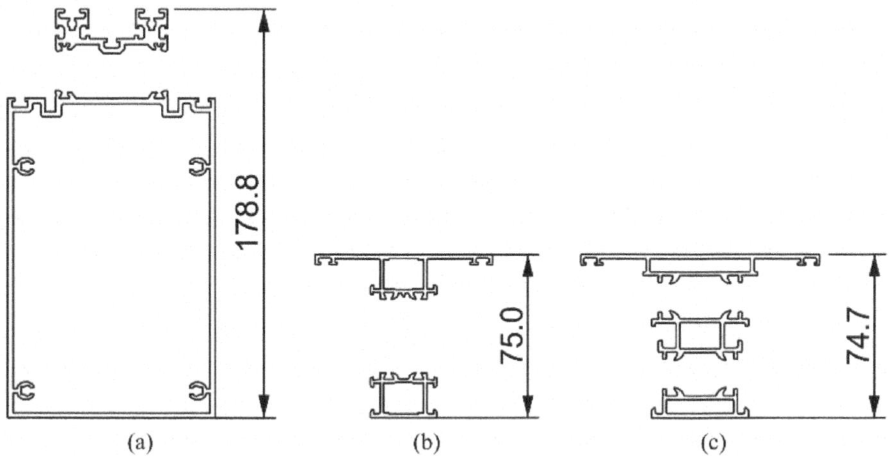

Fig. 2 Geometry of **a** façade profile (one thermal break layer) [9], **b** window profile (one thermal break layer), and **c** door profile (two thermal break layers). Dimensions in mm

as framed by the thermally broken profiles of the grid. Therefore, the profiles of the grid receive loading, e.g. wind pressure, from the tributary area of the corresponding IGU/infill. This area can be approximated as trapezoidal or triangular and therefore also a trapezoidal or triangular distribution of load can be considered on the thermally broken profile. However, for the simplicity of calculations, the load is frequently considered uniformly distributed along the profiles as this results in simpler analytical expressions of deflections and stresses. Other types of loading may also be exerted along the profiles e.g., point loads transferred by the transoms to the mullions. More precisely, in curtain wall applications, mullions are fixed in a statically determined manner to the building, and the transoms span between mullions and rest on them also in a statically determined manner i.e., they are simply supported and transfer point loads (reactions to wind loading) to the mullions.

1.2 Mechanical Behavior and Modelling of Sandwich Assemblies

The polymeric thermal breaks are sandwiched and crimped strongly enough into the adjacent metal profiles to allow for perfect shear transfer (i.e. no sliding at the metal-polymer interface, until a specific strength limit is achieved) and therefore a composite action is activated between metal profiles in the first steps of load application. The degree of composite action (*DCA*) of the thermally broken composite profile when subjected to flexural loads (e.g., wind loading) therefore depends on the shear stiffness (*G*-modulus) of the material of the thermal break, as well as on the shear area *A* of the sandwich assembly. The flexural response ranges between two limits: (i) the monolithic limit consisting of full composite action between the components due to shear-rigid thermal break ($GA \rightarrow \infty$) and (ii) the layered limit where no composite action is generated between the components due to shear-compliant thermal break ($GA \rightarrow 0$).

The sandwich construction provides the theoretical basis that underpins the mechanics and mathematics for these calculations. In this sense the sandwich theory by Stamm and Witte in the 70's [1], adapted by Feldmeier and Schmid in the 80's [2], made its place into the standardized method of EN14024 (Appendix C) for calculating stresses and deflections in thermally broken composite profiles for façade applications [3]. However, the geometry, equations, and loading scenarios presented in the standard have the following shortcomings: (i) cantilever configurations are absent, (ii) point loads are not considered, (iii) although solutions (stresses & deflections) are a good approximation for distributed loading case, they are only exact for sinusoidal loads, (iv) only 3-layer assemblies are considered. Furthermore, in earlier versions of the EN14024 standard, the approach in modelling the flexural response adopted a "deflection-effective" second moment of area I_{eff}, which corresponds to the second moment of area of an equivalent monolithic cross section to achieve the same mid-span deflection under uniformly distributed load and a simply supported configuration. Such approach allows the use of classical Euler–Bernoulli beam theory and formulae to calculate deflections by simply using $I = I_{eff}$.. The latest version (2023) of the standard clarifies that such approach, despite still valid, shall not be used to calculate stresses and it provides explicit formulae for the calculation of the stresses in the four critical cross-section points (top and bottom parts of the top and bottom metal profile extrusions, respectively) and mid-span deflections. This inconsistency demonstrates that I_{eff} is not an appropriate indicator for the study of multilayer beams as it does not provide the engineer with a closed-form solution for the mechanics of the beam. Based on the authors' experience, it is not unusual to see I_{eff} being inappropriately used in live projects beyond its original intent such as for direct modulus of resistance calculation or by using equivalent EI_{eff} within finite element models to verify stresses. Despite this general confusion, I_{eff} remains consolidated into day-to-day practice and it forms part of established jargon by professionals and for this reason it is likely to be kept being used moving forward. However, the authors consider that a theory providing a closed-form solution for the mechanics of the multilayer beam would be beneficial.

In this sense, a comprehensive and analytical sandwich theory developed by Allen [4] in the 60's shows potential to overcome the limitations of the standardized method described above. In this theory, the flexural response is explained as the superposition of two bending mechanisms: (i) a local mechanism in which the metal components bend about their own centroidal axis, and (ii) a global mechanism in which the metal components bend about the global centroidal axis of the composite profile. The model accounts for bending and shear deformations along the local and global centroidal axes in the through-thickness plane of the composite profile. The sandwich theory of Allen was initially developed for 3-layer composite sandwich panels and allows for exact solutions (under standard assumptions of linear elasticity) of deflections and stresses for a variety of loading conditions and geometries—including cantilevers. In recent years Pascual et al. have extended the theory to 5-layer sandwich panel configurations, i.e. allowing for two outer components, an inner core and polymeric adhesive connections between the inner and outer components. The theory of Allen and the extensions by Pascual et al. have been validated numerically and experimentally in several investigations during the last decade [5–8].

1.3 Study Objectives

The present work applies the sandwich theory of Allen to thermally broken composite profiles for façade applications exhibiting polymeric connections (i.e. thermal break profiles) with a height approximately one order of magnitude bigger than the connections evaluated in previous investigations [5–8]. Case studies analysed in this research are listed in Table 1. Firstly, the predictions of the model are compared with experimental results from large-scale tests on 3-layer mullions subjected to point loading (4-point bending, 4 PB). Then the model is compared with predictions of EN 14024 for cases contemplated in the standard (i.e. uniformly distributed load, UDL). Subsequently, the model is applied to a complex 5-layer mullion profile and results are compared with numerical analysis (FEM). Finally, the model is applied to a 3-layer mullion with cantilevers and subjected to point loading (three-point bending, 3 PB). In all cases the models based on Allen's theory show agreement with finite element analyses and with EN14024. This study bridges the gap between classical sandwich theory and modern façade applications, encouraging the development of new standards based on Allen's theory to support the design of optimised thermally broken profiles.

Table 1 Case study methodology to validate Allen's theory against EN14024, FEM and experimental results for simply supported beams

Case study	Beam configuration	Cantilevers	Loading	Allen's theory checked vs
1	3-layer, Fig. 2a	No	4 PB	Experimental data
2	3-layer, Fig. 2b	No	UDL	EN14024 & FEM
3	3-layer, Fig. 2b	No	4 PB	FEM
4	5-layer, Fig. 2c	No	3 PB	FEM
5	3-layer, Fig. 2b	Yes	3 PB	-

2 Methods and Materials

2.1 Methods

This research aims at applying Allen's theory (for 3-layer beams), and the extension by Pascual et al. (for 5-layer beams), to mechanically loaded and simply supported thermally broken profiles and validate its linear-elastic predictions by comparison with experimental data, finite element method (FEM) results and formulae in standard EN14024 (Table 1). In this regard, the comparison of Allen's theory versus EN14024 requires a further note about the relationship between the *G*-modulus and the elasticity constant *c* of every single thermal break profile. Both properties are intrinsically related by the cross-sectional geometry of the webs forming the thermal break according to Eq. (1):

$$G = c \cdot \frac{h}{b} \tag{1}$$

where *b* is the width and *h* is the height of the net shear area of the thermal break used for calculating *c* as schematically shown in Fig. 3. Therefore, when modelling using Allen's theory and FEM, appropriate values of *G* will be selected according to Eq. (1) based on specific values of *c*, *h* and *b*. If two identical thermal breaks of elasticity constant *c* are present, the model will present a shear area equal to 2*bh*. In this regard, *G* shall be interpreted as an "equivalent" shear modulus and not as the actual material property.

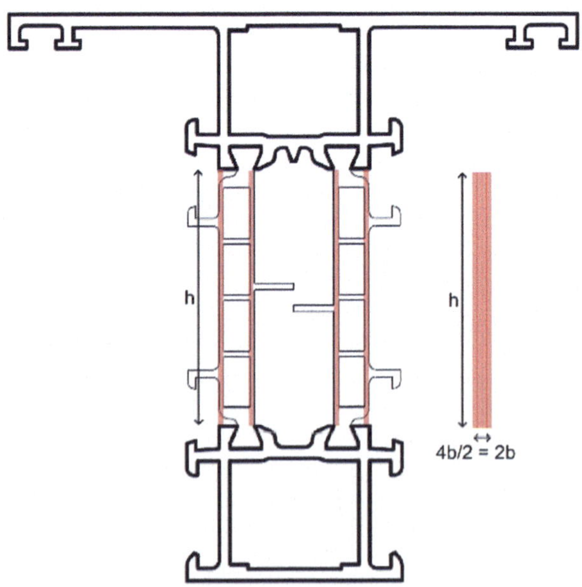

Fig. 3 Net shear area (in red) for the thermal break (left) and equivalent block for shear stiffness (right)

Detailed background on Allen's theory, and subsequent extensions by Pascual et al., can be found in references already cited above. However, the authors highlight here the importance of a^2 parameter, which governs the geometrical compatibility between the bending and shear deformations along the local and global centroidal axes in the through-thickness plane of the sandwich assembly. This mechanical parameter a^2 is given by Eq. (2):

$$a^2 = \frac{U}{D_{local} \cdot \left(1 - \frac{D_{local}}{D_{global}}\right)} \tag{2}$$

where U is the shear stiffness of the sandwich assembly (proportional to the G-modulus of the thermal break), and D_{local}, and D_{global}, are the local and global flexural rigidities, respectively. These parameters depend solely on geometric and material properties. Further information on how to calculate these parameters and on the analytical solutions for stresses and deflections, are given in [5]. Based on Allen's model, also the degree of composite action, DCA, can be calculated according to Eq. (3):

$$DCA = \frac{w_{layered} - w}{w_{layered} - w_{monolithic}} \cdot 100\% \tag{3}$$

where w are the deflections predicted by the model for the sandwich assembly (actual values of U and G), and $w_{layered}$ and $w_{monolithic}$ are the deflections predicted for the limit scenarios of $U,G \rightarrow 0$ and $U,G \rightarrow \infty$, respectively.

All models' assumptions and characteristics are as follows:

- All results presented below are generated with models (Allen's theory, EN 14024 formulae and FEM) that assume material linear elasticity, small displacements and uniaxial bending.
- Perfect shear connection at the metal-polymer interfaces is assumed in all models, i.e. insufficient mechanical bond or sliding at the interfaces is not considered in the following.
- The models based on Allen theory and EN 14024 disregard shear-lag effects, i.e., they neglect the decay of axial stresses in flanges.
- The models from Allen and EN14024 are based on advanced beam theory and are therefore unidimensional models (i.e. the differential equations that describe the mechanics have a single independent variable: the coordinate along the beam), however the models from FEM are three-dimensional models. FEM models are built in Strand7 R3.1.6 using brick elements (with 6 degrees of freedom per node) and congruent meshing i.e. nodes of adjacent materials are shared across the boundaries. If symmetry is present (i.e. symmetry of both loading and geometrical cross-section), half of the system was modelled and appropriate symmetry boundary conditions were applied, see Fig. 4.
- The Allen model presents the "antiplane core" hypothesis, i.e. the thermal breaks exhibit relevant shear stiffness but negligible bending stiffness.

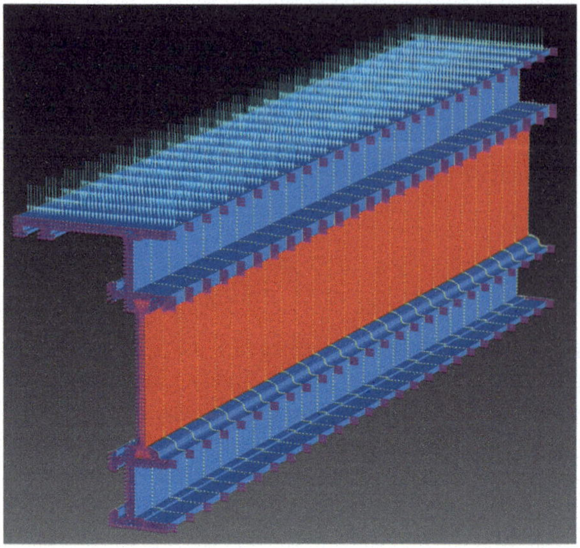

Fig. 4 Example of a Strand7 FEM model showing congruent meshing between aluminium (blue) and fiber-reinforced polyamide (red) materials and using translational restraints (magenta lines) to represent model symmetry

2.2 Materials and Cross-sections of Investigated Thermally Broken Profiles

Three geometries of thermally broken composite profiles, adapted to three different applications, are investigated here. The intended use of these profiles is as mullions that frame IGUs in: (1) façade profile (Fig. 2a), (2) window profile with one thermal break layer (Fig. 2b), and (3) door profile with two thermal break layers (Fig. 2c).

For all three mullions, the components of the sandwich assemblies are made of aluminium (metal parts) and fiber-reinforced polyamide (thermal breaks). The aluminium is considered linear elastic isotropic with an elastic modulus of E = 70 GPa and a Poisson ratio of $\nu = 0.33$.

The fiber reinforced polyamide can be considered as a linear elastic orthotropic material. However, for the purpose of this research, and for applying Allen's model and EN14024 formulae, the only relevant mechanical parameter is its elasticity constant *c* (or alternatively, the equivalent *G*-modulus as per Eq. (1)) in the plane of the webs which typically ranges from 30 to 80 MPa. In parametric analysis lower values (in the range 0–15 MPa) were also considered to account for reduced properties (e.g. due to ageing and elevated temperatures) and particularly to validate the analytical model (based on Allen theory and Pascual et al. developments) under conditions of higher shear deformations. Relevant properties for the three investigated mullions are provided in Table 2. The elastic modulus of the fiber-reinforced polyamide is mostly irrelevant and does not play a role in the equations of Allen's theory and EN14024 formulation as it is of an order of magnitude lower than that of aluminium and can therefore be disregarded, in line with the antiplane core hypothesis. However, when FEM is performed, the elastic modulus has to be implemented and is approximated

Table 2 Geometrical and mechanical properties of investigated profiles

Profile	Ref. figure	2b (mm)	h (mm)	*c* (MPa)	*G* (MPa)
Facade	2a	3.5	17.1	N/A	1089
Window	2b	3	34	0 – 80*	0 –1813*
Door	2c	3	12.4	15	124

*Despite in real-world applications the elasticity constant *c* typically ranges between 30 and 80 MPa, lower values were also considered to account for reduced properties and to validate the investigated analytical model under high shear deformations

from the *G*-modulus and Possion ratio ($v = 0.45$) applying classical isotropic linear elastic relation.

3 Modelling and Results

3.1 Case 1—Allen Theory (3-Layer) Versus Experimental Data (Loading: 4 PB)

Before delving into the case studies comparing Allen with FEM and the EN14024 standard, evidence is required to prove the ability of Allen's theory to predict experimental results on façade mullions subjected to flexural loads. Therefore, in the following, Allen's predictions are compared with experimental data reported by Huang [9]. The experiments correspond to four point bending tests performed on four façade mullions with different spans (2120 mm $\leq 2L \leq 3640$ mm). All mullions have identical cross sections made from two metal extrusions (aluminium) and one thermal break layer (fiber reinforced polyamide), see Fig. 2a. The mullions are all equipped with displacement transducers and strain gauges (Fig. 5).

The capability of Allen's theory to capture the linear elastic mechanical response of mullion composite profiles is demonstrated in Fig. 6. Clearly, the model predicts accurately the response in the linear elastic regime, i.e. it matches closely the initial tangent of every load–displacement diagram. As the mullions approach higher load levels, non-linearities arise due to possible yielding of the aluminium extrusions, sliding at the polymer-metal interfaces and deformations in the loading frames themselves, all producing a loss of stiffness in the load–displacement curves.

Similarly, Fig. 7 shows that Allen's theory predicts accurately cross-sectional deformation by matching closely the experimental axial strains (at a selected load level in the linear elastic regime of 5 kN). In particular, it can be noticed that for shorter spans Allen's theory correctly predicts a kink in the cross-sectional deformation (i.e. shearing of the thermal break, located at a depth between 140 and 160 mm approx.) and therefore proves inaccurate the classical Euler–Bernoulli assumption of cross-sectional planarity.

Fig. 5 Four point bending experiment on façade mullion as reported by Huang ([9])

Fig. 6 Force-midspan displacement in façade mullions of different spans subjected to bending as obtained experimentally [9] and analytically (Allen's theory)

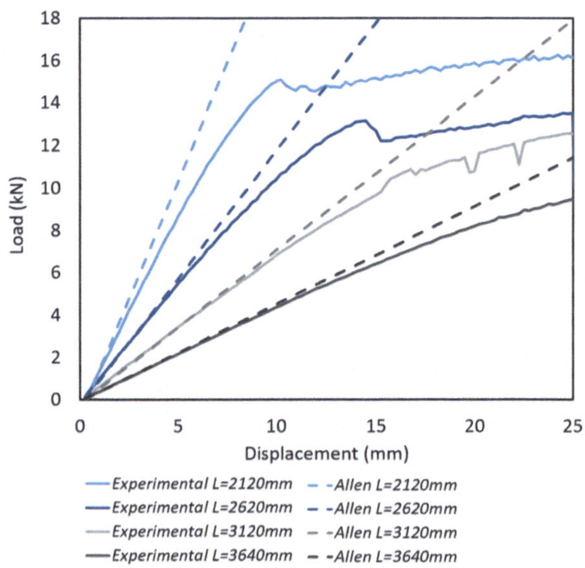

However, as the span (and therefore slenderness) increases, Allen's theory converges with the plane section deformation (i.e. no shearing) hypothesis since bending deformations govern, i.e. the relative effect of the shearing deformations of the thermal break profiles becomes irrelevant.

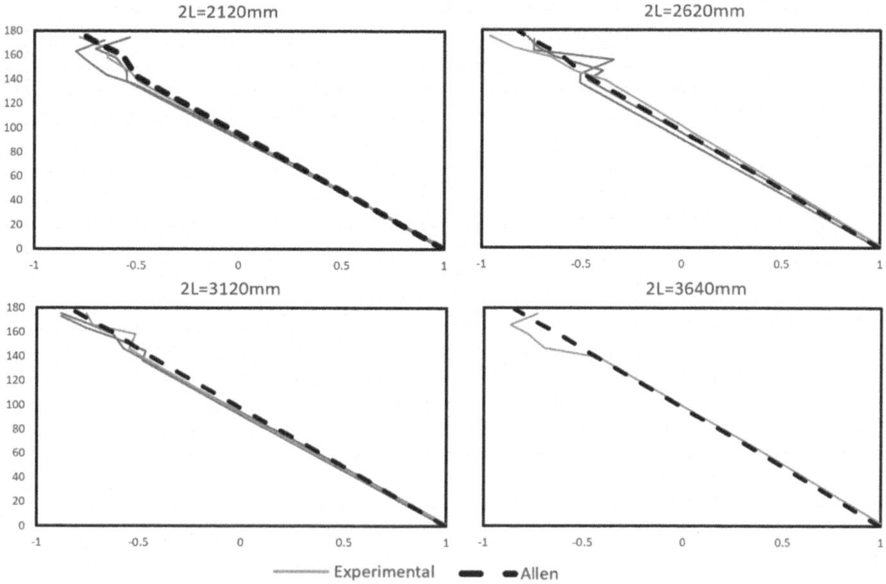

Fig. 7 Through-thickness normalised axial microstrains obtained experimentally [9] (solid lines) and analytically (dashed line) as per Allen theory [4] for different spans

3.2 *Case 2—Allen Theory (3-Layer) Versus EN14024 and FEM (Loading: UDL)*

In this case, a simply supported beam of span *2L* under uniformly distributed load q (Fig. 8) is modelled following:

- Allen theory
- EN14024 (appendix C of the standard)
- Finite Element Method (FEM)

The goal of this second case study is to validate both EN14024 and Allen theory against a full-fledged FEM model. The FEM was built as explained in Sect. 2. The cross section chosen for the analyses corresponds to an aluminium window profile as shown in Fig. 2b. Furthermore, a parametric study has been carried out to understand how deflection-effective second moment of area I_{eff} (Fig. 9) and axial stresses (Fig. 10) vary as a function of the following:

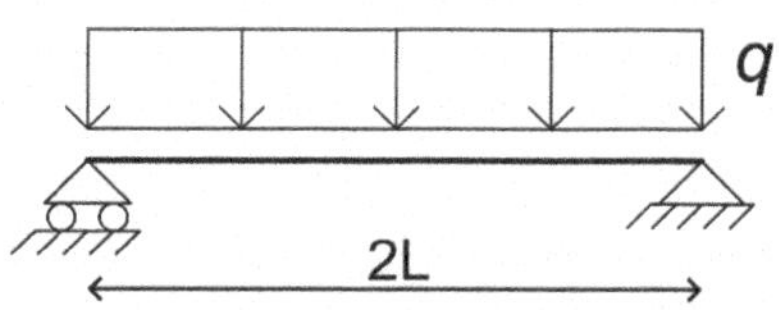

Fig. 8 Simply supported beam configuration for case study 2

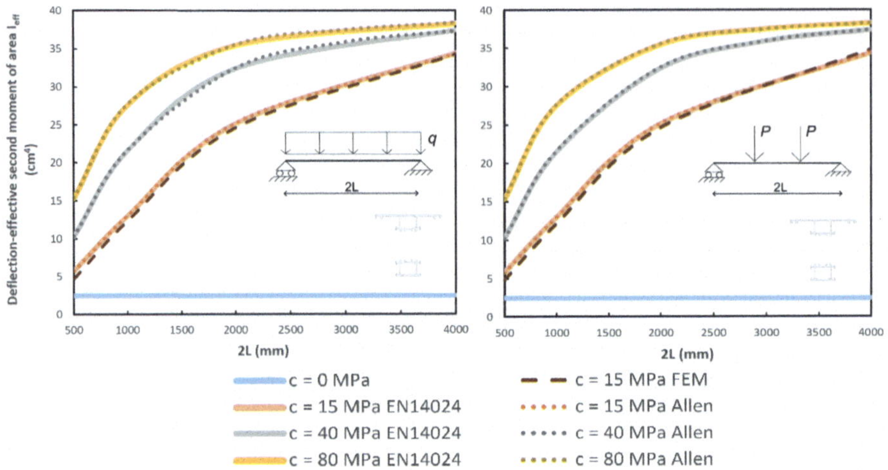

Fig. 9 Mid-span deflection-effective second moment of area of the investigated window profile under a simply supported configuration and UDL (left) or 4 PB (right) as a function of the span and thermal break elasticity constant *c*, as calculated per Allen, EN14024 and FEM. Note: I_{eff} under 4 PB as per EN14024 calculated by assuming an equivalent UDL equal to $q = P/(L_a + L_b)$

- $2L$ (mm) = [500, 1000, 2000, 3000, 4000]
- c (MPa) = [0, 15, 40, 80]
- q (N/mm) = [1] constant as models are linear with q.

The left chart of Fig. 9 shows the results from the investigation in terms of I_{eff} under the different configurations. It emerges that there is exact correspondence between FEM, EN14024 and Allen theory for $c = 15$ MPa as a function of the beam span $2L$, thus confirming the correctness of the two analytical modelling approaches compared to the numerical FEM. For $c = 0$ MPa, results show a constant I_{eff} as a function of $2L$ and it is equal to the layered second moment of area, i.e. the sum of the individual second moments of area of the two aluminium extrusions about their own centroidal axes.

Increasing the value of the elasticity constant *c* to values equal to 40 and 80 MPa increases the values of the deflection-effective second moments of area as *DCA* increases as a result of the stiffer-in-shear thermal break connection. No FEM was run for these two configurations, but results show that there is exact correspondence between EN14024 and Allen theory, thus confirming that the latter accounts for the various degrees of composite actions faithfully and in line with the EN14024.

In particular, it can be noticed that for the profile analyzed, the increase in *DCA* (and therefore I_{eff}) is significant for beams $2L < 2500$ mm, beyond which the benefit of the shear connector is not as evident. More precisely, it can be understood that as slenderness increases ($2L \rightarrow \infty$) the deflections of the composite converge to the monolithic and the contribution of shear deformations to deflections become progressively irrelevant in comparison with bending deformations.

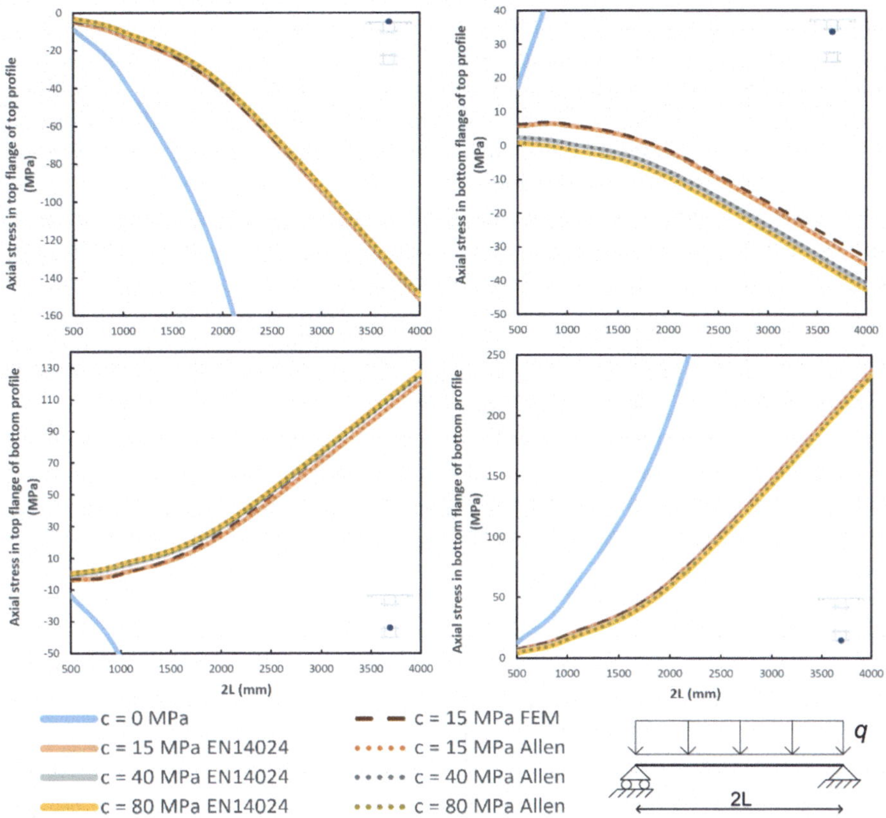

Fig. 10 Mid-span axial stresses of the investigated window profile under a simply supported configuration and UDL as a function of the beam span and thermal break elasticity constant *c*, as calculated per Allen, EN14024 and FEM

Figure 10 shows mid-span axial stresses resulting from the simply supported configuration for the three modelling approaches (EN14024 appendix C, Allen theory and FEM) as a function of varying beam spans, *2L*, and elasticity constant *c*. There are four charts each representing a critical position for evaluating the stresses: top flange of top profile, bottom flange of top profile, top flange of bottom profile, bottom flange of bottom profile).

Results show that a value of $c = 0$ MPa leads to a typical configuration of layered beams with the top flanges of both profiles being subjected to compression and the bottom profiles being under tension. As soon as some degree of composite action is introduced (due to $c > 0$ MPa), compression in the top flanges and tensions in the bottom flanges reduce in absolute values, indicating a more mechanically efficient cross-section. Higher values of the elasticity constants provide a more pronounced reduction in stresses. It can also be noticed that, depending on the span and the value of *c*, introducing composite action leads to an inversion in the sign of the stresses compared to the situation of $c = 0$ MPa. In particular, this leads to situations where all points in the top profile are subjected to compression and all points in the bottom profile are subjected to tension, similarly to a typical Euler–Bernoulli monolithic beam subjected to bending about its neutral axis. This behavior is a confirmation that increasing values of elasticity constants and/or beam lengths lead to the asymptotic configuration of fully monolithic beam.

It can also be noticed that there are not substantial differences in the mid-span axial stresses once the value of the elasticity constant assumes values different than zero. This is attributed to low shear effects in the central part of the beam, and particularly at mid-span the cross-section is subjected to pure bending. Although the next case study of a thermally broken beam under four point bending will shed more light on this aspect, one would typically expect that axial stresses increase as a function of the distance from the supports (Fig. 11).

However, if the beam is short and the value of the elasticity constant is small, the effect of the shearing due to the point load (reaction) is demonstrated to produce higher axial stresses closer to the supports than at mid-span (namely: left chart, bottom flange of top profile and top flange of bottom profile). As the beam becomes longer and the thermal break stiffer in shear, such effect becomes less pronounced since the beam starts to behave more monolithically in accordance with traditional Euler–Bernoulli theory.

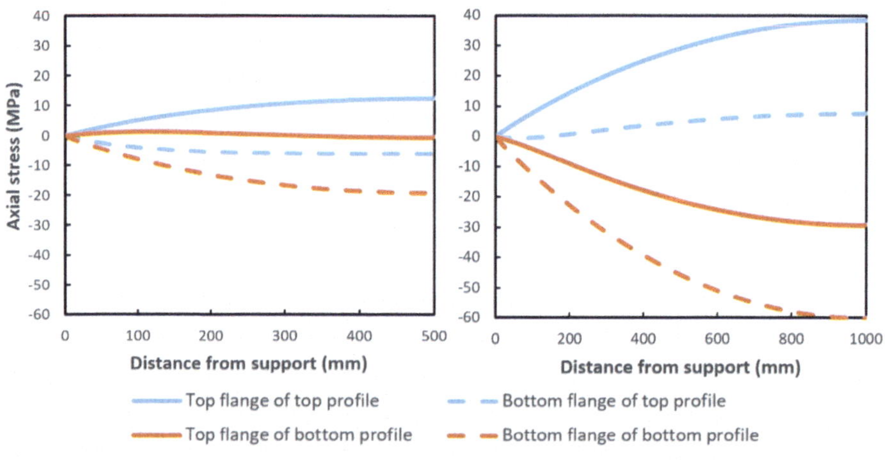

Fig. 11 Axial stresses as per Allen theory as a function of the span at the four characteristic points of the cross-section for a short beam with low elasticity constant (left: $2L = 1000$, $c = 15$ MPa) or for a longer beam with a higher elasticity constant (right: $2L = 2000$, $c = 40$ MPa)

3.3 Case 3—Allen Theory (3-layer) Versus FEM (Loading: 4 PB)

A similar scenario to Case 2 is now analysed under a different configuration, where the window profile shown in Fig. 2b is simply supported but under a four-point bending configuration. Such scenario is not contemplated by the EN14024 and therefore it constitutes an improvement to the standard. The configuration parameters for the beam are shown in Fig. 12 and they are equal to:

- $2(L_a + L_b)$ (mm) = [500, 1000, 2000, 4000]
- $L_a = 2L_b$ (arbitrary)
- c (MPa) = [0, 15, 40, 80]
- P (N) = [1000] as all models are linear with P

Furthermore, and for the sake of comparison with Allen and FEM, the deflection-effective second moment of area under this configuration was calculated as per EN14024 standard by assuming a simply supported configuration under fictitious UDL with equivalent load resultant as the 4 PB configuration:

$$q = \frac{P}{L_a + L_b} \tag{4}$$

Results from this configuration are shown in the right chart of Fig. 9. By comparing results with the same diagram obtained from the UDL configuration, it can be noticed that results in terms of I_{eff} are nearly identical between loading scenarios. This result is not obvious, as the definition of the effective second moment of area results from different Euler–Bernoulli formulations. Such evidence requires more analyses and further studies.

Once the value of the deflections and relevant I_{eff} are determined, the study of the stresses can be carried out. Figure 13 shows stresses as a function of the beam length for the configuration of a window profile with a $c = 15$ MPa. Results are plotted for both Allen theory and FEM modelling and they show that there is a very close correspondence between analytical and numerical findings. In particular, it is possible to notice that the effect of the increased length of the beam shifts the compression/tension states of the four critical cross-sectional points. By looking at the evolution of stresses across the cross section, it is possible to see that for shorter beams there is more shearing effect leading to the top flanges of both profiles to act in compression and the bottom flanges to act in tension, in a configuration similar to a layered one. As the span increases, and therefore also the slenderness of the beam,

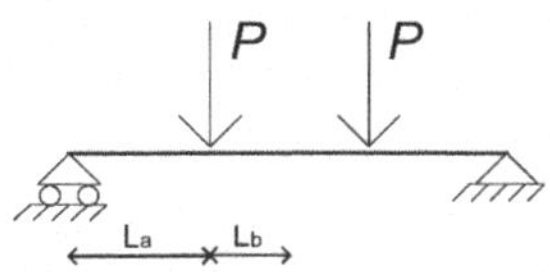

Fig. 12 Four-point bending beam configuration for case study 3

the cross section acquires more tendency to rotate about the global neutral axis as a result of the effect of the flexural behavior becoming predominant. At mid-span for a 4000 mm long beam, the top profile is entirely compressed while the bottom profile is entirely under tension, in a configuration which resembles more like a monolithic cross section under bending.

In terms of accuracy of Allen model compared to FEM results, it can be noticed that the model accurately captures the composite behavior in terms of stresses both away and in proximity of point loads. For shorter spans, while the Allen model correctly captures the theoretical value of the axial stress at supports (axial stresses are null), there is some discrepancy between the model and the FEM results, which sometimes display non-zero stress and not fully obvious trends close to the supports. This may be due to specificities of FEM: despite being able to capture local behaviors more effectively than analytical modelling, it suffers from the difficulty of modelling idealised configurations (such as point loads).

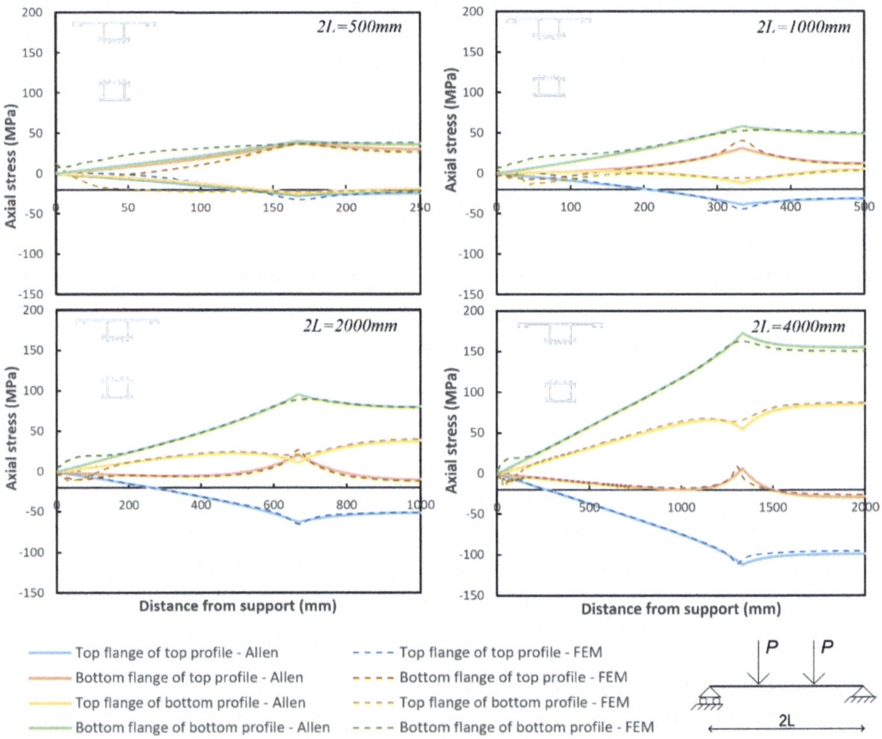

Fig. 13 Axial stresses for different beam lengths under 4 PB and $c = 15$ MPa (from top left to bottom right: 500, 1000, 2000, 4000 mm)

3.4 Case 4—Extended Allen Theory (5-layer) Versus FEM (Loading: 3 PB)

In this case study, we present an extension of Allen model as extended by Pascual et al. [5]. Under this configuration a cross section made from 5 layers (three aluminium profiles and two layers of intermediate thermal breaks) is used. The beam is studied under a three-point bending configuration, which represents a collapsed version of a 4 PB configuration with $L_b = 0$. The cross section of the chosen profile is shown in Fig. 2c while the parameters governing the beam configuration are shown in Fig. 14.

Under this case study, no parametric study was conducted and the value of the parameters are equal to:

- $2L$ (mm) = [1000]
- c (MPa) = [15]
- P (N) = [1000] as all models are linear with P

The deflections predicted by extended Allen theory (by Pascual et al.) and FEM are shown in Fig. 15. It emerges that the difference at mid-span between the two modelling approaches is only 3% for the scenario investigated.

Similarly, the comparison between axial stresses in the six characteristic points (top flange of top profile, bottom flange of bottom profile, top flange of mid profile, bottom flange of mid profile, top flange of bottom profile and bottom flange of bottom profile) is shown in Fig. 16. Despite the FEM results appear more convoluted, there still is close correspondence between the FEM results and the extended Allen approach. In this regard, it is important to note that although FEM has potential for high accuracy (e.g. due to the use of finite elements with 6 degrees of freedom), still suffers from modelling challenges such as stress concentrations produced by idealised loads (e.g., point-like loads and restrains). This is reflected in the peaks shown in Fig. 16 at point application, as well as in the non-zero axial stress at support locations.

Finally, the comparison of the shear stresses in the thermal breaks predicted by the modified Allen theory and FEM are shown in Fig. 17. The modified Allen theory considers pure shear deformation (no bending) in the thermal break, and a constant shear stress across the web is assumed – on the contrary FEM provides a more complex state of stress considering both bending and shear stiffness of the thermal break. However, it is important to note that the discrepancy is only in the order of 10%.

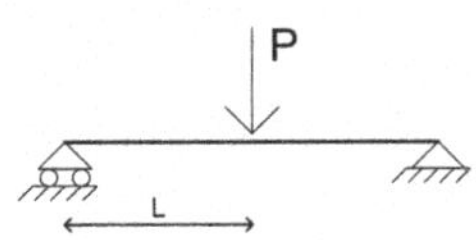

Fig. 14 Three point bending beam configuration for case study 4

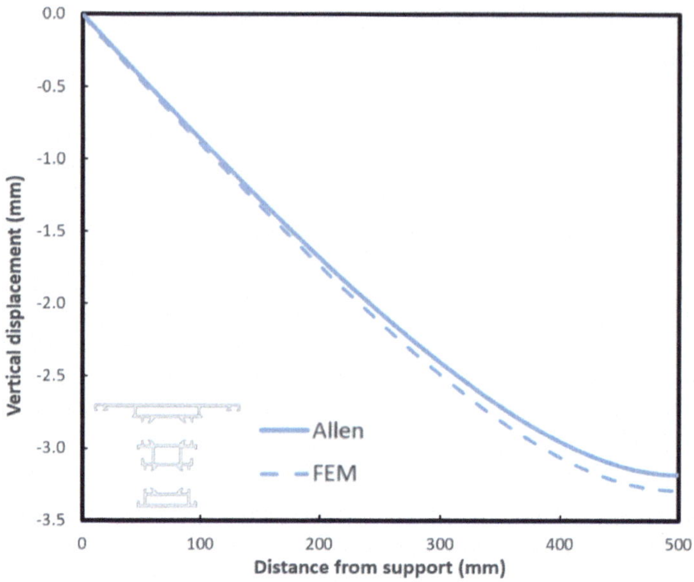

Fig. 15 Deflections for a 5-layered profile as predicted by FEM and extended Allen theory

3.5 Case 5—Allen Theory (3-layer) and Effect of Cantilevers (Loading: 3 PB)

In this case study, the effect of cantilevers on the beam deflection and stresses are assessed by using original Allen theory. It will be shown that Allen theory can be used to assess the beneficial effect of unloaded cantilevers in reducing deflections and stresses upon certain conditions. The parameters governing the configuration of the beam are shown in Fig. 18.

As an example, Fig. 19 shows a first situation in which a beam spanning 1000 mm between supports ($2L = 1000$ mm) is loaded with a central load of 1kN. The chosen cross-section is the window profile shown in Fig. 2b. Two parameters have been modified (ℓ, c) while the rest have been kept constant:

- $2L$ (mm) $= [1000]$
- ℓ(mm) $= [0, 500]$
- c (MPa) $= [15, 80]$
- P (N) $= [1000]$ as all models are linear with P

Under this configuration, it is possible to note that there is a beneficial effect of the cantilevers if the value of the c is not high (15 MPa). Please note that the cantilever is unloaded and that the only benefit results from the thermal break in the cantilever exerting shear stresses that increase the overall beam stiffness, if compared to a situation of $\ell = 0$.

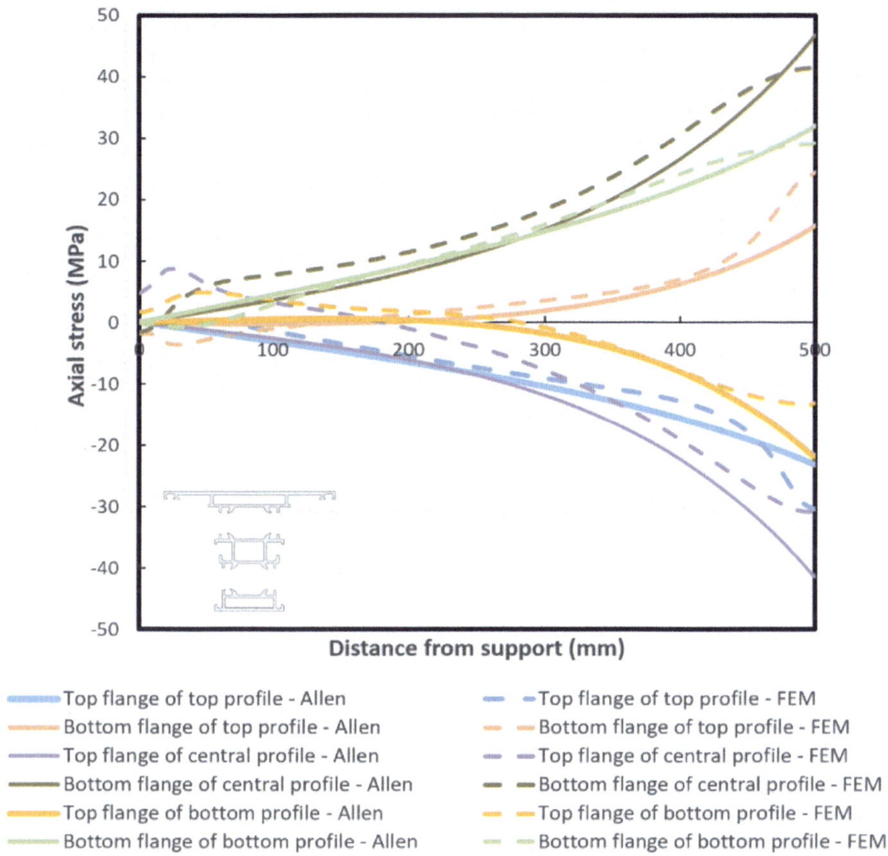

Fig. 16 Axial stresses for a 5-layered profile as predicted by FEM and extended Allen theory

For the scenario in which $c = 15$ MPa, adding 500 mm of cantilevers on both sides over a 1000 mm long beam decreases deflections by circa 11%.

Figure 20 shows the effect of adding cantilevers to beams of certain lengths for different elasticity constants c. The parameters that were modified for the study were:

- $2L$ (mm) = [1000,4000]
- ℓ (mm) = [100, 3200]
- c (MPa) = [1.5, 15]
- P (N) = [1000] as all models are linear with P

Results show that the presence of the cantilever is the major cause in reducing both deflections and stresses. A short cantilever ($\ell = 100$ mm) can produce a significant benefit if the beam is short ($2L = 1000$ mm) and if the elasticity constant is in the same order of magnitude as common practice ($c = 15$ MPa). This effect is more pronounced if the elasticity constant decreases, while the effect is negligible if the

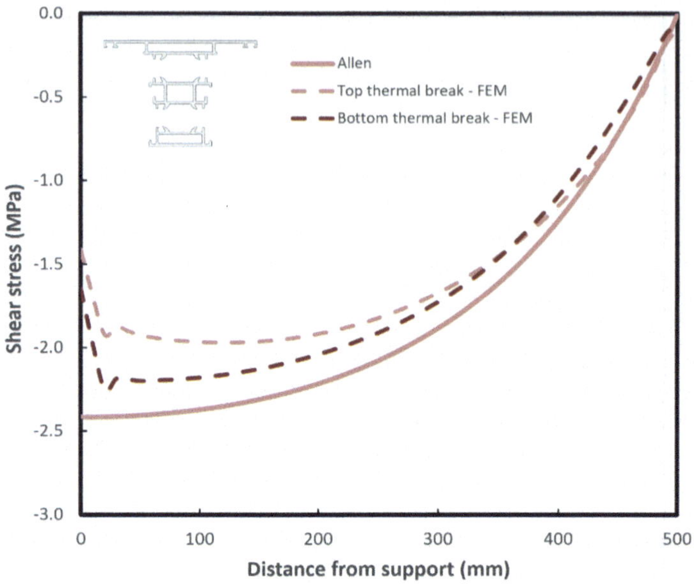

Fig. 17 Shear stresses for a 5-layered profile

Fig. 18 Three point bending (3 PB) with cantilevers of equal length for case study 5

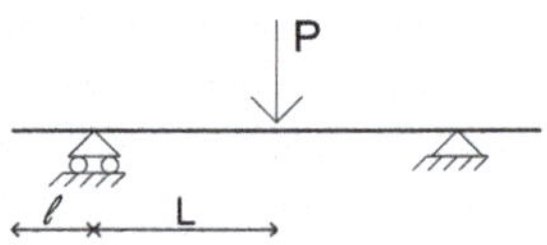

beam is already long ($2L = 4000$ mm).In terms of axial stresses, the benefits are less evident than in terms of deflections and, for the case study investigated, the value of the elasticity constant needs to be sufficiently low to achieve substantial benefits.

Finally, Fig. 20 shows how *DCA* varies as a function of c and ℓ. It emerges that for long enough spans (e.g., $2L = 4000$ mm) and for the profile investigated, the values c and ℓ become irrelevant as the very high slenderness of the profile produces that bending deformations govern over shear deformation – as a result the *DCA* approaches 1 (monolithic limit).

Conversely, for shorter spans (and in line with typical applications of the investigated profile), c and ℓ become important parameters that contribute to increase the *DCA* and therefore to reduce deflections and stresses.

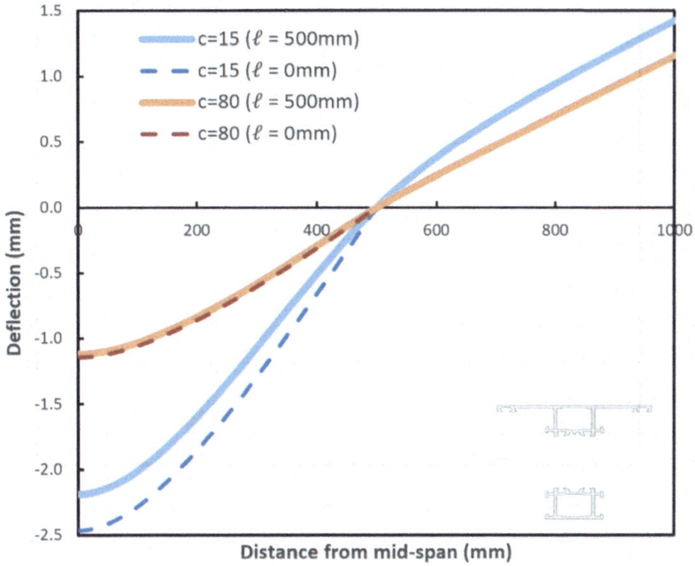

Fig. 19 Deflections for a simply supported beam of $2L = 1000$ mm without and with unloaded cantilevers ($\ell = 500$ mm) under 3 PB as a function of the elasticity constant c

4 Conclusions

This research has investigated the applicability of Allen's theory on sandwich structures and its extension by Pascual et al. to the analysis of deflections and stresses of thermally broken metal profiles for façades. The accuracy of the theory has been validated by comparison with finite element method (FEM) and with results from formulations included in EN14024.

From the results obtained in this study, it is demonstrated that Allen theory matches the predictions of EN14024, both in terms of deflections (via the parameter I_{eff}) and axial stresses. Allen theory also closely matches FEM results, especially in terms of deflections, with the accuracy on stresses (axial and shear) being variable depending on the modelling complexity, but with an overall good agreement.

In contrast to EN 14024, Allen's theory allows to model point loads perpendicular to the beam. As the theory is linear elastic and under the hypothesis of small displacements, any distribution of loading can be potentially modelled by superposition of point loads.

Allen theory also allows modelling beams with cantilevers. The effect of cantilevers has been demonstrated to be beneficial (reduction of deflections and stresses and increase of degree of composite action) especially when shear effects dominate i.e., shorter spans between supports and lower elasticity constants c. Cantilevers of about 100 mm length may be sufficient to produce significant benefits, based on the investigated case studies.

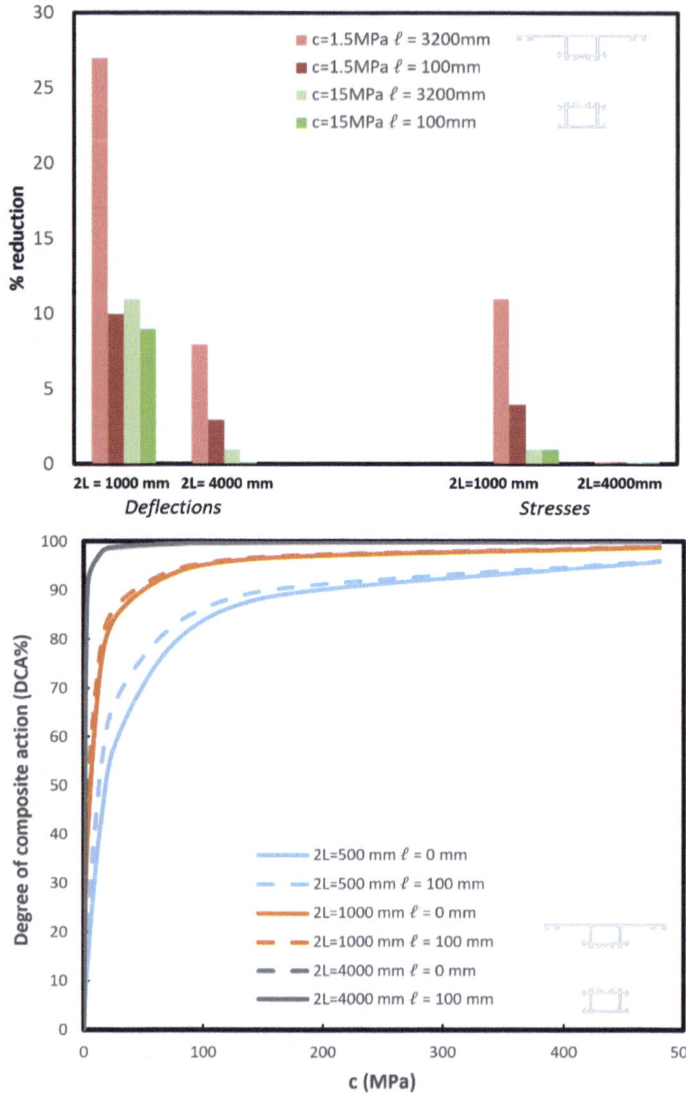

Fig. 20 Top: Reduction (%) in mid-span deflections (left) and axial stresses (right) due to the effect of cantilever lengths and elasticity constants c. Bottom: Degree of composite action, *DCA*, as per Eq. (3) for different elasticity constants and for varying spans between supports ($2L = 500$, 1000 and 4000 mm) without ($\ell = 0$ mm) and with cantilevers ($\ell = 100$ mm)

Allen's theory shows potential for the analysis of more complex multilayer components e.g., laminated glass with generic boundary conditions on the perimeter and for the analysis of other loading scenarios e.g., axial loading and buckling instabilities, as well as dynamic problems. Furthermore, Allen theory can be extended from beams to plates, thus paving the way for a unifying theory between a variety of sandwich-like systems currently used in the construction sector, such as thermally broken metal profiles, laminated glass, aluminium composite materials and sandwich panels, among others. Allen theory also has the potential to be extended to structural engineering problems such as concrete shear walls in towers or girders made from materials with different mechanical properties.

Due to the non-linear nature of such systems, the effect of material and geometry non-linearities should be assessed to better understand the post-elastic behavior both in terms of deflections and stresses [6].

While FEM approaches are the most established solution to demonstrate compliance at detailed design stages on real-world projects, they still remain very time-consuming and error-prone. The development of novel methodologies and tools to accurately achieve similar accuracy with less inputs and execution time can speed up the early involvement stage assessments, be they conceptual and/or tender, depending on the stakeholder. Such stages are known to commit and freeze the majority of future costs. Compared to a full-fledged FEM model, the amount of time required by using simpler, closed-form analytical solutions is lowered by orders of magnitude, thus allowing for more design iterations. This in turn leads to more sophisticated designs that optimise the use of material while retaining structural stiffness and strength, with subsequent cost and embodied carbon benefits.

5 Market Status, Software Readiness and Conflict of Interest

The authors are currently working on extensions of Allen theory to more generic geometries, boundary conditions and load distributions. This progress is implemented into a novel software named ASquared with the intention to support the design of optimized multilayer composites. Contrary to traditional FEM tools that approximate solutions by time consuming successive iterations, the software has an algorithm based on closed-form analytical formulations producing immediate results and includes a friendly graphical user interface (GUI) to draw and modify geometries. A free version of the software [10] with limited functionality is available online and a complete standalone application programming interface (API) is to be released shortly. A trial version of this API is operational and currently used by a major façade contractor, indicating a maturity level corresponding to Technology Readiness Level 9 (TRL 9).

The authors are jointly developing the software described above and share equal intellectual ownership and rights associated with its use. The authors declare no conflict of interest regarding the content of this document.

References

1. Stamm K, Witte H (1974) Sandwichkonstruktionen: berechnung, fertigung, ausführung. Springer-Verlag, Wien – New York
2. Feldmeier JS (1987) Statische Nachweise bei Metall-Kunststoff-Verbundprofilen. Fenster und Fassade
3. European Committee for Standardization (CEN) (2023) EN 14024:2023 Metal profiles with thermal barrier: Mechanical performance, requirements, proof and tests for assessment. Brussels
4. Allen HG (1969) Analysis and design of structural sandwich panels. Pergamon Press, Oxford
5. Pascual C, Montali J, Overend M (2017) Adhesively-bonded GFRP-glass sandwich components for structurally efficient glazing applications. Compos Struct 160:560–573. https://doi.org/10.1016/j.compstruct.2016.10.059
6. Pascual C, Nhamoinesu S, Overend M (2019) The flexural response of large scale steel-framed composite glazing panels. Glass Struct Eng 4(2):199–207
7. Pascual C, Overend M (2021) Buckling of sandwich struts with a particular application in composite multi-layer glazing. Eng Struct 230:111545
8. Pascual C, Hartwell R, Buskermolen P, Overend M (2025) Shear-lag effects in composite glazing panels. Eng Struct 342:120863
9. Huang SS (2014) Investigation of composite façade mullions. PhD thesis, University of Technology Sydney, Sydney
10. ASquared Zero (free version), https://www.food4rhino.com/en/app/asquared-zero. Last accessed 13 April 2026

Performance Requirements for Curtain Walls in Super-High-Rise Buildings—Effective Waterproofing Tests for Skyscraper Facades

Naofumi Isomoto and Akira Kudo

Abstract In Japan, curtain wall systems are required to satisfy a broad range of performance criteria, including resistance to wind pressure, watertightness, inter-story displacement adaptability, air tightness, acoustic performance, thermal insulation, fire resistance, durability, and resistance to thermal differentials. Among these, wind pressure resistance, watertightness, and inter-story displacement adaptability constitute the fundamental performance requirements, with particularly high performance demanded for curtain walls applied to super-high-rise buildings. These stringent requirements arise from Japan's severe natural environment. The country is among the world's most rainfall-intensive regions, experiencing prolonged rainy seasons and intense storms associated with typhoons, while also being subject to frequent seismic activity, with some urban areas recording more than 100 earthquakes annually. As a result, curtain wall systems must demonstrate high levels of weather resistance, seismic performance, and long-term durability, with watertightness representing a particularly critical evaluation criterion. Performance verification is therefore essential. Watertightness is commonly evaluated using chamber testing methods that reproduce severe wind and rain conditions representative of the Japanese climate. A distinctive feature of Japanese practice is that watertightness testing is frequently conducted in combination with other performance tests—such as wind pressure resistance and inter-story displacement tests—allowing staged evaluation under continuous load histories. Within this context, several Japanese curtain wall manufacturers have established in-house testing facilities accredited to ISO/IEC 17025, enabling reliable performance verification, including watertightness assessment. This paper presents the testing facilities of Fujisash Co., Ltd. as a representative example and describes practical approaches to watertightness testing for curtain wall systems used in super-high-rise buildings.

Keyword Curtain wall · Watertightness · High-rise building

N. Isomoto
Fujisash Co., Ltd., Tokyo, Japan

A. Kudo (✉)
Japan Sash Manufacturers Association, Tokyo, Japan
e-mail: kudou@jsma.or.jp

A. Mejorin et al. (eds.), *Performance-Based Façade Design*, Springer Tracts in Civil Engineering, https://doi.org/10.1007/978-3-032-28835-6_3

1 Introduction

Contemporary interpretations of resilience in architecture and engineering frequently emphasize robustness, redundancy, and resistance to external forces. Within this framework, building envelopes are commonly evaluated according to their capacity to withstand extreme loading conditions through strength, rigidity, and material durability. While these characteristics are undeniably important, such an interpretation does not fully encompass the conceptual foundations of resilience as they have been understood and practiced in Japan over many centuries.

A distinctive feature of Japanese architectural philosophy is the recognition that the built environment exists in continuous interaction with natural forces, including wind, rain, seismic activity, and biological processes. Rather than pursuing absolute resistance to these forces, traditional Japanese architecture has historically prioritized continuity through adaptability, regenerative capacity, and acceptance of change. This perspective is exemplified by the Shikinen Sengū ritual of the Ise Grand Shrine, which has been perpetuated for more than 1300 years. In this ritual, the shrine buildings are dismantled and reconstructed every twenty years. Architectural durability is thus achieved not through the permanent preservation of material form, but through an intentional cycle of destruction and creation, enabling the transmission of construction knowledge, techniques, craftsmanship, and cultural values across generations.

This understanding of resilience as a form of dynamic equilibrium contrasts with interpretations based primarily on rigidity or material permanence, often likened to the protective exoskeleton of an insect. In the Japanese context, resilience may instead be understood as a system-level characteristic, whereby the functionality and meaning of the whole are preserved while individual components are permitted—and indeed expected—to undergo deformation, ageing, and renewal. This conceptual approach has influenced not only traditional timber architecture but also contemporary engineering practice, particularly in fields where buildings are exposed to recurrent natural hazards.

In the context of modern skyscrapers, Japan presents a distinctive convergence of severe environmental conditions and advanced construction technologies. Frequent typhoons generate intense wind-driven rain, prolonged rainfall events challenge long-term watertightness performance, and persistent seismic activity repeatedly imposes inter-story displacement demands on façade systems. Under such conditions, building envelopes cannot be designed solely to resist peak actions associated with isolated events; instead, they must sustain functional performance throughout a continuous loading history over extended service lives.

Curtain wall systems, as the primary interface between interior spaces and the external environment, are central to this challenge. Their performance is determined not only by initial resistance to wind pressure and water ingress, but also by their capacity to accommodate deformation, recover following loading, and retain watertightness after repeated mechanical and environmental stresses. In this respect, curtain wall performance assessment in Japan represents a contemporary engineering manifestation of the broader philosophical concept of resilience as continuity through change.

Accordingly, Japanese curtain wall testing methodologies place particular emphasis on sequential and combined performance verification. Watertightness is not evaluated solely under initial or idealized conditions, but is repeatedly assessed following wind pressure resistance tests, inter-story displacement histories, and simulated ageing effects. This approach reflects an understanding that façade performance degradation often results from cumulative actions rather than instantaneous failure, and that resilience lies in the ability to maintain acceptable performance under evolving conditions.

This paper situates contemporary watertightness testing methods for curtain wall systems within this broader conceptual framework. By describing the performance requirements and testing methodologies applied to high-rise buildings in Japan, it demonstrates how a culturally rooted understanding of resilience—characterized by adaptability, recovery, and dynamic equilibrium—has been translated into rigorous, performance-based engineering practice. The insights presented are relevant not only to domestic applications in Japan, but also to international efforts aimed at enhancing the resilience of building envelopes in an era of increasing climatic and environmental uncertainty.

2 Japan's Super-High-Rise Buildings

As of 2024, Japan has 1,253 buildings exceeding 100 m in height, the majority concentrated in the Greater Tokyo Area. Figure 1. shows the geographical distribution of these skyscrapers, while Fig. 2. illustrates the relationship between building height and year of completion.

Japan's first skyscraper was completed in 1968. Since the 1990s, the number of buildings around 200 m tall has steadily increased. As building heights continue to rise, the performance requirements imposed on curtain wall systems have become stricter, particularly regarding watertightness.

This emphasis is closely linked to Japan's climatic conditions. The country experiences prolonged rainfall during the rainy season, coupled with intense precipitation associated with typhoons. In recent years, the frequency of short-duration torrential downpours exceeding 50 mm/hour has increased significantly. Statistical data from 1976 to 2023 indicates that the number of occurrences over the most recent decade reached approximately 1.5 times that of the initial ten-year period (See Fig. 3). Such extreme rainfall events frequently cause widespread damage.

Furthermore, Japan's high seismic activity necessitates curtain walls possessing sufficient seismic resistance and inter-story displacement adaptability. Considering the long service life expected of façade systems under these complex environmental conditions, the reliable maintenance of watertightness is critically important.

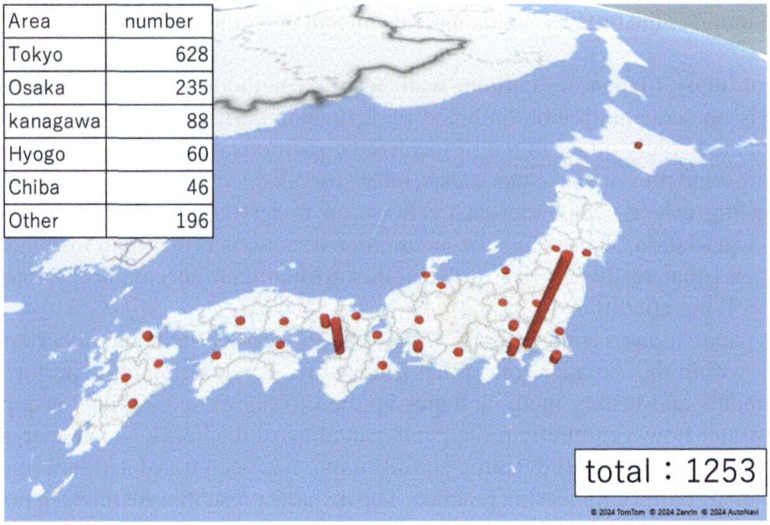

Fig. 1 Areas and numbers of built skyscrapers

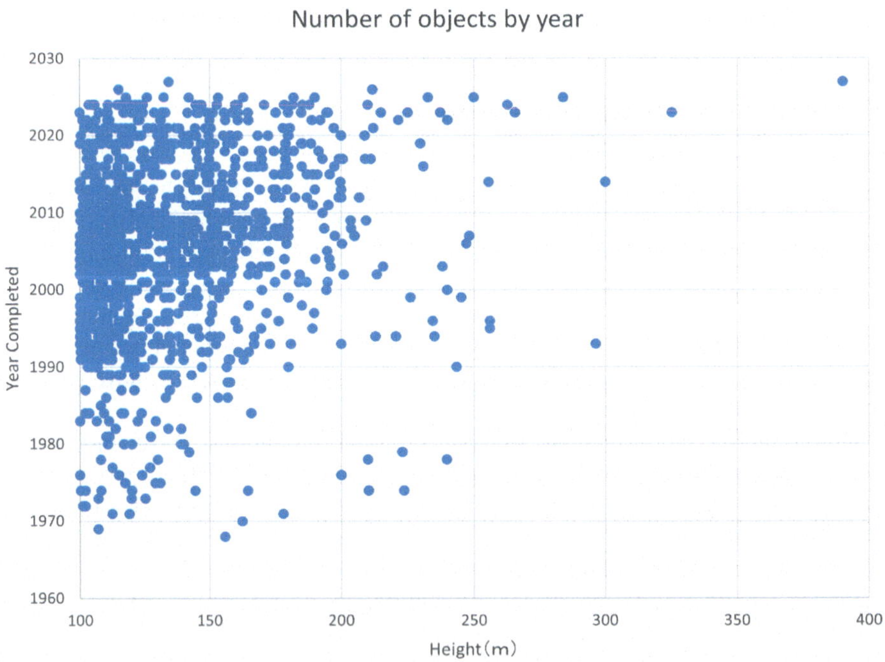

Fig. 2 Relationship between building height and year of completion

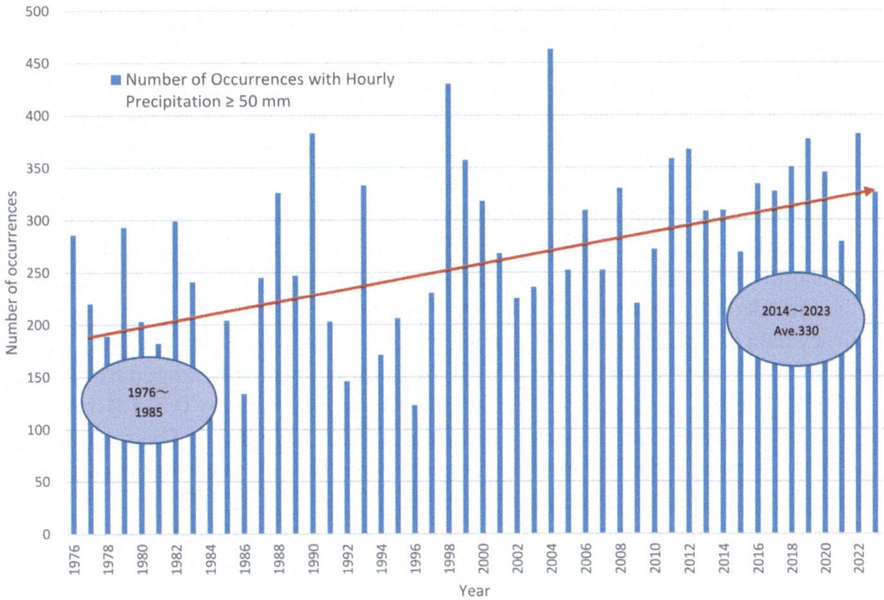

Fig. 3 Occurrences of hourly rainfall ≥ 50 mm (per year)

3 Performance Requirements for Curtain Walls

In Japan, the performance requirements for curtain wall systems are primarily defined by the industry standard Curtain Wall Performance Standards [1]. These standards specify a comprehensive set of performance items that curtain walls must satisfy to ensure safety, serviceability, and durability throughout the entire lifecycle of a building. The required performance items include:

- Inter-story displacement accommodation
- Wind pressure resistance
- Watertightness
- Airtightness
- Sound insulation
- Thermal insulation
- Fire resistance
- Durability
- Resistance to temperature differentials

Each performance item is associated with a defined evaluation method and, where applicable, graded performance levels. Among these items, watertightness, wind pressure resistance, and inter-story displacement accommodation are widely recognized as the most critical for super-high-rise buildings, as deficiencies in these areas can directly compromise occupant safety, building functionality, and long-term serviceability.

Watertightness performance is classified into multiple grades according to increasing resistance to wind-driven rain under elevated pressure differentials. For curtain wall systems applied to buildings exceeding 60 m in height, the standard recommends Grade 5 performance (See Table 1). It should be noted, however, that these recommended grades do not constitute mandatory limits. In practice, required performance levels are determined on a project-specific basis, taking into account site wind conditions, building height (See Table 2) and geometry, façade configuration, and the intended design service life of the structure.

Survey data collected up to 2003 indicate that watertightness performance values for curtain walls installed in Japanese high-rise buildings typically ranged from 2,000 to 4,000 Pa (See Fig. 4). These values correspond to approximately 0.5–1.0 times the design wind pressure, equivalent to wind speeds of 57.3–81.0 m/s. The relationship between wind speed and wind pressure is expressed by Eq. (1). These figures illustrate the stringent performance demands imposed on façade systems in Japan.

Unlike regions that adopt nationally prescribed façade performance criteria, Japan employs a responsibility-sharing framework for performance assurance. Within this framework, designers specify required performance levels based on regulatory provisions and project-specific conditions; general contractors verify compliance through testing and inspection; and manufacturers provide performance guarantees for the supplied curtain wall systems. Performance testing therefore plays a central role in demonstrating conformity with stipulated requirements.

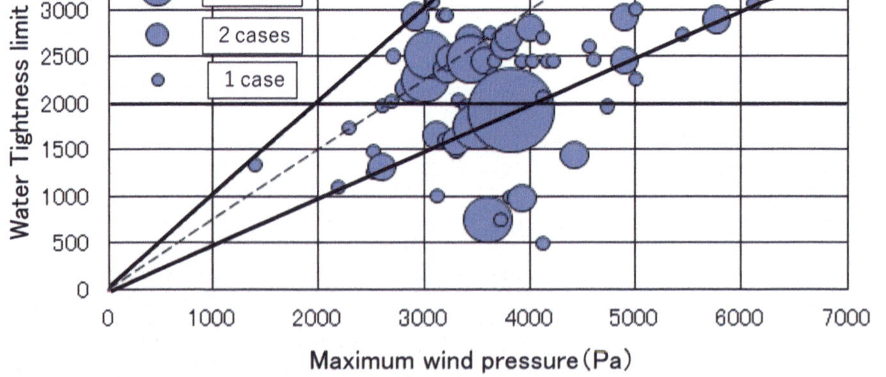

Fig. 4 Watertightness and wind pressure

In recent years, many curtain wall manufacturers in Japan have established ISO/IEC 17025[2]-accredited laboratories, enabling in-house performance testing with technical reliability equivalent to that of independent third-party laboratories. This development has facilitated more flexible and timely performance verification, particularly for large-scale and technically demanding super-high-rise projects. The testing facilities of Fujisash Co., Ltd., discussed later in this chapter, represent a typical example of this trend.

$$V = \sqrt{W/0.61} \tag{1}$$

V: Wind velocity (m/s) *W:* Wind pressure (Pa)

From an international perspective, Japan's approach differs in emphasis from regulatory frameworks commonly applied in Europe and North America, such as EN-based systems [3, 4] and ASTM-based systems [5]. While these frameworks often rely on standardized test procedures and uniform acceptance criteria, Japanese practice places greater emphasis on project-specific performance verification and phased testing. This distinction is particularly significant for high-rise buildings, where façade behavior is strongly influenced by building-specific characteristics.

The method for calculating design wind pressure in Japan was established through amendments to the Building Standards Act enacted in 2000. Under this framework, notifications specify the standard wind speed for each construction site and define ground surface roughness categories based on the surrounding conditions of the building location. This enables the calculation of site-specific wind pressure values. The standard wind speed is generally defined as corresponding to a recurrence interval of 50–100 years, indicating the statistical frequency with which wind speeds exceeding the calculated value are expected to occur. It should be noted that this recurrence interval does not represent the maximum wind speed that may occur during that period, but rather a probabilistic basis for design.

4 Test Apparatus

In Japan, waterproofing tests are generally conducted using the chamber method, similar to that employed for windows and doors. This method involves applying a pressure differential equivalent to the specified wind speed while spraying water onto the outer surface of the test specimen.

Fujisash Co., Ltd.'s test system (See Fig. 5) comprises a pressure chamber (See Fig. 6) and an earthquake loading device capable of inducing inter-story displacement. Two types of test apparatus are available: the F-type apparatus for flat façade test specimens and the L-type apparatus for corner configurations (See Figs. 7, 8). The maximum test specimen dimensions are 10 m × 13 m for the F-type and 8.6 m × 13 m × 3.6 m for the L-type, enabling testing of full-scale curtain wall assemblies under realistic conditions (See Table 3).

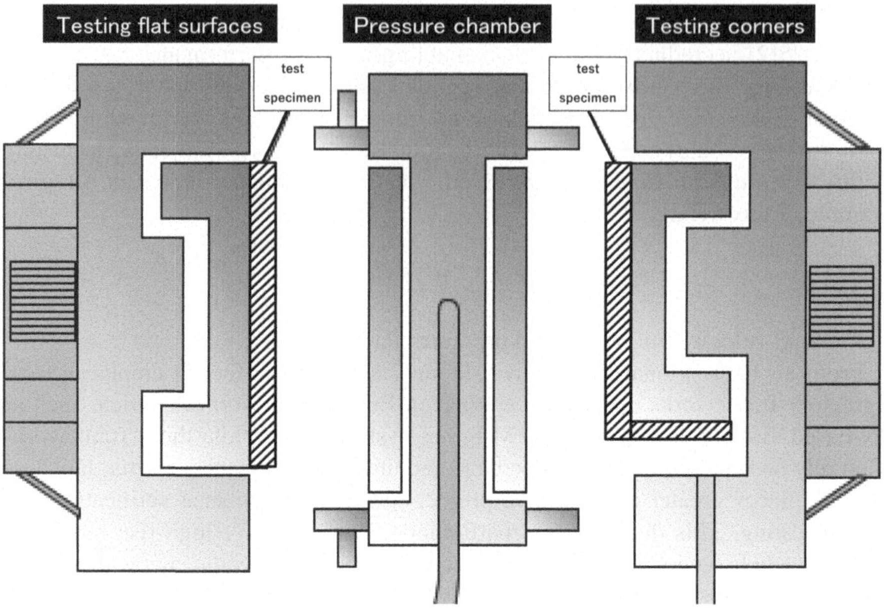

Fig. 5 Outline figure of test apparatus

The maximum permissible specimen weight is 15 tons for the F-type rig and 40 tons for the L-type rig, enabling testing from lightweight metal curtain walls to heavyweight precast concrete systems. Wind pressures up to ± 15,000 Pa can be applied, meeting the increasingly stringent wind resistance requirements for modern high-rise buildings.

For waterproofing evaluation, three water supply methods are available: water jet, water film spray, and mist spray. This enables the reproduction of conditions such as heavy rain, rainwater runoff from façade surfaces, and light rain during the rainy season. All test equipment is installed indoors, ensuring a stable test environment independent of external weather conditions. The facility holds ISO/IEC 17,025 accreditation, confirming its capability equivalent to that of a third-party testing organization.

5 Test Specimens and Waterproofing Test Methods

5.1 Test Specimen Configuration

The test specimen shall be designed to reproduce the representative façade configuration of the target building as accurately as practicable. In general, the specimen is fabricated at or near full scale, with typical plan dimensions of approximately

Fig. 6 Pressure chamber

10 m × 10 m, and incorporates the principal components of the curtain wall system, including the framing system, glazing units, joints, anchorage details, and sealant configurations (See Fig. 9). This approach enables realistic assessment of system behavior under simulated service conditions.

The support and boundary conditions of the test specimen shall be configured to replicate the interface between the curtain wall and the primary building structure. Anchorage locations and restraint conditions shall reflect those assumed in the design stage, and particular attention shall be given to ensuring that the global deformation behavior of the specimen approximates that expected under actual loading conditions.

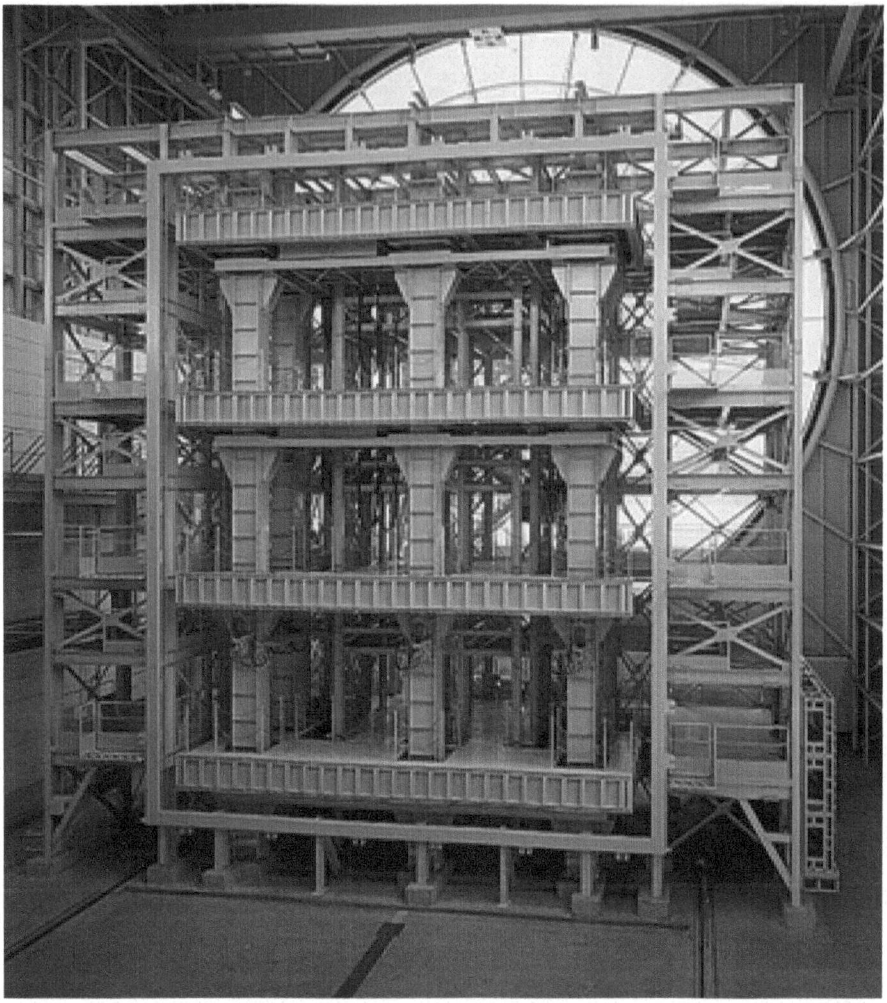

Fig. 7 F-type test apparatus

This includes consideration of load transfer mechanisms, stiffness distribution, and compatibility between the curtain wall and the supporting structure.

Typically, the test specimen comprises three vertical stories or layers. This configuration is adopted because the upper and lower layers are affected by boundary conditions imposed by the test apparatus, such as support frames, reaction points, and load introduction systems. As a result, the central layer most closely represents the deformation and loading conditions experienced by curtain wall panels in actual buildings.

Accordingly, the primary evaluation of inter-story displacement accommodation, joint behavior, and watertightness performance is conducted on the central layer of the

Fig. 8 L-type test apparatus

Table 1 Grades of watertightness

Grade	1	2	3	4	5
Pressure difference P (Pa) Curtain wall section	Less than 975	975	1500	P × 0.5 And Minimum value 1500	P × 0.5 And Minimum value 2250
Pressure difference P (Pa) Window section	Less than 525	525	750	1000	1500

Table 2 Building heights and recommended performance grades

Building height $h(m)$	Recommended performance grades
$h \leqq 31$	1, 2, 3
$31 < h \leqq 60$	4
$h < 60$	5

Table 3 Specifications of test apparatus

		F-type		L-type	
Test specimen	Size(mm)	W × H= 10,000 × 13,080		W1 × H × W2= 8600 × 13,080 × 3600	
	Depth (mm)	Min: 200 Max: 1000		Min: 200 Max: 1000	
	Weight (ton)	Max 15(3 layers)		Max 40(3 layers)	
Wind pressure	Max (Pa)	15,000		15,000	
Water sprinkle	Jet	2–10 L/(m^2 min)		2–10 L/(m^2 min)	
	Film	20–40 L/(m min)		20–40 L/(m min)	
	Mist	4 L/(m^2 · min) (200–400μ)		4 L/ (m^2 · min) (200–400μ)	
Inter-story displacement (mm)	Method	Independent slabs control		Coupled (master-follower) slabs control	
	In-plane	Static H/80= ± 150	Dynamic (0.5 Hz) H/150 = ± 80	Static H/80= ± 150	Dynamic (0.5 Hz) H/150 = ± 80
	Out-plane	Static H/80= ± 150	Dynamic (0.5 Hz) H/150 = ± 80	Static H/80= ± 150	Dynamic (0.5 Hz) H/150 = ± 80
	Doglegged-shape deformation	Static H/80= ± 150	Dynamic (0.5 Hz) H/150 = ± 27	Unavailable	
	Upward	W/200 = 50		Unavailable	
	Wind Swing	Max. Wind Pressure ± 2,500 Pa	Dynamic in-plane H/240 = ± 50	Max. Wind Pressure ± 2,500 Pa	Dynamic in-plane H/240 = ± 50

specimen. Observations of the upper and lower layers may be used as supplementary data to assess the influence of boundary effects and to confirm the overall validity and consistency of the test results.

Fig. 9 Test specimen installation

5.2 Waterproofing Test Method

Watertightness testing is conducted using the chamber method (See Fig. 10), in which water is applied to the exterior surface of the test specimen while a controlled pressure differential (See Fig. 11) is imposed across the façade. This method is intended to reproduce wind-driven rain conditions representative of severe storms and typhoons (See Fig. 12).

Prior to the commencement of water application, a pre-pressurization step is performed. This step serves two primary purposes: first, to induce initial deformation in the test specimen and stabilize its structural response; and second, to confirm that the specimen can safely withstand the target pressure level before water injection begins. By introducing this pre-loading process, the influence of initial settlement or adjustment effects on subsequent test results is minimized.

The design pressure value, P, is typically selected within the range of 2,000–4,000 Pa. The precise value is determined on a project-specific basis, taking into account factors such as building height, site wind conditions, and the design wind speed. Pressurization is generally applied in three to four incremental stages, progressing until the specified maximum pressure is reached.

Pulse pressurization is employed with a cycle period of 2 s to simulate fluctuating wind pressures. For each stage, the pressure range is defined as follows:

- For a central pressure $P \leq 1{,}500$ Pa:

  upper limit = 1.5P, lower limit = 0.5P

- For a central pressure $P > 1{,}500$ Pa:

  upper limit = P + 750 Pa, lower limit = P − 750 Pa

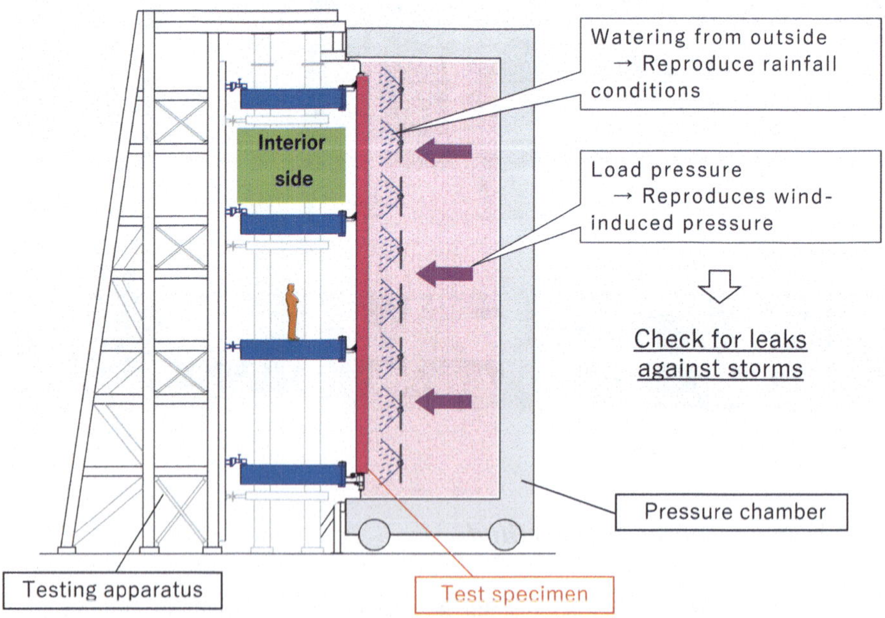

Fig. 10 Outline diagram of watertightness performance test method

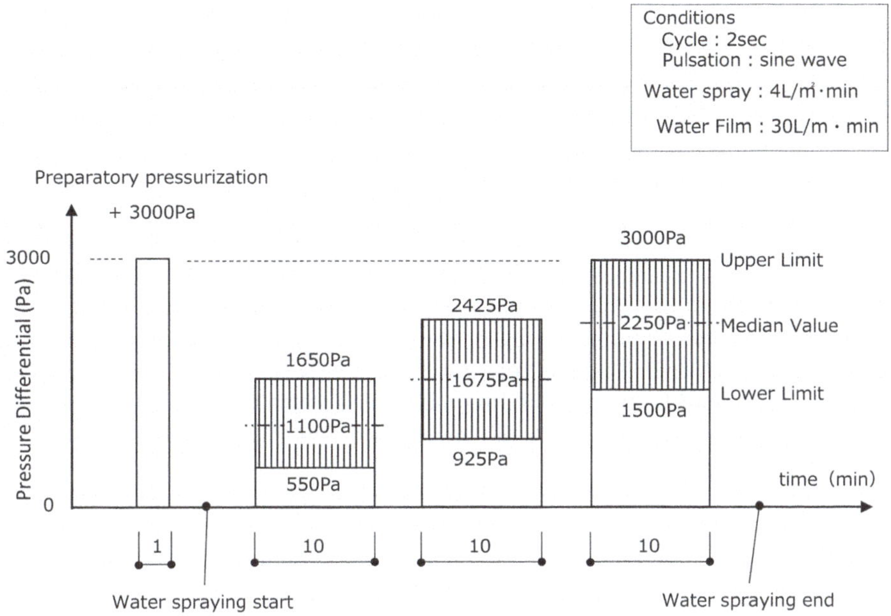

Fig. 11 Watertightness Performance Test: Pressurization Process

Fig. 12 Watertightness test

Water application conditions are selected to represent heavy rainfall. The standard sprinkling rate is 4 L/(m^2·min), while water film spraying is conducted at approximately 30 L/(m·min), corresponding to a rainfall intensity of approximately 40 mm over 10 min. In addition, mist spraying tests may be conducted to simulate prolonged drizzle conditions typical of the Japanese rainy season.

Each test stage is maintained for a duration of 10 min. During this period, changes in watertightness performance in response to pressure fluctuations are carefully monitored. The primary evaluation criterion is the presence or absence of water penetration on the interior side of the test specimen, as determined by visual inspection. When leakage is observed, its location, onset, and progression are recorded in detail.

As an optional procedure, prolonged watertightness tests extending over several hours may be conducted to evaluate performance under sustained rainfall conditions. Furthermore, water film tests simulating rainwater flowing over the façade surface may be applied to assess the effects of concentrated water flow on joints, interfaces, and drainage paths.

A water application rate of 4 L/(m^2·min) corresponds to a rainfall intensity of approximately 240 mm/h. This value, adopted in various testing standards, represents severe rainfall conditions acting on curtain walls and window systems during strong winds, based on the statistical distribution of rainfall events observed in Japan. In addition, the pulsating pressure applied during testing simulates wind gust effects under natural wind conditions. A cycle period of 2 s was selected to enable practical and repeatable evaluation of façade response.

6 Performance Test Procedure

This section describes a comprehensive performance test procedure applicable to curtain wall systems used in high-rise and super-high-rise buildings in Japan. A defining characteristic of this procedure is the sequential combination of multiple performance tests, with watertightness repeatedly evaluated after major loading histories. This approach reflects the need to verify façade performance not only in its initial condition, but also after exposure to wind loads, inter-story displacement, and simulated ageing effects that occur over the service life of a building.

Unlike test protocols that assess watertightness under a single, isolated condition, the Japanese approach emphasizes performance continuity. By sequentially applying specified mechanical actions to the test specimen and verifying watertightness at each stage, potential degradation mechanisms—such as sealant damage, joint opening, loss of deformation-following capability, and changes in drainage behavior—can be systematically identified (See Fig. 13).

A typical performance test procedure is outlined below. The sequence and content correspond directly to the original test protocol and are presented without omission.

(1) Initial Watertightness Performance Test

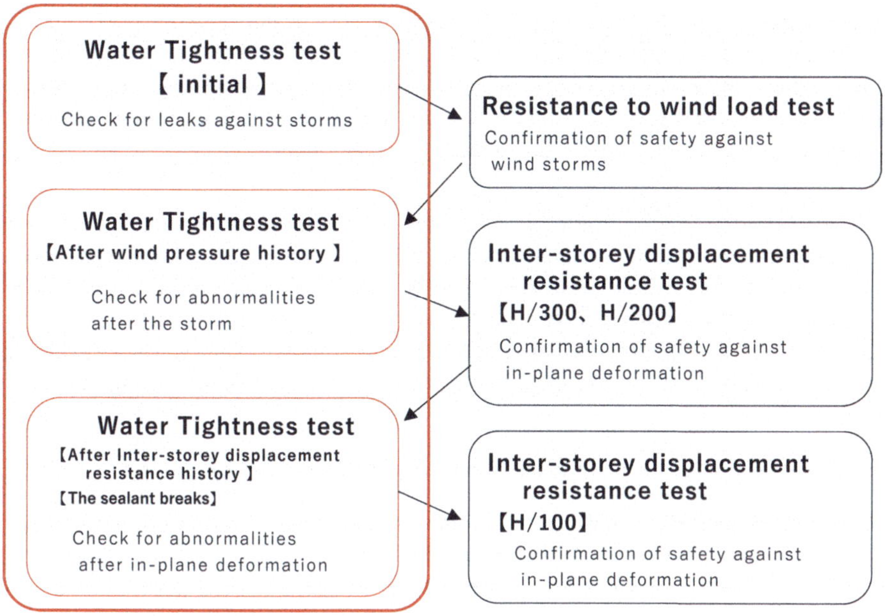

Fig. 13 General test procedure

This test is conducted immediately after installation of the test specimen. Its primary purpose is to confirm that the curtain wall system satisfies the specified watertightness performance in its as-installed condition, without prior mechanical or environmental loading. This step establishes a baseline against which subsequent performance changes can be evaluated. The presence or absence of water penetration to the interior side of the specimen is determined by visual inspection.

(2) Wind Pressure Resistance Test

Following confirmation of initial watertightness, a wind pressure resistance test is conducted. The test specimen is pressurized up to the maximum design wind pressure to simulate severe storm conditions. The objective of this test is to verify structural safety, including the integrity of framing members, anchorage systems, and glazing components under extreme wind loads.

Performance under design wind pressure is evaluated primarily in terms of deflection. Where the span (H) is 4.0 m or less, the deflection of primary members is typically limited to H/150 or less, with an absolute maximum of 20 mm. Where the span (H) exceeds 4.0 m, the allowable deflection is generally limited to H/200 or less. These criteria also take into account psychological considerations, as larger visible deflections at greater spans may cause discomfort or concern to building occupants and observers.

(3) Watertightness Test Following Wind Pressure Resistance Test

After completion of the wind pressure resistance test, watertightness performance is re-evaluated. This test examines whether the curtain wall system maintains its watertightness after experiencing deformation induced by maximum wind pressure. Particular attention is given to joints, seals, interfaces, and drainage components that may have been affected by pressure-induced displacement.

(4) Inter-Story Displacement Test (Small Deformation)

This test simulates relatively small inter-story displacements associated with wind-induced building sway or moderate seismic events. The objective is to verify the curtain wall system's ability to accommodate repetitive in-plane and out-of-plane movements without loss of functional performance. During this stage, the deformation-following behavior of joints, connections, and sealing components is carefully observed.

(5) Watertightness Test After Inter-Story Displacement History

Following the small-deformation inter-story displacement test, watertightness performance is again evaluated. This step assesses whether the accumulated displacement history has caused performance degradation, such as the formation of micro-gaps, sealant fatigue, or reduced contact pressure at interfaces. The results provide insight into the resilience of the watertightness system under service-level displacement conditions.

(6) Watertightness Test After Partial Sealant Removal

To simulate ageing-related deterioration and potential sealant damage associated with long-term use, sealant is intentionally and partially removed from designated locations. A subsequent watertightness test is then conducted to evaluate the capacity of the curtain wall system against localized loss of sealing performance. This step reflects realistic deterioration scenarios that may result from material ageing, workmanship variability, or maintenance deficiencies.

(7) Inter-Story Displacement Test (Large Deformation)

The test specimen is next subjected to inter-story displacement corresponding to extreme seismic events. This test verifies the ultimate deformation-following capability and safety of the curtain wall system under conditions approaching design limits. Structural integrity, residual functionality, and the absence of critical damage are assessed to confirm that the system can withstand large deformation demands without catastrophic failure.

(8) Drainage Path Verification Test

Finally, a drainage path verification test is conducted to confirm that water entering the curtain wall interior—either as condensation or as controlled penetration during watertightness testing—is properly discharged through the designed drainage pathways. Colored water is injected into internal zones intended to collect condensation, as well as into areas where rainwater is permitted to penetrate during watertightness tests. The internal flow of water and its discharge from drainage outlets are visually observed to identify any blockages, stagnation, or backflow. This test verifies the validity and reliability of the drainage design.

Through this eight-stage procedure, watertightness performance is repeatedly verified following major loading scenarios. This methodology enables a comprehensive evaluation of curtain wall performance from the perspective of continuous use and cumulative effects, rather than a single-point compliance check. Such an approach characterizes Japanese curtain wall performance assessment practice and is particularly suited to super-high-rise buildings exposed to the combined influences of wind, rain, and seismic actions.

7 Discussion

7.1 Implications for Facade Design Based on Performance Standards

Traditionally, watertightness design for curtain wall systems has been governed by regulatory frameworks that prescribe uniform testing parameters, such as water spray volume, pressurization conditions, and acceptance criteria for leakage. Within such frameworks, performance is evaluated primarily through compliance with standardized test methods. While this approach is effective in ensuring minimum levels of watertightness, it does not always adequately reflect the diversity of actual wind and rain conditions or the variation in external force demands arising from differences in building height, geometry, and façade configuration.

In contrast, a performance-based framework defines watertightness objectives in terms of preventing rainwater ingress under design wind pressure conditions. Test conditions and evaluation methods are then established on a project-specific basis, taking into account anticipated wind and rain exposure, building height, and façade detailing. This approach enables more rational and context-sensitive watertightness design and performance evaluation. At the same time, it requires a high level of technical expertise to establish appropriate test conditions and to verify the validity of evaluation results. Consequently, in curtain wall watertightness design, it is important to introduce performance-based approaches progressively, while continuing to benefit from the consistency and reliability offered by regulation-based standard evaluation methods.

The testing methodology described in this study exemplifies the performance-based approach that has been increasingly adopted in Japanese façade engineering practice. Performance-based design emphasizes verification through experimentally demonstrated behavior under realistic loading conditions, rather than reliance solely on prescriptive requirements. This approach is particularly well suited to super-high-rise buildings, where façade systems are subjected to complex combinations of wind pressure, wind-driven rain, seismic actions, and long-term environmental degradation.

By sequentially integrating watertightness testing with wind pressure resistance testing and inter-story displacement testing, the methodology enables evaluation of both initial performance and residual performance following mechanical loading. This reflects the understanding that façade performance degradation often arises not from insufficient initial capacity, but from cumulative damage caused by repeated deformation or extreme loading events. Verifying watertightness at key stages within the load history therefore provides valuable insight into the robustness and resilience of curtain wall systems.

The knowledge obtained through such testing extends beyond performance verification for individual projects. Observations of factors contributing to reduced watertightness performance, as well as deformation-following behavior under inter-story displacement, can be incorporated into future façade designs. Specifically, these findings can inform decisions regarding joint configuration, sealant specification, anchorage design, and support conditions in projects with comparable environmental and structural demands. In this way, the systematic accumulation and feedback of project-specific test results contribute to improving the reliability, reproducibility, and transparency of performance-based curtain wall design.

7.2 Role of Manufacturer-Owned Accredited Testing Laboratories

The establishment of manufacturer-owned testing laboratories accredited in accordance with ISO/IEC 17,025 represents a significant institutional development within the Japanese façade industry. Such facilities enable manufacturers to conduct project-specific performance testing efficiently while maintaining technical credibility equivalent to that of independent third-party laboratories. This capability is particularly advantageous for super-high-rise projects, which often involve bespoke façade systems, complex performance requirements, and constrained construction schedules.

At the same time, the operation of an in-house accredited testing laboratory requires rigorous governance to ensure impartiality, traceability, and compliance with recognized testing standards. Transparent documentation of test procedures, calibration records, and test results is essential to maintain confidence among designers, contractors, and building owners. When these requirements are properly satisfied, manufacturer-owned laboratories can contribute significantly to improving the overall quality, consistency, and reliability of façade performance verification.

7.3 International Context and Standardization

Although the testing methodologies described in this chapter are rooted in Japanese practice, many of the underlying challenges are shared internationally. Climate change has increased the frequency and intensity of extreme weather events in many regions, leading to greater demands on façade systems to resist wind-driven rain. At the same time, ongoing urban densification has driven the construction of increasingly tall buildings, amplifying wind pressure effects on building envelopes.

From this perspective, Japan's approach of phased and composite performance testing offers potentially valuable insights for international façade standardization. While regional climatic conditions, regulatory frameworks, and construction practices differ, the fundamental principle of verifying watertightness performance after sequential mechanical loading is broadly applicable. Enhancing the compatibility of such performance-based methodologies with international standards may contribute to greater global consistency in façade performance evaluation and support the development of more resilient building envelope systems worldwide.

8 Limitations and Future Research

This study focuses primarily on watertightness performance testing conducted under controlled laboratory conditions. While chamber-based testing provides a high level of reproducibility and enables systematic investigation of key performance parameters, it cannot fully reproduce all aspects of in-service conditions experienced by curtain wall systems. Factors such as construction tolerances, variability in site workmanship, differences in maintenance practices, and long-term material degradation may significantly influence actual performance over a building's lifecycle.

To address these limitations, future research should incorporate field-based performance monitoring of curtain wall systems installed in high-rise and super-high-rise buildings. Long-term observational data on leakage occurrences, maintenance interventions, component ageing, and repair histories would enable more accurate calibration of laboratory test conditions and acceptance criteria. Such data would also support the development of performance thresholds that more closely reflect in-service behavior.

In addition, the integration of numerical simulation techniques with experimental testing represents an important direction for future research. Computational fluid dynamics analyses may be used to better characterize wind-driven rain exposure and pressure distributions, while structural and hygrothermal simulations can assist in predicting deformation behavior, moisture transport, and long-term degradation mechanisms. Combined experimental–numerical approaches have the potential to enhance predictive capability and reduce uncertainty in façade performance assessment.

Another key area for future investigation is the harmonization of watertightness performance metrics across different regions. Comparative studies examining Japanese testing methodologies alongside those applied in Europe, North America, and other Asian regions would help to identify common principles as well as region-specific adaptations. Such research could support the development of internationally applicable performance-based frameworks, contributing to greater consistency and reliability in curtain wall performance evaluation worldwide.

9 Conclusion

This paper has examined the watertightness performance requirements for curtain wall systems applied to high-rise buildings in Japan and has described the testing methodologies used to verify compliance with these requirements. Owing to exposure to typhoons, intense rainfall, and frequent seismic activity, façade systems in Japan are required to satisfy stringent and multifaceted performance criteria. As a result, watertightness testing is not conducted in isolation, but is systematically integrated with wind pressure resistance and inter-story displacement tests to evaluate performance under realistic and continuous loading histories.

The testing facilities and procedures operated by Fujisash Co., Ltd. illustrate how manufacturer-owned laboratories accredited in accordance with ISO/IEC 17025 can support reliable, project-specific performance verification. By enabling comprehensive assessment of curtain wall systems—including the effects of deformation history and simulated ageing—such facilities contribute to improving the reliability of façade performance in demanding applications.

As buildings continue to increase in height and complexity and as climatic conditions evolve, the importance of rigorous watertightness performance evaluation is expected to grow further. Key challenges for future practice include closer integration between performance standards, façade design, and experimental verification; the systematic accumulation, analysis, and application of test data; and continued alignment with international standards and evaluation frameworks. Addressing these challenges will support the development of safer, more resilient, and more durable building envelope systems in Japan and beyond.

References

1. Japan Building Openings Association: Curtain Wall Performance Standards. Tokyo (2013)
2. International Organization for Standardization: ISO/IEC 17025: General requirements for the competence of testing and calibration laboratories. Geneva (2017)
3. European Committee for Standardization: EN 13830: Curtain walling – Product standard. Brussels (2015)
4. European Committee for Standardization: EN 12154: Curtain walling – Watertightness – Performance requirements and classification. Brussels (2016)

5. ASTM International: ASTM E331-00(2016): Standard Test Method for Water Penetration of Exterior Windows, Skylights, Doors, and Curtain Walls by Uniform Static Air Pressure Difference. West Conshohocken (2013)

Wind-Borne Debris Impact Resistance of Façades: A Workflow for Defining Performance-Based Impact Tests According to ISO 16316:2024

Angela Mejorin

Abstract The escalating impact of climate change is ushering in a new era of extreme weather events, characterized by heightened frequency and intensity, resulting in unprecedented monetary losses and fatalities. Previously unaffected regions are now grappling with the urgent need to enhance urban resilience against these growing threats. Among the myriad challenges posed by extreme weather, wind-borne debris emerges as an urgent danger, posing a significant risk to the built environment and the safety of its inhabitants. The May 22nd, 2011 Joplin tornado serves as a stark reminder of the havoc wrought by wind-borne debris, with shattered windows leaving a hospital vulnerable and compromising critical services. In response to such risks, codes and standards have evolved to fortify building envelopes against wind-borne debris impacts, particularly in cyclone and hurricane-prone regions. This review chapter presents current code and standard requirements for wind-borne debris impact resistance of the building envelope and synthesizes recent performance-based design directions. It describes a generic workflow to define alternative impact test criteria, consistent with emerging standardization pathways, enabling project-specific testing strategies within established best practices. This chapter combines a critical review of existing standards with a synthesis of recent performance-based methodologies now entering international standardization and can serve as a guideline for addressing the urgency of mitigating the adverse impacts of windstorms, thereby bolstering urban resilience sustainably.

Keywords Wind-borne debris · Façade design · Impact test · Façade resilience · Performance-based design · Hurricanes · Tornadoes · Damage mitigation

Nomenclature/Abbreviations

ASCE/SEI American Society of Civil Engineers/Structural Engineering Institute

A. Mejorin (✉)
Performance-Based Façade Design, Dorsoduro, Venice, Italy
e-mail: amejorin@performancebasedfacadedesign.org

A. Mejorin et al. (eds.), *Performance-Based Façade Design*, Springer Tracts in Civil Engineering, https://doi.org/10.1007/978-3-032-28835-6_4

AS/NZS	Australia/New Zealand Standard
ASTM	ASTM International
C&C	Components and Cladding
CFD	Computational Fluid Dynamics
FBC	Florida Building Code
FEMA	Federal Emergency Management Agency
ICC	International Code Council
IPCC	Intergovernmental Panel on Climate Change
ISO	International Organization for Standardization
MWFRS	Main Wind Force Resisting System
NIST	National Institute of Standards and Technology
NSSA	National Storm Shelter Association
PBD	Performance-Based Design
PBWD	Performance-Based Wind Design

1 Introduction

The evidence of climate change has become increasingly undeniable, with its impacts vividly manifesting in the increasing intensity and frequency of extreme weather events worldwide [1–4]. This trend creates a concerning rise in natural disasters, marked by significant economic losses and tragic human casualties [5–8]. Extreme weather events rank among the top risks globally in the World Economic Forum's Global Risks Perception Survey 2021–2022, emphasizing the urgent need for proactive climate action to mitigate their devastating consequences [9].

Across the spectrum of extreme weather phenomena, wind-related events, including hurricanes, typhoons, tornadoes, and severe wind gusts, pose substantial threats to life, ecosystems, and infrastructure on a global scale. The repercussions of climate action failure, compounded by the failure to adopt effective adaptation and mitigation measures, exacerbate these risks, underscoring the imperative for decisive action [9]. In regions not traditionally governed by hurricane/tornado design provisions (e.g., much of Europe) a concerning uptick in severe wind events has been observed in recent years [10, 11]. This trend underscores the need for comprehensive research and enhanced building resilience to withstand the onslaught of extreme winds and their associated debris impacts. Debris propelled by extreme winds poses a significant threat to building envelopes, with projectiles ranging from roof gravel to façade elements and antennas. The vulnerability of buildings to wind-borne debris underscores the critical importance of fortifying the building envelope as the primary line of defense against such hazards [12, 13]. The emergence of international initiatives, such as the ongoing efforts by the International Organization for Standardization, underscores a growing recognition of the need for standardized testing requirements to evaluate building envelope resilience against wind-borne debris impacts

[14, 15]. This collective endeavor aligns with the broader goals of sustainable development, aiming to foster resilient infrastructure and sustainable urban development [16].

The catastrophic May 22nd, 2011, Joplin tornado serves as a poignant reminder of the devastating toll wind-borne debris can exact on communities and infrastructure. The harrowing accounts of structural damage and human casualties underscore the urgent imperative for robust design strategies to mitigate such risks [17]. Despite decades of research and the development of codes and standards to address wind-induced structural failures, current regulations in North America and Europe often overlook the specific requirements for mitigating wind-borne debris impacts on building envelopes. However, some countries [18–20] have made strides in developing testing and certification systems to enhance the resilience of building components against wind-borne debris [21].

This review chapter aims to open the discussion on performance-based design strategies to mitigate extreme wind events' devastating effects, focusing on wind-borne debris impact resistance. By examining existing standards, ongoing research initiatives, and emerging technologies, this chapter endeavors to offer insights and guidance for enhancing the resilience of building envelopes in the face of extreme wind events. In civil engineering, ensuring the safety and resilience of buildings against extreme weather events, particularly wind-borne debris impacts, stands as a critical imperative. This review chapter embarks on a comprehensive exploration of three pivotal documents central to fortifying structures against wind-borne debris: the ASCE 7-22 [19] for minimum design requirements for buildings and other structures, the International Code Council ICC/NSSA 500 [22], which focuses on storm shelters, and the reference standards ASTM E1886 [23] and ASTM E1996 [24] for reader convenience. This review delves into the intricacies of ASCE 7-22 [19], which addresses wind-borne debris impacts for both regular synoptic windstorms (covered in chapters 26–30) and tornadoes (chapter 32). It elucidates the performance-based design objectives delineated by ASCE [25], considering wind-borne debris with ad-hoc considerations, with the ultimate goal of optimizing structural resilience and safety.

Through examination of its provisions concerning debris impacts, this chapter aims to unveil the symbiotic relationship between structural design, occupant safety, and debris mitigation strategies as enshrined within the standard [19]. Performance-based design strategies can be implemented using the design framework developed by Mejorin [26], which is included in Annex G of ISO 16316 [15] to identify performance-based design requirements, following the directions highlighted by ASCE [25] and by Scott et al. [27]. This review chapter highlights the best practices for wind-borne debris impacts in structural design, providing insights for engineers, architects, policymakers, and stakeholders involved in ensuring the safety and resilience of built environments against the ravages of extreme weather events. This critical review chapter presents the best available practices for synoptic and non-synoptic winds, considering the latest directions and issues [27]. As discussed above, one of the major causes of building damage in extreme winds is due to wind-borne debris impact [28].

2 Current Wind-Borne Debris Impact Requirements

Governing bodies in areas that are prone to extreme weather events have implemented measures to mitigate the effect of wind on the built environment. Studies and field observations have been conducted on the consequences of major hurricanes, typhoons, and tornadoes on the built environment and human safety [29, 30]. Such measures have one single goal: preventing injury to people by ensuring structures are safe. Such an objective is sought by adopting two sets of measures: on the one hand, reducing the likelihood of the wind-borne debris phenomenon by improving the resistance of building elements to the wind; on the other hand, mitigating the consequences of wind-borne debris by setting design guidelines and testing methodologies to ensure the resistance of the building envelope to the debris impact. These measures apply to buildings in areas prone to hurricane and cyclonic winds: less demanding target performances are requested for houses, while essential facilities (such as hospitals, emergency services, schools, etc.) are required to meet more demanding standards, especially for the building envelope [19, 20].

Wind-borne debris resistance of façades is currently widely assessed in the United States' wind-borne debris regions [31] and Australian wind regions C and D [20]. The impact testing is conducted using two different impactor typologies to be shot through an air cannon either on the spandrel or the glass panel. The U.S. façade test procedures are currently the international benchmark requirements for façade impact resistance to wind-borne debris caused by windstorm events [32].

Differently from the United States testing, there is no provision for cyclic load testing post-debris impact for façades in AS/NZS 1170.2 [20]. According to Australian Standard's requirements, the building envelope components are required to be flying debris-resistant for a height of only 25 m from the ground level. This last criterion is based on research, fixing the upper limit for flight initiation at 20 m [33], but the small projectile (2-g steel ball) has been chosen because it is representative of roof gravel; thus, it could be higher than 25 m.

2.1 ASCE/SEI 7-22

In ASCE/SEI 7-22 [19], the requirements for protection against wind-borne debris in hurricane and tornado-prone regions are delineated in Chapter 26 and Chapter 32, respectively. These chapters establish crucial guidelines to fortify structures against wind-borne debris impacts, ensuring structural integrity and occupant safety.

In hurricane-prone regions, glazed openings in buildings categorized as Risk Category II, III, or IV are mandated to be safeguarded. These requirements are contingent upon specific conditions, including proximity to high-water lines and the prevailing basic wind speed. Glazing within defined proximity to the mean high-water line in areas subject to high wind speeds must either be impact resistant or be equipped with an impact-protective system. Furthermore, glazing located above certain heights may be exempt from protection requirements, provided they meet specified conditions to mitigate risks associated with wind-borne debris. Compliance with ASTM [23, 24] standards for missile tests and cyclic pressure differential tests is obligatory to ensure the efficacy of impact-protective systems and glazing.

For existing buildings in regions prone to wind-borne debris, non-compliant glass windows must either be replaced with impact-resistant alternatives or retrofitted with protective systems. Failure to meet these requirements necessitates the assumption that existing non-compliant elements are susceptible to penetration by debris, necessitating recalculations of structural loads. Additionally, when replacing roofs, careful consideration must be given to the uplift resistance of both the roof deck and supporting structures to enhance structural resilience.

In contrast, tornado-prone regions pose unique challenges due to the greater wind-borne debris hazard associated with tornadoes. Tornadoes, characterized by strong updrafts, can propel debris at higher speeds and quantities to higher heights, increasing the vulnerability of unprotected windows. Assemblies that resist penetration by specified test missiles offer varying levels of protection against wind-borne debris generated by tornadoes. Furthermore, specialized glazing assemblies tested in accordance with prescribed standards provide enhanced protection against more intense tornadoes.

Research into the detailed use and flight distances of various wind-borne debris elements as tools for estimating tornado wind speeds has been somewhat limited. Wills et al. [34] developed a model describing building damage caused by wind-borne debris and noted that loose debris is more likely to roll than to become airborne. Kopp et al. [35] conducted a failure analysis of complete roofs, using the observed flight distances of these roofs to estimate the wind speeds responsible for their detachment and subsequent flight. However, both models assume a straight-line wind field, overlooking the impact of the three-dimensional tornado wind field on the angle of attack affecting large, compact objects.

ASCE 7-22 standard [19] introduces stringent requirements for buildings and other structures classified as Risk Category III or IV and located in tornado-prone regions. These structures, including the main wind force resisting system (MWFRS) and all components and cladding (C&C), must be designed to withstand the greater of either tornado loads, as specified in Chapter 32, or wind loads, as outlined in Chapters 26 through 31, using the load combinations provided in Chapter 2.

For the MWFRS, tornado loads must be determined using one or more procedures specified in Chapter 32 [19]. For C&C, tornado loads must be determined using analytical procedures specified in Chapter 30 or the Wind Tunnel Procedure in Chapter 31. ASCE 7-22 [19] also permits performance-based tornado design procedures, subject to approval by the Authority Having Jurisdiction. These procedures must conform to Section 1.3.1.3 and be documented accordingly. Additionally, the standard underscores the importance of protecting glazed openings from wind-borne debris, which is more severe in tornadoes compared to hurricanes. Tornadoes can loft debris higher, resulting in more and faster traveling debris. Failure of unprotected glazing during tornadoes can lead to significant interior damage and hazards to life safety. Consequently, glazed openings in essential facilities and buildings intended to remain operational during a tornado should be specified to resist wind-borne debris impacts.

The ASCE 7-22 standard [19] draws a clear distinction between the requirements for tornado shelters and conventional buildings. Tornado shelters and safe rooms are designed to withstand extreme hazard levels, including wind-borne debris impacts from a 15 lb (6.8 kg) sawn lumber 2 × 4 missile traveling at speeds of 53–100 mi/h (23.7–44.7 m/s). These shelters must provide near-absolute protection, as outlined in ICC 500 [20] and FEMA P-361 [36], which use deterministic tornado speed maps based on historical tornado data from 1950 to 2006. In contrast, the tornado speed maps in ASCE 7-22 [19] are probabilistic, based on return periods of 1700 and 3000 years for Risk Category III and IV buildings, respectively.

In summary, while studies on wind-borne debris in general are relatively few, research specifically focusing on tornadoes is even scarcer. Notable exceptions include Twisdale et al. [37], who developed a methodology for simulating the initial release and subsequent motion of objects transported by tornadoes, and Huo et al. [38, 39], who performed CFD simulations of tornado-like vortices to estimate debris trajectories and flight initialization. Zhao et al. [40] provided a comprehensive summary of previous analytical and experimental approaches to studying the dynamics of wind-borne debris in both straight-line and tornadic wind fields. The ASCE 7-22 [19] requirements further emphasize the need for robust design and construction practices to ensure the resilience of structures in tornado-prone areas.

Storm shelters and safe rooms, designed to withstand extreme hazards, adhere to stringent criteria established by standards such as ICC/NSSA 500 [22]. These shelters are engineered to provide maximum protection from wind and wind-borne debris, even in rare and catastrophic storm events. Compliance with ICC/NSSA 500 [22] standards ensures the construction of shelters capable of safeguarding lives and mitigating the impacts of severe weather events.

Overall, ASCE/SEI 7-22 [19] lays out comprehensive requirements to enhance the resilience of structures against wind-borne debris impacts in hurricane and tornado-prone regions. By adhering to these guidelines and standards, stakeholders can mitigate risks, safeguard occupants, and bolster the structural integrity of buildings in the face of extreme weather events.

2.2 ASTM E1886 and ASTM E1996

According to ASCE 7-22 [19], the wind zone map is shown to identify the wind-borne debris regions and the boundary for hurricane-prone regions. ASTM E1886 [23] and ASTM E1996 [24] requirements, or local standards requirements, whichever is more stringent, must be followed by buildings constructed in these United States' areas. The testing missile weights and speeds vary on the level of importance of the building and on the wind zone location of the building, ranging from Wind Zone 1 to Wind Zone 4 (highest wind speeds) (Tables 1 and 2).

Table 1 Missile types according to ASTM E1996 [24]

Missile type	Material and weight	Impact speed (m/s)
A	(2 ± 0.1) g (small steel ball)	39.62 (130 f/s)
B	(0.91 ± 0.1) kg (2 × 4 in small lumber)	15.25 (50 f/s)
C	(2.05 ± 0.1) kg (2 × 4 in small lumber)	12.19 (40 f/s)
D	(4.1 ± 0.1) kg (2 × 4 in medium lumber)	15.25 (50 f/s)
E	(4.1 ± 0.1) kg (2 × 4 in medium lumber)	24.38 (80 f/s)

Table 2 Missile for testing classification according to ASTM E1996 [24]

Level of protection	Essential facilities		Basic protection		Unprotected	
H (m)	≤ 9.1 m	> 9.1 m	≤ 9.1 m	> 9.1 m	≤ 9.1 m	> 9.1 m
Wind Zone 1	D	D	C	A	None	None
Wind Zone 2	D	D	C	A	None	None
Wind Zone 3	E	D	D	A	None	None
Wind Zone 4	E	D	D	A	None	None

Wind Zone 1: 130 mph (58 m/s) < basic wind speed < 140 mph (63 m/s)
Wind Zone 2: 140 mph (63 m/s) < basic wind speed < 150 mph (67 m/s)
Wind Zone 3: 150 mph (67 m/s) < basic wind speed or 140 mph (63 m/s) < basic wind speed within 1.6 km (1 mile) of the coastline
Wind Zone 4: 170 mph (76 m/s) < basic wind speed

The impact locations include the center of each type of infill; 150 mm from supporting members at the corner; 150 mm from supporting members at a diagonally opposite corner; integral mullion; meeting stile; combination mullion; meeting rail/check rail. The impact testing procedure shall be conducted on one single specimen if the building envelope component is designed to be installed as a single element. The test should be repeated for a testing assembly composed of three units if the building envelope component is designed to be installed as a façade system. This method considers both lumber and steel-ball missile impacts. The pressure cycling to conduct on the façade after the successful impact tests are based on the design wind pressure (inward and outward) from the building code, based on an intact building envelope. A total of 4500 positive and 4500 negative pressure cycles has to be conducted. When it comes to window or curtain wall assemblies, at least three panels should be tested together and that has been shown to be an important parameter to classify the façade products. The test specimen shall consist of the entire fenestration or impact protective system assembly and contain all devices used to resist wind and wind-borne debris. Test specimens for a large fenestration element shall be one panel unless otherwise specified. Where it is impractical to test the entire fenestration assembly such as curtain walling and heavy commercial assemblies, usually the largest size of each type of panel is tested, to qualify the entire assembly.

Individual building envelope components such as windows, doors, curtain walling, or windstorm protective systems should be tested separately. When these building envelope systems are intended to be installed in combination, they shall be tested by joining at least three lites into one mounting frame, separated only by the mullions. These mullions should be representative of the mullions of the building envelope to be tested. Windstorm protective systems should be tested also to verify their deflection after the testing, to verify what should be the minimum design distance of these building components installation from the internal building envelope to be protected and/or from the internal content.

ASTM standard leaves the possibility open to identify any other representative projectiles with different mass, size, shape, and impact speed determined by engineering analysis to conduct tests on the building envelope components, considering the design reference wind speed of the wind zone location. From ASTM E1886 [23]: "other representative missile with mass, size, shape, and impact speed as a function of basic wind speed determined by engineering analysis" can be adopted.

2.3 ICC/NSSA 500-2023

The ICC/NSSA 500 [22] Standard for the Design and Construction of Storm Shelters was developed by the International Code Council (ICC) and the National Storm Shelter Association (NSSA) to establish minimum requirements for storm shelters' design and construction. The standard aims to ensure that storm shelters provide safe refuge during high-wind events such as hurricanes and tornadoes. These storms generate intense wind pressures and flying debris that exceed typical building design standards. Storm shelters must be capable of withstanding these extreme conditions to protect occupants' lives and health.

2.4 Hurricane Requirements

The missile testing for all components of the storm shelter envelope of hurricane shelters shall be a 9-pound (4.1 kg) sawn lumber 2 by 4. The speed of the test missile impacting vertical storm shelter surfaces shall be a minimum of 0.50 times the design wind speed. The speed of the test missile impacting horizontal surfaces shall be 0.10 times the design wind speed.

Storm shelter envelope components meeting impact test requirements for tornado shelters at the 250 mph (111.8 m/s) design tornado speed shall be considered acceptable for the impact test requirements for hurricane shelters provided the components meet the structural design load requirements for hurricane shelters.

Table 3 Missile speed for tornado shelters [19]

Design tornado speed	Missile speed and impact surface
58.12 m/s (130 mph)	35.76 m/s (80 mph) vertical surfaces 23.69 m/s (53 mph) horizontal surfaces
71.53 m/s (160 mph)	37.55 m/s (84 mph) vertical surfaces 25.03 m/s (56 mph) horizontal surfaces
89.41 m/s (200 mph)	40.23 m/s (90 mph) vertical surfaces 26.82 m/s (60 mph) horizontal surfaces
111.76 m/s (250 mph)	44.70 m/s (100 mph) vertical surfaces 29.95 m/s (67 mph) horizontal surfaces

Exceptions

1. Window assemblies and other glazed openings where the opening is protected on the exterior side by an impact-protective system are not required to be tested for impact
2. Window assemblies and other glazed openings where the opening is protected on the interior side by an impact-protective system are not required to be tested for impact and cyclic pressure
3. Non operable, permanently affixed shields or cowlings designed to resist the design tornado pressure or design wind pressures are not required to be tested for static and cyclic pressure

2.5 Tornado Requirements

The missile testing for all components of the storm shelter envelope of tornado shelters shall be a 15-pound (6.8 kg) sawn lumber 2 by 4 traveling at the speeds shown in Table 3.

2.6 Impact Testing

Impact testing apparatus, calibration, and procedure follow ASTM E1886 [23]. ICC/NSSA 500 [22], however, gives additional requirements in terms of impact locations and number of impacts to conduct on the specimens. These requirements are characterized by differences, based on panel/wall/window typologies. The additional requirements are:

- Panel or framed wall assemblies and roof assemblies;
- Solid wall assemblies and roof assemblies of concrete or other materials;
- Masonry unit wall assemblies and roof assemblies;
- Door assemblies;
- Window assemblies and other glazed openings.

2.7 ISO 16316 [15]

With major references into ASTM E1886 [23] and ASTM E1996 [24] standards, ISO 16316 [15] has been developed, to introduce international requirements for wind-borne debris impact performance of façades. For the same wind regions [23, 24], impact test requirements are respected, but ISO 16316 [15] introduces advancements in the standardization of wind-borne debris impact testing, refining both the methodologies and practical applications to address modern construction challenges.

- **Introduction of Second-Test Requirement for Smaller Panels**

  One notable addition is the requirement for additional testing for smaller panels when they exceed a 50% reduction in their individual length size (e.g., height or width) compared to the originally tested panel. This second test ensures that the increased stiffness of smaller components does not compromise their impact performance under wind-borne debris simulation. This requirement addresses the potential for discrepancies in stiffness and energy absorption, which may alter the results of impact tests. By mandating a second evaluation for these cases, ISO 16316 [15] ensures that all panel sizes, regardless of dimension, adhere to consistent and reliable safety standards.

- **Substitution Criteria for Fenestration Assemblies (Annex E)**

  Annex E [15] introduces a flexible substitution framework, enabling modifications in fenestration assemblies while maintaining compliance with performance benchmarks. Key features include: Automatic Substitution (allowed for assemblies meeting size reduction and performance equivalence criteria without additional testing); Engineering Analysis (permits substitution based on material properties and documented evidence); Single Specimen Testing (required for substitutions not qualifying under automatic allowances). This annex minimizes redundant testing, reduces costs, and enhances the adaptability of construction materials and designs while safeguarding performance integrity.

- **Performance-Based Design for Impact Testing (Annex G)**

  Annex G [15] shifts the focus toward performance-based design [26], enabling impact testing tailored to localized environmental risks. Highlights include: Localized Impact Tests (derived from engineering analyses of debris characteristics, flight trajectories, and kinetic energy); Numerical Simulations (incorporates models such as Monte Carlo simulations to predict debris behavior and estimate impact probabilities); Adaptable Testing Framework (aligns testing requirements with specific wind-borne debris features and building typologies, including curtain walls, double-skin façades, and more). This methodology enhances urban resilience, ensuring building envelopes can withstand region-specific challenges, from wind-borne debris to severe storms.

The changes introduced in ISO 16316 [15] set a new standard for precision, adaptability, and reliability in wind-borne debris impact testing. By addressing size-dependent stiffness, enabling flexible substitutions, and embracing performance-based designs, the standard provides a robust framework for safer, more resilient building practices. These advancements mark a significant evolution in how construction standards address the complexities of climate resilience and urban safety, with a focus on wind-borne debris.

3 Performance-Based Wind Impact Design

In the realm of performance-based design and wind-borne debris impact resistance, this review chapter embarks on an examination of the provisions outlined within the standards and codes, aiming to illuminate the intricate interplay between structural design, occupant safety, and debris mitigation strategies. Central to this endeavor is the elucidation of the symbiotic relationship inherent within the standards, wherein robust structural design not only ensures the resilience of buildings but also safeguards the well-being of occupants against the onslaught of wind-borne debris.

To operationalize this approach, this chapter advocates for the utilization of the design framework which considers the details around the facility being designed. This framework serves as a guiding reference, facilitating the identification of performance-based design requirements embedded within the standards.

3.1 ASCE/SEI Prestandard for Performance-Based Wind Design

The Prestandard for Performance-Based Wind Design aims to enhance building wind design by enabling performance-based design (PBD) and improving building envelope performance [19]. It offers alternatives to prescribed procedures in building codes. The Prestandard's recommendations apply to engineered buildings' wind-loading resistance design and the building envelope, with a focus on buildings in hurricane-prone regions or categorized as Risk Category III or IV. The Prestandard identifies the ICC/NSSA 500 [22] as a suitable standard for performance-based design requirements when it comes to hurricane and tornado-prone regions.

The acceptance criteria for building envelope systems consider factors like wind pressure, wind-borne debris, and wind-driven rain. Specific criteria are provided for roof systems in both hurricane-prone and non-hurricane-prone regions, with guidelines for enhanced protection against water leakage and wind-borne debris. This prestandard's recommendations for design and detailing apply to the wind-loading resistance design of engineered buildings, and the building envelope, and select internal systems that are desired to have enhanced performance. This prestandard shall be permitted to be used for MWFRS design, for the building envelope, or for both as determined by the project stakeholders. All projects using performance-based wind design (PBWD) for the building must use wind tunnel data for the MWFRS. In addition, the prestandard shall be permitted to be applied to the envelope if the MWFRS is designed according to prescriptive standards.

The building types that shall be subject to the mandatory inclusion of the enhanced criteria described within are:

- Risk Category III or IV buildings; and
- Buildings in hurricane-prone regions.

The requirements for mandatory enhanced criteria are minimum requirements. The owner and project designers shall weigh the benefits of the inclusion of the enhanced criteria for buildings that fall outside of the minimum criteria.

Wind-borne debris shedding from the roof can damage the building's façade or lower-level roofs. Debris can either be from the roof itself or it can be items such as loose access panels on rooftop equipment, as well as various types of debris that were left on the roof before a hurricane. Debris shedding is particularly problematic at hospitals and hurricane shelters, wherein people approaching the building during a hurricane are vulnerable to injury from the debris. To avoid debris shedding, the following are recommended.

Do not specify brittle roof coverings such as lightweight pavers, slate, or tile unless testing indicates that the covering is not susceptible to blow-off after impact testing. These types of roof coverings can be easily broken by low-momentum wind-borne debris. Once broken, these types of roof coverings are susceptible to shedding a significant amount of debris.

Specify a tall parapet. Storm damage research has shown that tall parapets often prevent debris from blowing off a roof. An appropriate parapet height likely depends on the basic wind speed, exposure, and roof height. Laboratory research has not been conducted to provide design criteria for the parapet height needed to retain debris on the roof. However, parapet height alone is not sufficient to prevent debris from leaving the roofing surface [41]. Armoring guidance is provided in FEMA P-424 [42].

This prestandard adopts a fenestration impact standard for opaque wall assemblies. Glazing systems shall comply with the requirements of ASTM E1886 [23]. The prestandard offers comprehensive guidance for performance-based wind design, particularly focusing on building envelope systems' resilience against hurricane and tornado winds. The goal is to enhance the safety, sustainability, and durability of buildings in wind-prone regions.

3.2 Scott et al. [27]

The 2023 workshop identified a diverse array of research and development endeavors aimed at advancing the application of Performance-Based Wind Design (PBWD) to mitigate the effects of severe wind events. Its objective was to evaluate the current status of PBWD and facilitate the creation of a roadmap for advancing knowledge and practical implementation in this field [27]. It is crucial to assess techniques or strategies to enhance building resilience and reduce uncertainty in high wind conditions, addressing factors like wind-borne debris impact, water infiltration, and structural damage. Concerns have been raised within the structural engineering community regarding the efficacy of envelope systems in protecting buildings from wind-induced damage, prompting the need for long-term improvements in this area.

The techniques and strategies may include enhancements to building codes, enhanced testing, enhanced detailing or toughness, enhanced performance objectives, and leveraging of databases or datasets not presently in common use. As of early 2023, the structural engineering community is concerned that the level of damage protection offered by envelope systems is not achieving the expected performance implied by ASCE/SEI 7-22 target reliabilities [19]. Consequently, too many buildings and facilities are taken offline, damaged, and in some cases destroyed because of the envelope system permitting wind-borne debris damage in wind events less severe than those prescribed by ASCE/SEI 7-22 [19]. These observations form the basis for the need for long-term building envelope system improvement.

The envelope system is central to the performance of a building subject to extreme winds. However, this system is scrutinized less than the structural system. Moving to a PBWD paradigm necessitates the holistic treatment of the structural and the envelope systems, which constitutes a two-way coupled system (feedback system) that experiences progressive damage during extreme winds.

One of the priority research needs (nr. 2, [27]) focuses on enhancing the reliability of components and cladding by reassessing testing methods. It aims to determine if existing wind-borne debris impact tests accurately reflect real impact risks and achieve desired performance levels in real-world scenarios. The current methods [22–24] employ standard projectiles and impact velocities, assuming complete failure of typical balloon frame construction and roof aggregate. However, these protocols do not consider the aerodynamics of wind-borne debris in local environments, nor do they utilize a database with debris speeds in various windstorm intensities.

Current testing methods for wind-borne debris impact performance adopt standard "missiles" and impact velocities. These consider the complete failure of typical balloon frame construction—the flight of structural members (large missiles representative of the 2 × 4 in. section wood frame construction) and the roof aggregate (small missiles)—to conduct impact tests on glazed assemblies. The current testing protocols, therefore, also assume that wind-borne objects have these two features. Thus, the impact energies are not based on the aerodynamics of wind-borne debris in local environments, nor is a database with wind-borne debris speeds in windstorms of various intensities available. The weights and velocity of the testing projectiles change according to the building's wind zone and level of protection.

Various methods are available to estimate the failure wind speeds of an object. Potential debris sources in the surroundings include surrounding buildings' roof tiles, roof aggregate, roof pavers, and other building envelope components.

3.3 Mejorin [26]

Mejorin [26] contributed to the identification of performance-based design impact test requirements for wind-borne debris by developing an organized design tool to estimate façade impact performances in specific environments (Fig. 1). The final objective of the design framework is enhancing building envelope resilience. It synthesizes advanced research in building technology and wind engineering, presenting a performance-based design framework.

In building technology, the framework focuses on failure capacities of building components and proposes a probabilistic approach to estimate real-world failure capacities, focusing on roofing components. These are the same building components identified as priority research objectives by Scott et al. [27]. In wind engineering, it analyzes studies to estimate wind-borne debris failure and flight trajectories, considering real-world objects and building aerodynamics. The framework can improve the resilience of urban habitats to extreme winds, particularly for essential facilities. Stakeholders such as building professionals, owners/managers, and insurance companies can benefit from the research findings to assess risks and implement measures to reduce them. The design framework's impact on alternative wind-borne debris façade impact test requirements is significant, potentially influencing new testing standards at both voluntary and regulatory levels.

The design framework can be taken as a reference to develop recommendations for debris impact requirements for building envelope systems which considers realistic scenarios for tornadoes for common debris elements from adjacent structures, considering the approach used in ANSI/ICC 500 [22]. This performance-based design framework [26] provides a method to consider the immediate surroundings of the building being designed including the potential trajectories considering tornado wind field models.

- **Wind-Borne Debris Failure and Trajectory Analysis**

Kordi et al. [43] highlighted that the resultant distributions and maximum values in the debris flight and the flight distances depend on the mode of flight of the debris element. Considering debris trajectories, Grayson et al. [44] developed a model that can provide relevant debris trajectory information (e.g., linear and rotational position, velocity, and acceleration) needed to track wind-borne debris and assess any impacts on the building envelope that may occur due to the debris flight. Previously conducted wind-borne debris trajectory analyses [45–47] considered debris elements that are

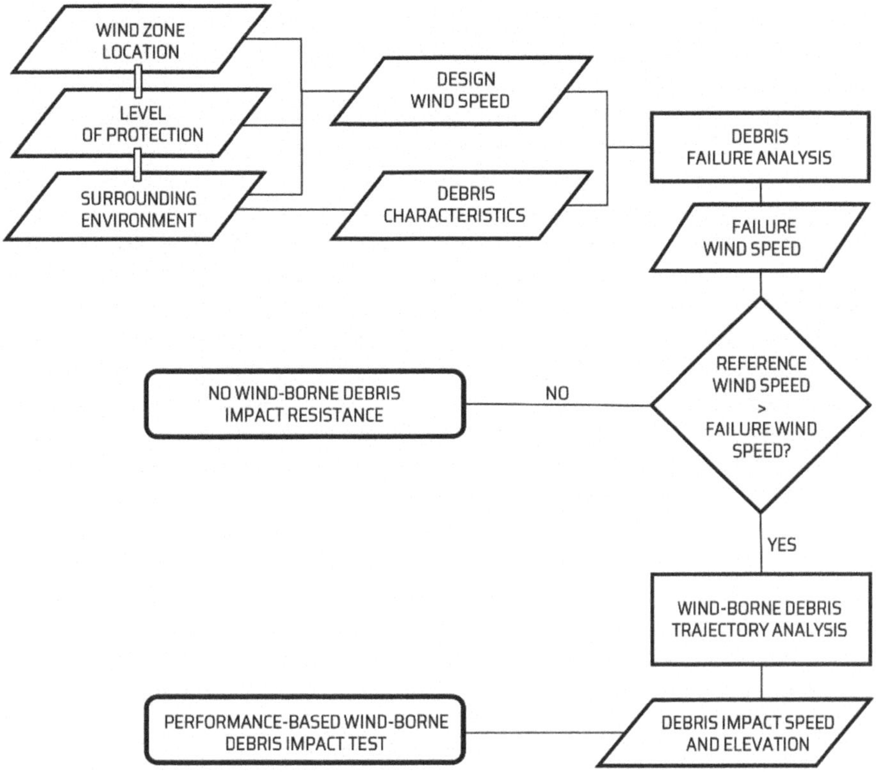

Fig. 1 Flowchart of the design framework for the identification of performance-based impact tests for exterior windows, curtain walls, doors, and impact protective systems impacted by wind-borne debris [26]

free to fly in a uniform flow, thus, Kordi [48] improved the representation of real-world conditions by introducing appropriate boundary and initial conditions. The experimental database that has been developed at Western University represents a unique example of data collection on the specific research topic of wind-borne roof tiles' flight analysis. Based on the wind tunnel experiments, Kordi et al. [43] defined a numerical model to estimate debris trajectories and velocities with sufficient accuracy, to capture the effects of initial conditions in terms of turbulence and the building wake. They found that the 3-s gust failure wind speed represents a practical and reasonable upper-bound wind speed to estimate the upper-bound flight trajectory, but that it is overestimated the mean trajectory. Somehow, debris elements can miss the wind gust and, consequently, when the failure occurs, they accelerate and fly in "lower wind speeds environments" [49]. This has been explained by the effect of a negative vertical component of the wind velocity, which is not considered in the numerical calculations [43]. This vertical component is linked to the wake region behind the building, which reduces the maximum below the uniform, smooth flow trajectory results for debris. For plate-like debris [34], it has been found [43, 49] that the lower bound wind speed to estimate debris' trajectories, is the average 10-min wind speed at source building roof height.

Mejorin [26] developed an alternative design framework for wind-borne debris-impact-resistance of façades. It recognizes that the current best practices in terms of impact testing to simulate wind-borne debris in windstorm events are based on the ASTM standard procedures [23, 24]. These adopt standard projectiles and impact velocities: the lumber projectile (large missile) and the steel ball projectile (small missile), representative, respectively, of a piece of the balloon frame structure, and roof gravel used on flat-roof constructions. The impact speeds used to conduct these tests have been developed through an empirical approach and the performance-based design method shown in this document (Fig. 1) sets a new approach to establishing façade impact testing performance, to provide site-specific wind-borne debris impact resistance of building envelopes, based on building aerodynamics, wind-borne debris trajectory analysis, and risk assessments. Mejorin [26] offers façade designers a solution to relate to local wind-borne debris types and specific design goals. The impact energies on target façades are, accordingly, evaluated by estimating the trajectory and the velocity of specific debris elements after the failure occurs. The ASTM standard requirements [23, 24] already give façade designers the chance to go through engineering assumptions to develop ad-hoc wind-borne debris impact tests for their projects, adopting an "other missile" for the impact test. A façade design tool for wind-borne debris resilience of the building envelope is generalized in Mejorin [26]. The alternative design strategy lays its foundations in the study of the aerodynamics of wind-borne debris, to guide designers in evaluating site-specific situations, and to achieve building envelope resilience to local exogenous objects. The overall result is, accordingly, the development of a design tool for the identification of alternative testing requirements, which uses the impact design inputs achieved through aerodynamics and building technology considerations, to reach the objective of improving built environment resilience to wind-borne debris impact. The impact of the presented design framework to identify alternative wind-borne debris façade impact test requirements is significant in the development of new testing standards, to be introduced in the standardization regulation framework, at the voluntary level and regulatory level.

- **Quantification of Source-Building Aerodynamics Effects on Debris Trajectories**

Recent work by Mejorin and Kopp [50] further advances the understanding of wind-borne debris trajectories by explicitly quantifying the role of source-building aerodynamics in debris flight initiation and development. While earlier analytical approaches typically assumed uniform flow conditions, the study demonstrates that neglecting roof-level flow modification, wake effects, and turbulence-debris interaction can lead to systematic overestimation of both mean and upper-bound debris flight distances.

Building upon experimental datasets developed at Western University for roof covering elements released from low-rise buildings [43, 48, 49], the study introduces the concept of an equivalent uniform wind speed, U_{debris}, defined as the uniform flow velocity that reproduces experimentally observed debris trajectories when used within classical debris-flight equations [50]. This approach preserves the analytical structure of established trajectory models while embedding corrections that account for realistic flow conditions at the debris source.

- **Debris Response Time, Gust Factors, and Trajectory Scaling**

A key contribution of Mejorin and Kopp [50] is the introduction of a debris response time, t^*, representing the characteristic time required for a debris element to accelerate from rest to a significant fraction of its asymptotic flight velocity. This parameter captures the physical tendency of debris elements to either fully engage with, or partially miss, peak gusts responsible for failure.

By linking t^* to gust statistics, a debris-specific gust factor G is derived, allowing the governing wind speed for debris flight to be expressed as Eq. 1:

$$U_{debris} = G\overline{U}F_R \tag{1}$$

where $\overline{U}$ is the mean failure wind speed and F_R is a roof/wake factor accounting for local flow modification due to roof geometry, position on the roof, wind direction, and surrounding buildings [50]. This formulation provides a physically grounded explanation for previously observed discrepancies between experimental debris trajectories and analytical predictions based solely on peak gust wind speeds.

Monte Carlo simulations calibrated against experimental datasets show that the proposed formulation significantly improves agreement with observed mean trajectory distributions, while highlighting inherent limitations in predicting extreme outlier trajectories using simplified analytical models.

- **Implications for Performance-Based Impact Testing**

The findings of Mejorin and Kopp [50] have direct implications for performance-based façade design and the definition of alternative impact test requirements. By enabling estimation of debris impact velocities that are consistent with realistic source-building aerodynamics, the framework supports the rational selection of non-standard missiles (mass, geometry, and impact speed) aligned with site-specific debris hazards.

This approach provides a quantitative underpinning for the performance-based provisions outlined in Annex G of ISO 16316 [15], where engineering analysis is permitted to justify alternative impact testing scenarios beyond standardized projectiles. In particular, the use of U_{debris} allows designers to bridge the gap between debris generation mechanisms at the source building and impact demand on the target façade, strengthening the scientific basis for resilience-driven design decisions.

The trajectory-based methodology introduced by Mejorin [26] is further refined and quantitatively validated in Mejorin and Kopp [50], where debris response time, gust statistics, and roof-level flow effects are explicitly integrated into a unified framework for debris flight assessment.

4 Research Findings and Needs

To clarify the main limitations of current wind-borne debris impact requirements and to identify priority directions for improvement, Table 4 synthesizes the principal gaps across the main reference standards and regulatory approaches discussed in this chapter.

There is a lack of information regarding the impact performance of façades for projectiles beyond those currently standardized. Moreover, existing data on standardized projectiles is limited, particularly concerning tests involving debris elements impacting façades at various angles of incidence. To thoroughly investigate the performance of building components, it is essential to focus on their mechanical behavior under a wide range of impact angles. Varying the impact angle is expected to alter the mechanical resistance of the system and/or the mechanical properties of the constituent materials [51]. This results in a probability density assessment for failure that is not currently accounted for in existing testing procedures.

In the field of wind engineering, the limitations of wind-borne debris impact characteristics primarily stem from restricted trajectory analyses and the paucity of available databases (e.g., [49]). Existing datasets for validating numerical debris trajectory calculations are limited, with most studies failing to consider real-world scenarios [45] where debris originates from a source building. Developing a comprehensive database on various debris types, categorized by Tachikawa numbers [46], is necessary to associate these with building components or urban objects susceptible to failure in extreme winds. Investigating the source building is crucial, requiring studies on the failure and trajectory of debris elements from both low-rise and high-rise buildings, encompassing various roofing solutions and roofing shapes. Tall buildings still need to be experimentally tested for wind-borne debris generation in wind tunnel experiments.

Recent work by Mejorin and Kopp [50] partially addresses this gap for low-rise buildings by proposing a calibrated approach to debris flight assessment that explicitly accounts for source-building aerodynamics, debris response time, and roof/wake effects. In particular, the definition of an equivalent uniform wind speed, U_{debris}, provides a practical pathway to reconcile analytical debris-flight models with experimental datasets in which both failure initiation and post-failure flight occur within the flow field modified by the source building. This contribution highlights a broader research need: developing comparable calibration methodologies and datasets for additional debris typologies (including compact and rod-like debris [34]) and for a wider set of source-building geometries and urban contexts.

The current ASCE 7-22 [19] requirements specify geographic areas susceptible to wind-borne debris without a thorough wind engineering analysis. The author suggests that these regions should be updated through detailed scenario analyses to determine the potential for debris to travel longer distances from coastal areas.

A related implication emerging from Mejorin and Kopp [50] is that scenario-based updates of debris-prone regions would benefit from integrating physics-informed

Table 4 Synthesis of current limitations and areas for improvement in wind-borne debris impact requirements

Code/standard	Strength	Main limitations	Areas for improvement
ASCE/SEI 7-22 [19]	Establishes design requirements for hurricane- and tornado-prone regions; distinguishes between risk categories and hazard types; links protection requirements to code-based design practice	Wind-borne debris regions and protection requirements remain largely prescriptive and geographically delimited, without detailed project-specific trajectory analysis or local debris-source characterization	Introduce more explicit performance-based pathways for debris hazard characterization, including site-specific debris-source assessment, trajectory analysis, and scenario-based regional updates
ASTM E1886 [23]	Provides established test procedures for impact and post-impact cyclic pressure performance of façade and fenestration systems; internationally recognized benchmark method	Test methodology is based on a limited number of standardized missiles and nominal impact conditions, which may not represent local debris typologies, incidence angles, or realistic aerodynamic conditions	Expand the framework for alternative missiles through engineering analysis, including non-standard debris geometries, impact angles, and velocity distributions derived from wind engineering assessments
ASTM E1996 [24]	Defines missile types and impact velocities according to wind zones and protection levels; provides a practical classification structure for U.S. applications	Classification is linked to a small family of conventional projectile assumptions (steel ball and lumber missile), with limited flexibility to represent other common debris sources such as roof coverings, façade fragments, or urban objects	Support broader classification options for project-specific debris hazards, including calibrated “other missile” criteria and guidance for selecting representative local debris types
ICC/NSSA 500 [22]	Provides robust impact requirements for storm shelters and safe rooms, with strong life-safety orientation and severe missile criteria for tornado conditions	Focuses on near-absolute protection for shelters rather than on the broader performance differentiation needed for conventional buildings and façades	Develop complementary performance-based guidance for non-shelter buildings that bridges the gap between conventional façade design and shelter-level protection philosophies

(continued)

Table 4 (continued)

Code/standard	Strength	Main limitations	Areas for improvement
AS/NZS 1170.2 [20]	Includes debris-impact-related provisions for cyclone-prone regions and recognizes façade vulnerability in severe wind climates	Does not require cyclic pressure testing after debris impact for façades; debris resistance provisions are limited in vertical extent and debris representation	Integrate post-impact performance verification and reassess debris typologies and height-related exposure assumptions, particularly for smaller dense projectiles and taller buildings
ISO 16316:2024 [15]	Provides an international framework for wind-borne debris impact testing; introduces substitution criteria and Annex G performance-based design directions	Although Annex G opens the door to performance-based design, practical implementation still depends on limited databases, few validated trajectory models, and scarce calibration data for real debris types and urban conditions	Strengthen implementation guidance for Annex G through validated case studies, debris databases, probabilistic tools, and examples of alternative impact-test derivation for practice
Scott et al. [27]	Increasing recognition of the need for PBWD and resilience-driven envelope design	Market adoption remains strongly prescriptive; façade testing is often disconnected from surrounding-site debris risk, source-building failure mechanisms, and resilience-based project objectives	Promote integration between façade engineering, wind engineering, and risk assessment so that testing criteria can reflect realistic site exposure, asset criticality, and resilience targets
Mejorin [26]	Framework for PBD identification of alternative impact test protocols for building enclosures to enhance their resilience to wind-borne debris	Limited experimental databases exist for real debris generation, source-building effects, non-uniform wind fields, impact angle variation, and non-synoptic wind scenarios	Prioritize experimental and numerical research on debris generation, flight, impact energy, response time, and source-building aerodynamics to support future standard revisions and performance-based test methods

trajectory scaling rather than relying solely on conservative uniform-flow assumptions. Specifically, where current delineations implicitly assume debris flight is driven by peak gusts acting uniformly over the duration of flight, the response-time concept suggests that different debris elements may experience markedly different effective wind environments after failure, influencing both travel distance distributions and expected impact speeds at target façades.

Turbulence, associated with instantaneous wind velocity peaks, significantly influences both uplift loads and debris flight post-failure. The design framework presented by Mejorin [26] addresses turbulence effects with a statistical approach, providing a range of results instead of a deterministic output. This framework simplifies the analysis compared to real-world scenarios. For debris flight assessment, this design framework can verify current requirements for steel ball (representing roof gravel) and lumber (representing structural members of balloon frame construction) projectiles, in simplified problems. The ASCE 7-22 [19] requirements extend wind-borne debris impact testing to 1.6 km from the coastline, necessitating assessment of whether rod-like debris can travel greater distances based on source location. Expanding assessments to include roof components, frequently observed to fail and become airborne in extreme winds, should cover different roof tile types and asphalt shingles.

In this context, the response-time framework introduced in Mejorin and Kopp [50] also points to a critical need for debris-type-specific gust characterization for both synoptic and non-synoptic winds. For debris elements with short response times, localized speed-up effects and near-roof flow structures may dominate the effective flight wind speed, whereas heavier or slower-responding elements may "miss" peak gusts and be governed by lower effective wind speeds. Expanding experimental campaigns to systematically quantify response times across common roof-cover typologies would therefore support more defensible impact-energy estimates and reduce uncertainty in performance-based "other missile" test definitions.

Trajectory calculations for wind-borne debris face limitations due to analytical models not considering the actual wind field at the target building location, particularly in complex urban environments. Computational fluid dynamics (CFD) methods are widely used to assess design parameters for target buildings and urban environments, including flow-generated noise, thermal, and ventilation analyses. In extensively studied urban environments, CFD models have been validated through experimental campaigns. Integrating current best practices for urban design and CFD modeling with wind-borne debris generation, flight, and impact analyses on urban building envelopes would be beneficial. Accordingly, CFD software can be used for debris trajectory simulation after a significant dataset has been analyzed to validate the software analytically and identify correct boundary conditions for debris failure and flight analysis.

The calibration strategy presented in Mejorin and Kopp [50] can also be interpreted as a bridge between simplified analytical trajectory models and higher-fidelity CFD approaches: it provides a structured method to embed the net effects of roof boundary layer development and wake interference into reduced-order calculations. A priority research direction is to generalize this approach by coupling validated CFD flow fields (for representative urban morphologies and building typologies) with probabilistic debris initialization and flight modeling, enabling the development of transferable correction factors (or surrogate models) that can be used in practice without resorting to full CFD for every project.

Finally, research needs include the fact that extreme wind events can be both synoptic and non-synoptic. There is currently no available experimental database for failure and trajectory analysis from a source construction and subsequent flight for the three debris types identified by Wills et al. [34], for non-synoptic winds.

5 Conclusions

While both hurricane and tornado wind-borne debris impact test requirements emphasize protection, differences exist in the standards and methodologies. Hurricane requirements focus on glazing protection in Risk Category buildings within designated regions, with testing standards based on ASTM E1996 [24]. Tornado requirements, on the other hand, emphasize the need for absolute protection in storm shelters and safe rooms, with standards based on ICC/NSSA 500 [22]/FEMA P-361 [36]. From a performance-based design standpoint, a key limitation of today's standardized missile tests is that they encode only a small number of debris typologies and assumed impact speeds, without explicitly representing the source-building flow mechanisms that govern realistic debris acceleration and flight. The calibrated trajectory methodology proposed in Mejorin and Kopp [50] reinforces that the wind environment relevant to debris flight is often not well represented by uniform peak-gust assumptions, particularly for roof-cover debris originating on low-rise buildings where roof boundary-layer effects and wakes are influential. In conclusion, effective protection against wind-borne debris in both hurricane and tornado-prone regions requires adherence to specific building codes and standards tailored to the respective hazards, ensuring the safety and integrity of structures and occupants during severe wind events. The ICC/NSSA 500 [22] provides comprehensive guidelines for designing storm shelters capable of withstanding the extreme wind pressures and debris hazards associated with hurricanes and tornadoes. Adherence to these standards is essential to ensure the safety and effectiveness of storm shelters in protecting occupants during severe weather events. This standard can be taken as a reference, together with the design framework developed by Mejorin [26] and reported in Annex G of ISO 16316 [15], to come out with performance-based design requirements for tornado and hurricane-prone buildings. Latest performance-based design directions highlighted that there is an urgent necessity [27] for enhanced impact requirements of the building envelope, which is highly affected by wind-borne debris impact in windstorm events. The current best practices in terms of impact testing to simulate wind-borne debris in windstorm events are based on the ASTM standard procedures [23, 24]. The design framework [26] reported in this document (Fig. 1) sets a new approach to establishing façade impact testing performance, to provide site-specific wind-borne debris impact resistance of building envelopes, based on building aerodynamics, wind-borne debris trajectory analysis, and risk assessments. This research offers façade designers a solution to relate to local wind-borne debris types and specific design goals. In particular, by introducing U_{debris} and debris response time t^*, Mejorin and Kopp [50] offer a quantitative pathway for translating experimentally observed debris flight behavior into engineering-design inputs (impact velocity distributions, likely incidence conditions, and kinetic energy ranges) that are compatible with the "other missile" provisions already allowed in ASTM E1886 [23]/ASTM E1996 [24] and with the performance-based direction formalized in ISO 16316 [15], Annex G. This supports a more explicit linkage between (i) local debris sources and failure mechanisms, (ii) trajectory-informed impact demand on façades, and (iii) defensible alternative impact test criteria, thereby strengthening the technical basis for resilience-driven façade design beyond prescriptive baseline requirements.

References

1. Intergovernmental Panel on Climate Change (IPCC) (2022) Climate change 2022: mitigation of climate change. In: Working group III contribution to the sixth assessment report. IPCC
2. Munich Re (2022) Hurricanes, cold waves, tornadoes: weather disasters in USA dominate natural disaster losses in 2021. Munich Re
3. Gergis J (2019) The terrible truth of climate change. The Monthly, Aug 2019
4. Sisco MR, Bosetti V, Weber EU (2017) When do extreme weather events generate attention to climate change? Clim Change 143(1–2):227–241
5. Munich Re (2021) Record hurricane season and major wildfires—the natural disaster figures for 2020. Munich Re
6. Lu S, Bai X, Zhang X, Li W, Tang Y (2019) The impact of climate change on the sustainable development of regional economy. J Clean Prod 233:1387–1395
7. Brody SD, Zahran S, Vedlitz A, Grover H (2008) Examining the relationship between physical vulnerability and public perceptions of global climate change in the United States. Environ Behav 40(1):72–95
8. Eckstein D, Künzel V, Schäfer L (2021) Global climate risk index 2021: who suffers most from extreme weather events? Germanwatch
9. World Economic Forum (2022) The global risks report 2022, 17th edn. World Economic Forum
10. Chronis T, Papadopoulos V, Nikolopoulos EI (2011) QuickSCAT observations of extreme wind events over the Mediterranean and Black Seas during 2000–2008. Int J Climatol 31:2068–2077
11. European Severe Storms Laboratory (ESSL) (2020) Annual report 2019. ESSL
12. National Institute of Standards and Technology (NIST) (2014) Measurement science R&D roadmap for windstorm and coastal inundation impact reduction. NIST GCR 14-973-13. NIST, Gaithersburg
13. Butler K, Kareem A (2012) Anatomy of glass damage in urban areas during hurricanes. In: Advances in hurricane engineering: learning from our past. ASCE
14. International Organization for Standardization (ISO) (2018) Contributing to the UN sustainable development goals with ISO standards. ISO, Geneva
15. International Organization for Standardization (ISO) (2024) ISO 16316:2024 windows, doors and curtain walling—impacted by wind-borne debris in windstorms—test method and classification. ISO, Geneva
16. United Nations (UN) (2022) The sustainable development goals report 2022. UN
17. Kuligowski ED, Lombardo FT, Phan L, Levitan ML, Jorgensen DP (2014) Technical investigation of the May 22, 2011 tornado in Joplin, Missouri. NIST NCSTAR report. NIST
18. International Code Council (ICC) (2020) Florida building code: building. ICC

19. ASCE/SEI (2022) Minimum design loads and associated criteria for buildings and other structures (ASCE/SEI 7-22). ASCE
20. Standards Australia, Standards New Zealand (2021) AS/NZS 1170.2: structural design actions—part 2: wind actions. Standards Australia/Standards New Zealand
21. Laboy S, Smith D, Fernandez G, Masters F (2012) Residential fenestration vulnerability to wind-borne debris. In: Advances in hurricane engineering: learning from our past. ASCE
22. ICC/NSSA (2023) ICC/NSSA 500 standard for the design and construction of storm shelters. ICC
23. ASTM International (2019) ASTM E1886-19 standard test method for performance of exterior windows, curtain walls, doors, and impact protective systems impacted by missile(s) and exposed to cyclic pressure differentials. ASTM International
24. ASTM International (2023) ASTM E1996-23 standard specification for performance of exterior windows, curtain walls, doors and storm shutters impacted by wind-borne debris in hurricanes. ASTM International
25. ASCE/SEI (2019) Prestandard for performance-based wind design. ASCE
26. Mejorin A (2022) Wind-borne debris resistance of façades: identification of alternative impact test requirements. Ph.D. thesis, Western University and Università Iuav di Venezia
27. Scott D, Goupil J, Burton M, Denoon R, Larsen R, Spence SMJ, Wu T (2023) Advancement in performance-based wind design: workshop report. NIST GCR 23-045-upd1. NIST
28. Minor JE (1994) Wind-borne debris and the building envelope. J Wind Eng Ind Aerodyn 53:207–227
29. FEMA (1993) Building performance: hurricane Andrew in Florida. FEMA
30. Cyclone Testing Station (CTS) (2017) Technical report No. 63: tropical cyclone Debbie damage to buildings. CTS
31. ASCE (2016) Wind engineering for natural hazards. ASCE
32. Trabucco D, Mejorin A, Miranda W, Nakada R, Troska C, Stelzer I (2017) Cyclone-resistant glazing solutions in the Asia-Pacific region. In: GPD 2017 proceedings, pp 64–69
33. Moghim F, Caracoglia L (2012) A numerical model for wind-borne compact debris trajectory estimation. Part 1: probabilistic analysis. Eng Struct 38:153–162
34. Wills JAB, Lee BE, Wyatt TA (2002) A model of wind-borne debris damage. J Wind Eng Ind Aerodyn 90:555–565
35. Kopp GA, Morrison MJ, Kordi B, Miller C (2011) A method to assess peak storm wind speeds using detailed damage surveys. Eng Struct 33(1):90–98
36. FEMA (2021) FEMA P-361 safe rooms for tornadoes and hurricanes. FEMA
37. Twisdale LA, Dunn WL, Davis TL (1979) Tornado missile transport analysis. Nucl Eng Des 51(2):295–308
38. Huo S, Hemida H, Sterling M (2020) Numerical study of debris flight in a tornado-like vortex. J Fluids Struct 99:103–134
39. Huo S, Hemida H, Sterling M (2023) A numerical investigation of debris initialisation in tornado-like wind fields. J Fluids Struct 121
40. Zhao J, Yan G, Han D (2021) A review of approaches to simulate wind-borne debris dynamics in wind fields. J Wind Eng Ind Aerodyn 212
41. Doddipatla LS, Kopp GA (2019) A review of critical scouring velocity of compact roof aggregate. J Wind Eng Ind Aerodyn 188:110–124
42. FEMA (2010) FEMA P-424 design guide for improving school safety in earthquakes, floods, and high winds. FEMA
43. Kordi B, Traczuk G, Kopp GA (2010) Effects of wind direction on the flight trajectories of roof sheathing panels under high winds. Wind Struct 13(2):145–167
44. Grayson M, Pang WC, Schiff S (2012) Three-dimensional probabilistic wind-borne debris trajectory model for building envelope impact risk assessment. J Wind Eng Ind Aerodyn 102:22–35
45. Tachikawa M (1983) Trajectories of flat plates in uniform flow with application to wind-generated missiles. J Wind Eng Ind Aerodyn 14:443–453

46. Holmes JD, Baker CJ, Tamura Y (2006) Tachikawa number: a proposal. J Wind Eng Ind Aerodyn 94:41–47
47. Lin N, Letchford C, Holmes JD (2006) Investigation of plate-type wind-borne debris. Part I: experiments in wind tunnel and full scale. J Wind Eng Ind Aerodyn 94:51–76
48. Kordi B (2009) Aerodynamics of wind-borne plate debris. Ph.D. thesis, University of Western Ontario
49. Kordi B, Kopp GA (2011) Effects of initial conditions on the flight of wind-borne plate debris. J Wind Eng Ind Aerodyn 99:601–614
50. Mejorin A, Kopp GA (2024) Estimation of debris flight trajectories of roof cover from low-rise buildings. Front Built Environ 10:1428693
51. Mejorin A, Trabucco D, Stelzer I (2019) Cyclone-resistant façades: best practices in Australia, Hong Kong, Japan, and the Philippines. CTBUH research report 160. Images Publishing

Development—Methods, Systems, and Innovation

Computer Simulation for Wind-Induced Noise Analysis on Building Facade

Alessandro Gambale, Sampath Kumar Raghunathan Srikumar, and Gabriele Mosca

Abstract Wind-induced noise on building facades presents a growing challenge to occupant comfort in dense urban environments, particularly with the increasing use of complex, functional, or decorative cladding. Traditional wind tunnel testing, while accurate, is often too costly and slow for early-stage design iterations. This paper demonstrates the efficacy of Computational Fluid Dynamics (CFD) as a robust numerical alternative for predicting aeroacoustic interactions. We propose a cost-effective hybrid workflow that combines micro-scale aerodynamic characterization of individual facade elements with macro-scale, full-building simulations. By integrating steady-state RANS turbulence statistics with Proudman's acoustic analogy, the method bridges the gap between geometric detail and computational efficiency. Validated against experimental data with a precision of $\pm$ 3 dB, this approach was applied to the KORN 16 high-rise project. The analysis successfully identified high-risk noise zones in the initial mesh design and guided the transition to a silent, optimized grating. The study concludes that this digital framework allows architects to visualize the risk of wind-induced noise early in the design process, ensuring a balance between aesthetic vision and environmental acoustic comfort.

Keywords Aeroacoustics · Computational Fluid Dynamics (CFD) · Building facade · Wind-induced noise · Building envelope · Building design

Nomenclature/Abbreviations

ABL	Atmospheric Boundary Layer
CFD	Computational Fluid Dynamics
dB	Decibel
DNS	Direct Numerical Simulation
LES	Large Eddy Simulation

A. Gambale (✉) · S. K. R. Srikumar · G. Mosca
BuildWind SRL, Rue Bara 175, 1070 Brussels, Belgium
e-mail: info@buildwind.net

A. Mejorin et al. (eds.), *Performance-Based Façade Design*, Springer Tracts in Civil Engineering, https://doi.org/10.1007/978-3-032-28835-6_5

RANS	Reynolds-Averaged Navier–Stokes
SPL	Sound Pressure Level
SST	Shear Stress Transport
SWL	Sound Power Level
TRL	Technology Readiness Level

1 Introduction

The rapid urbanization and the increasing height of skyscrapers have introduced new challenges in building envelope design. While structural integrity and thermal performance have long been primary concerns, the acoustic performance of building facades under wind loads has emerged as a critical comfort parameter.

Wind flowing over complex facade elements, such as sun shading louvers, perforated screens, and decorative fins, can generate significant aeroacoustic noise. This noise can penetrate the building envelope, causing annoyance to occupants and potentially violating local noise regulations. Notable historical failures highlight the importance of this analysis. For instance, the Beetham Tower in Manchester (171 m) became infamous for the "hum" generated by its glass plates [10]. Similarly, the Het Strijkijzer in The Hague (130 m) [11] and Cityspire Center in New York [3] experienced noise issues linked to steel grids and decorative domes, respectively.

Traditional wind tunnel testing, while accurate, is often prohibitively expensive and time-consuming due to the iterative nature of the early design phase. Consequently, architects and engineers are increasingly turning to numerical simulation. This paper explores the application of Computational Fluid Dynamics (CFD) combined with acoustic analogies to predict wind-induced noise. By adopting a hybrid computational approach, we demonstrate how potential noise sources can be identified and mitigated virtually, long before physical construction begins.

2 Theoretical Framework and Background

Noise is an inherently unsteady phenomenon: sound is generated by time-varying pressure fluctuations. When wind flows past a building facade, it is the time-dependent swirling of air that generates these fluctuations. If the flow were perfectly smooth and unchanging (steady), it would be silent. Therefore, the generation of sound is inextricably linked to the 'unsteadiness' or turbulence intensity of the wind.

Standard aerodynamic analysis of buildings typically relies on steady-state simulations to determine structural loads and pedestrian comfort. These methods are designed to calculate average conditions over time. However, this approach presents a challenge for acoustics. Because steady-state simulations mathematically average out the flow fluctuations to save computational time, they effectively remove the very data needed to predict sound directly. While they are excellent for predicting mean wind speeds, they do not inherently capture the rapid, time-varying pressure pulses that our ears detect as noise.

To address this, we present a methodology that assesses wind-induced noise using the turbulence statistics provided by steady-state Reynolds-Averaged Navier–Stokes (RANS) models. By correlating these steady metrics with Lighthill's aeroacoustic analogy [8], we predict where noise is likely to occur without the heavy computational cost of transient simulations, making this method a viable tool for rapid architectural design iterations.

2.1 *Aeroacoustic Mechanisms*

The generation of sound by fluid flow is governed by the interaction between flow turbulence and solid boundaries. In the context of building facades, noise is often primarily caused by vortex shedding.

Vortex shedding is an oscillating flow phenomenon that occurs when a fluid flows past a bluff (i.e., non-streamlined) body. As wind encounters a facade element, the flow separates from the surface of the object, and the separation creates a repeating pattern of swirling vortices. Each time a vortex sheds, it creates fluctuating pressure zones that act as a sound source if the frequency of this shedding lies within the audible range (20 Hz–20 kHz). The frequency of the sound generated by vortex shedding is governed by the geometry of the object and the speed of the wind, defined by the dimensionless Strouhal number:

$$f = \frac{St \cdot U}{d} \tag{1}$$

where

- f is s the frequency of vortex shedding (in Hz),
- St is the Strouhal (dimensionless) number,
- U is the flow velocity (e.g., wind speed in m/s),
- d is the characteristic dimension (diameter or thickness of the element in m).

The formula implies that as wind speed U increases, the pitch of the noise increases linearly. This is why a building facade might emit noise at a higher pitch during a gust. Smaller elements (i.e., lower d values) shed vortices at higher frequencies, while larger cross-sections lead to slower vortex separation, lowering the pitch of the aerodynamic noise.

In addition to vortex shedding, which produces coherent tonal noise, an important secondary aeroacoustic source in building facades is turbulence-generated noise. Unlike vortex shedding, which arises from organized, periodic flow structures, turbulence-generated noise originates from incoherent, stochastic turbulent fluctuations within the flow field. As wind interacts with facade surfaces, boundary layers transition to turbulence and separate at edges, corners, surface roughness elements, joints, and geometric discontinuities. The resulting turbulent eddies of varying scales generate unsteady, spatially distributed pressure fluctuations on solid surfaces, which act as broadband acoustic sources.

This form of noise is broadband in nature, lacking discrete tonal components, and its spectral content is governed by the intensity of turbulence and the characteristic length and time scales of the turbulent structures. Turbulence-generated noise is particularly important in facade aerodynamics because it is ubiquitous across large surface areas, does not require bluff-body geometries, and can dominate the overall acoustic signature of complex building envelopes, especially in high-wind conditions and in densely built urban environments.

The numerical prediction of aerodynamically generated sound is built upon the foundational work of Michael J. Lighthill, specifically his general theory of aerodynamic noise [8]. Lighthill's acoustic analogy reformulates the exact conservation equations of mass and momentum into a single inhomogeneous wave equation. The analogy is exact, but its application requires high-fidelity, time-resolved data of the entire flow field, typically only provided by Direct Numerical Simulation (DNS) or Large Eddy Simulation (LES), which is computationally expensive for urban-scale building aerodynamics. Therefore, the direct application of Lighthill's analogy in a practical engineering context requires simplifying assumptions to link the sound power to time-averaged, readily available turbulence quantities.

This is the critical step where Proudman's formula [12] is introduced. Proudman was one of the first to apply Lighthill's theory to the specific case of isotropic turbulence. Isotropic turbulence, while a theoretical idealization, allows for a powerful simplification. Proudman demonstrated that the total acoustic power generated per unit volume in a low-Mach-number flow can be approximated as being proportional to the turbulence intensity and the integral length scales of the turbulence. For the purposes of RANS simulation, this formulation is expressed in terms of the turbulent kinetic energy and its dissipation rate, which are the primary output variables of turbulence closure models.

This equation allows the use of computationally efficient, steady-state CFD results to infer the total acoustic energy. While the Proudman analogy cannot provide the frequency spectrum of the noise (i.e., the pitch), it provides a robust prediction of the overall sound power level (SWL), which is essential for assessing the magnitude of the noise nuisance.

Reynolds-Averaged Navier–Stokes (RANS) models are designed to solve for the time-averaged flow quantities, while modeling the effects of turbulence using additional transport equations. The omega Shear Stress Transport (SST) model is the specific turbulence closure chosen for the full-building simulations. The SST model is a hybrid formulation that blends the *k-omega* model near the wall (where viscous effects are dominant and the flow is critical for boundary-layer-based noise) with the *k-epsilon* model in the free stream. This provides robustness and accuracy for flows involving adverse pressure gradients and flow separation, which are characteristic features of wind flow around complex building facades.

The SST model directly computes the two key variables required for Proudman's formula: the turbulent kinetic energy k and the turbulent dissipation rate epsilon. k represents the average kinetic energy per unit mass associated with turbulent fluctuations, and *epsilon* is the rate at which this energy is dissipated into heat.

The aeroacoustic coupling step is a post-processing procedure: the converged steady-state flow field is extracted and used as the input to the regression model, which is based on the Proudman analogy.

3 Methodology and Approach

The approach adopted for this study aimed at performing accurate noise simulation at relatively low computational time and cost, so that it can be effectively integrated within the building design process workflow. The method is composed of two phases: Phase 1, consisting in the characterization of selected facade elements, and Phase 2, where the wind flow and noise generation on the whole building are simulated.

3.1 Phase 1: Facade Element Simulation

Before simulating the building of interest, the specific facade elements under investigation must be characterized. A detailed geometric model of the element or a section of it is simulated using Computational Fluid Dynamics (CFD), as if it was tested in a virtual wind tunnel. Simulations are performed for multiple inlet velocities (e.g. 5, 10, 15, 20 m/s), directions, and turbulence levels (e.g. 1, 5, 10, 15, 20%). The generated noise levels are then extracted, and a correlation is established between the local wind velocity, direction, turbulence level, and the generated Sound Pressure Level (*SPL*).

In order to compute the noise from steady-state wind data, Proudman's formula [12] is utilized, which was derived based on the works of Lighthill [8]. It approximates the sound intensity of isotropic turbulence for low Mach numbers in terms of steady-state turbulence quantities (turbulent kinetic energy k, dissipation rate ε). The formula is given as:

$$P_V = \alpha \rho \varepsilon \left(\sqrt{2k}/c \right)^5 \tag{2}$$

where P_V is the acoustic power in W/m^3, c is the speed of sound in air, ρ is the air density, and α is the model constant set to 0.1.

$$P = \frac{{p_{rms}}^2 A}{\rho c} \tag{3}$$

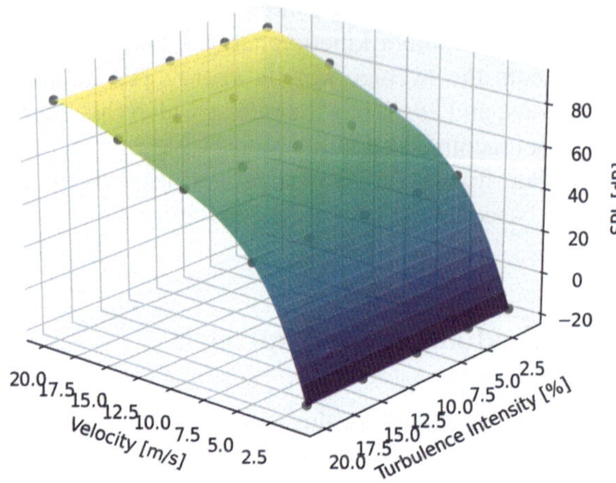

Fig. 1 Example of a regression model to estimate the noise generated by a section of the mesh cladding in *SPL* for varying inlet velocity and turbulence intensity. Figure created by the authors

$$SPL = 20\log_{10}\frac{p_{rms}}{p_{ref}} \tag{4}$$

The relationship between the acoustic power P in W and the RMS sound pressure (p_{rms}) is given in Eq. (3), where A is the area of the surface encompassing the sound source. Finally, the acoustic sound in terms of Sound Pressure Levels (SPL) can be obtained in decibels (dB) according to Eq. (4), with the reference pressure (p_{ref}) taken as 20 μPa.

Utilizing Eqs. (2), (3) and (4), wind-induced noise generated by elements on the building facade is computed for each inlet condition, and a regression model is built, which gives a relationship between *SPL*, inlet velocity, and turbulent intensity. An example is shown in Fig. 1.

3.2 Phase 2: Building Simulation

Once the regression models for wind-induced noise prediction have been established, the final step consists of their integration into full-scale building simulations. These full-scale simulations are conducted using steady-state, incompressible Reynolds-Averaged Navier–Stokes (RANS) formulations, employing customized atmospheric boundary layer (ABL) profiles for the inlet velocity and turbulence quantities, as well as for near-wall boundary conditions, in accordance with established methodologies [2, 13]. Turbulence closure is achieved through the use of the k–ω shear stress transport (SST) model, selected for its robustness in capturing near-wall effects and adverse pressure gradients relevant to building aerodynamics.

The computational domain is dimensioned to minimize blockage and artificial boundary influences, following the recommendations of the COST guidelines [5, 6]. In practice, this corresponds to a domain extent of at least 5H in the vertical direction,

5H upstream of the building, 15H downstream, and 5H in the lateral directions, where H denotes the tallest building height in the domain.

From the converged CFD solutions, local flow velocities and turbulence intensities are extracted in regions adjacent to the facade elements. These flow-derived quantities are subsequently provided as inputs to the regression model, which yields a spatially resolved distribution of wind-induced noise, expressed in terms of sound pressure level (*SPL*), across the facade.

3.3 Validation

The wind-induced noise assessment was validated by replicating the experimental study carried out by Feng [4]. The test setup consisted of a pressurized anechoic chamber connected by an opening to a reverberating chamber (Fig. 2) qualified by standard EN ISO 3741:2010. Several perforated panels measuring 0.4 m × 0.4 m, with a thickness of 1.5 mm, and with different circular hole sizes and distribution were placed in front of the opening, and noise sound power levels for different air speeds were measured.

In order to obtain a correct estimation of the acoustic sound power, an appropriate mesh resolution is required in the region of noise production. Therefore, a grid convergence study was conducted, where meshes with several cell refinements were tested and their corresponding noise results compared with the experimental data. As shown in Fig. 3, a cell refinement lower than or equal to 2 mm is necessary to get sound power in the ± 3 dB range (threshold of accuracy where variances are negligible to the human ear) of the experimental values.

Once the optimal mesh was selected, steady-state RANS simulations were performed with k–ω SST turbulence model for different hole sizes, and Proudman's

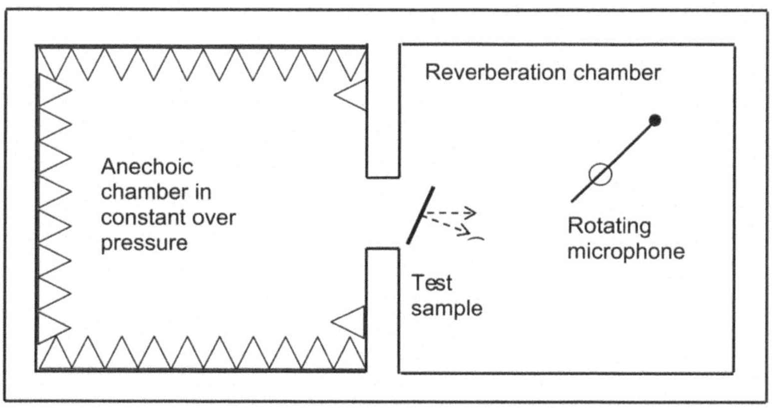

Fig. 2 Experimental test setup from Feng [4] used to validate the methodology

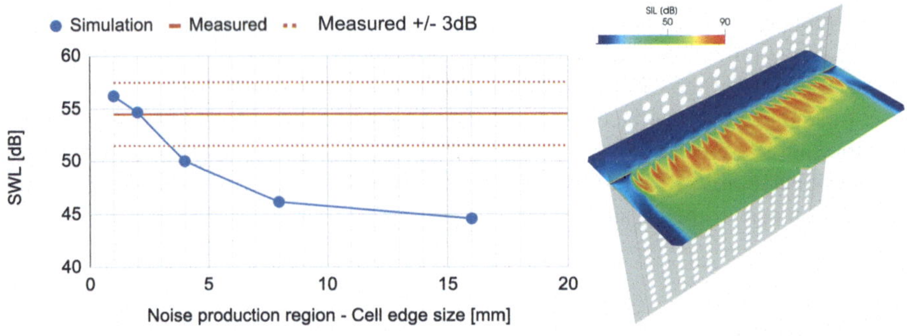

(a) (b)

Fig. 3 Comparison between sound power level (*SWL*) from simulations with different cell refinement and experimental measure (**a**). Sound intensity level (*SIL*) on a slice of the perforated plate using 1 mm cell refinement (**b**). Figure created by the authors

formula (Eq. 4) was utilized to compute the total noise generated for each case. A comparison of the results obtained from the CFD methodology with the experimental data is shown in Fig. 4. The noise production determined from the numerical simulations is within $\pm$ 3 dB of the experimental measurement. However, the simulation data also show that the current methodology is comparatively less sensitive to the variation in hole sizes, highlighting one of the main drawbacks of the current methodology.

Fig. 4 Sound power level (*SWL*) from simulations compared to experimental measures for different hole sizes of the perforated plate. Figure created by the authors

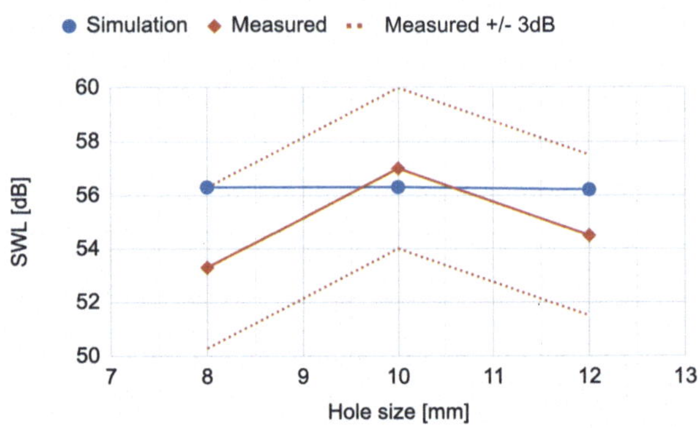

4 Results

4.1 Case Study: KORN 16

The KORN 16 project is a development located in Hasselt, Belgium. The building under investigation stands 65 m tall with a footprint radius of 17.5 m. Designed by *A-tract architecture* [1], the tower features extensive use of mesh cladding panels on the facade. Given the building's height and exposure, the client raised valid concerns about potential wind-induced noise nuisances for future occupants and neighbors (Fig. 5).

The KORN 16 building facade features geometrically complex mesh cladding. Explicitly meshing every individual wire for the full-building simulation would result in an extremely large computational mesh of approximately 2 billion cells. This would make wind and noise simulation computationally expensive and slow down the design process consisting in testing multiple candidate panels. This difficulty is overcome by using the approach described in Chap. 3: first, the facade elements are simulated to extract their properties that are then applied to the full simulation. In this case, a porous media model was employed to replicate the inertial and viscous effects of the panel openings through the addition of sink terms to the governing equations of fluid motion [9]. This way, instead of refining the perforation geometry explicitly, their effect is taken into account at a fraction of an additional cost: the resulting mesh size is approximately 60 million, which corresponds to about 3% of the computational cost (and time with equivalent resources) that would be required by fully representing the mesh claddings in the computational mesh.

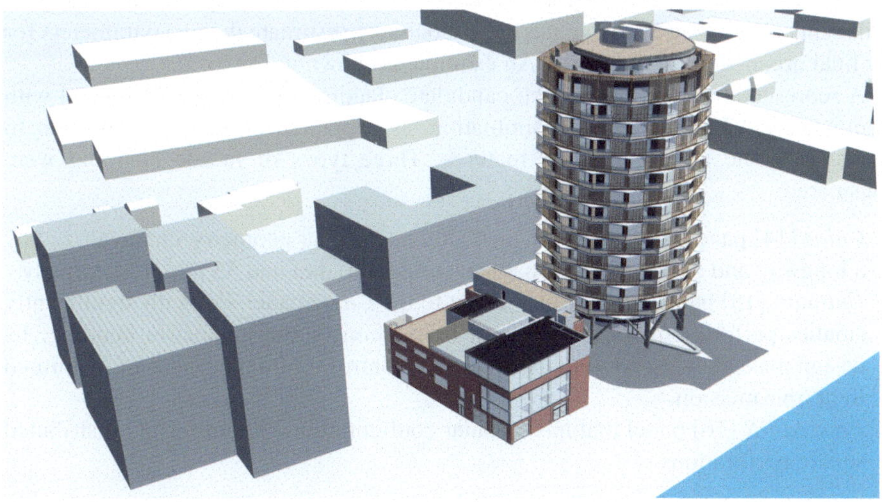

Fig. 5 KORN 16 project. Figure created by the authors, based on the architectural model provided by A-tract architecture [1]

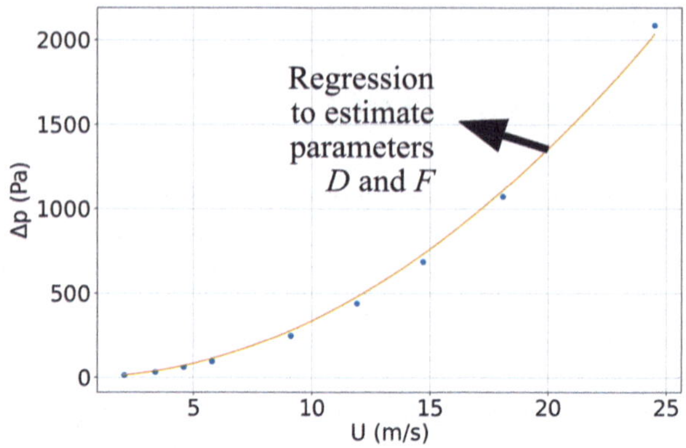

Fig. 6 Example of the regression performed to estimate the two parameters *D* and *F* of the Darcy–Forchheimer equation from pressure drop values for multiple CFD simulations with varying inlet speeds. Figure created by the authors

The Darcy–Forchheimer equation models a sink term added to the momentum equation as a pressure drop [7], as shown in Eq. (5), where μ and ρ are the dynamic viscosity and density of the air, respectively, U is the velocity field, and D and F are the model constants tuned for the porous media. The equation is characterized by the presence of a linear dependence on U, which is the Darcy contribution, and a quadratic dependence on U, which is the Forchheimer contribution.

$$\nabla p = -\left(\mu D U + \frac{1}{2}\rho F U^2\right) \tag{5}$$

To find the two parameters D and F of the porous media model, a portion of the entire mesh cladding was explicitly modeled, and multiple CFD simulations with different inlet speeds were performed. Based on the results, a regression model was built using the pressure drop values and velocities to estimate the two parameters for the final urban-scale simulations. An example is shown in Fig. 6.

A representative section of each candidate cladding system was examined with Computational Fluid Dynamics simulation, as explained in Sect. 3.1, in order to characterize the associated flow features. Three types of facade cladding were considered:

- *Chloe* [14] panel, that exhibits a distinctly non-planar geometry, characterized by a longway and shortway diagonal measuring 150 mm and 55 mm, respectively.
- *Cullinan* [15] panel, similar in concept to the *Chloe* panel, but with significantly smaller perforations and a different orientation of the geometric bending. Its design makes it very good to use in a solar control solution because of its limited light transmission.
- *Pressed RR* [16] panel that has a planar configuration with uniformly distributed square perforations.

All of them are commonly employed in residential building applications [14–16]. Figure 7 presents an image of the cladding panels and their relevant geometrical characteristics.

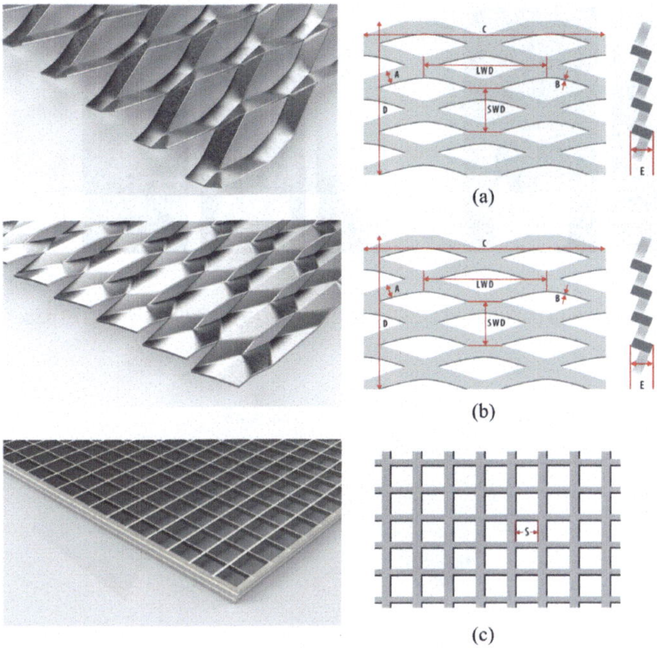

LWD	150 mm
SWD	55 mm
A	20 mm
B	1.5, 2.0 & 3.0 mm
E	21.6 mm

LWD	76 mm
SWD	24 mm
A	11 mm
B	2.0 mm
E	8.7 mm

S	33 mm

Fig. 7 Image and geometrical characteristics of the *Chloe* (**a**) [14], *Cullinan* (**b**) [15] and *Pressed RR* (**c**) [16] facade panel

Figure 8 illustrates the velocity magnitude distributions in the wake of both panels as obtained from the virtual wind tunnel simulations. Owing to its smaller perforation size and three-dimensional geometry, the *Cullinan* panel induces a pronounced acceleration of the flow and redirects it upward in the wake region. This strong acceleration is accompanied by a substantial increase in turbulence intensity and the formation of coherent vortical structures downstream of the panel. This behavior contrasts with that observed for the *Chloe* panel, for which the associated acceleration is significantly weaker, resulting in comparatively lower turbulence levels. In comparison, the *Pressed RR* panel exhibits negligible flow acceleration and correspondingly limited turbulence generation, consistent with its uniform, planar configuration.

Utilizing Eqs. (3), (4) and (5), wind-induced noise generated by each panel type was computed for each inlet condition, and a regression model was built, which gives a relationship between *SPL*, inlet velocity, and turbulent intensity (Fig. 9).

Subsequently, local velocity and turbulence quantities were extracted in regions adjacent to the building facades at locations corresponding to the prospective cladding installation. These flow parameters were then used as inputs to the previously described noise regression model. This procedure was repeated for all considered wind speeds and wind directions, and, when combined with the corresponding meteorological wind statistics, enabled the construction of a probabilistic map of wind-induced noise occurrence during the year.

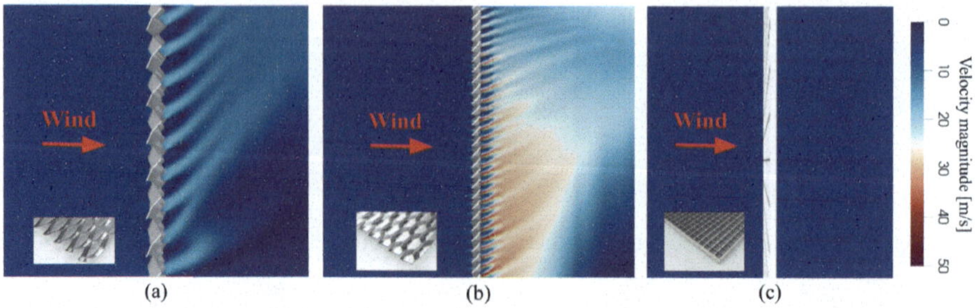

Fig. 8 Velocity magnitude contours in the wake of *Chloe* (**a**), *Cullinan* (**b**) and *Pressed RR* (**c**) panel. Figure created by the authors

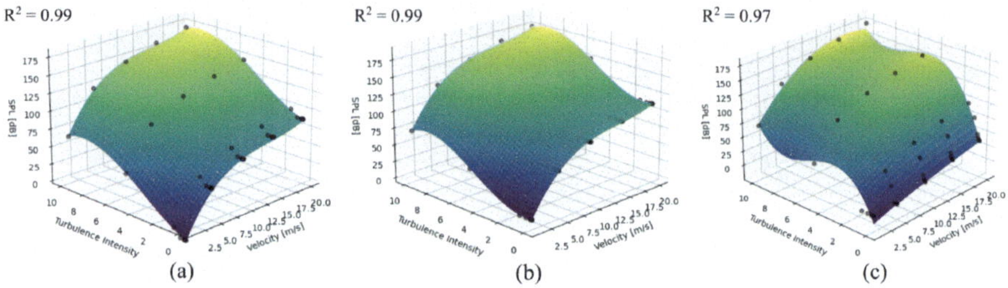

Fig. 9 Regression model to compute *SPL* for *Chloe* (**a**), *Cullinan* (**b**) and *Pressed RR* (**c**) panels. Figure created by the authors

Figure 10 presents representative probability maps of wind-induced noise for configurations employing the *Chloe* mesh cladding, evaluated for two sound pressure level (*SPL*) thresholds: 40 dB, corresponding to a quiet acoustic environment, and 60 dB, representative of typical conversational noise levels. For the 40 dB threshold, elevated probability regions are observed across the facades over the full building height, with the highest probabilities occurring near the rooftop. In these regions, the probability of exceeding the threshold is greater than 25%. When the higher threshold of 60 dB is considered, the majority of the facade exhibits probabilities below 5%, while the maximum values are again concentrated near the rooftop, reaching approximately 5–10%. These results indicate that, for the *Chloe* panel configuration, wind-induced noise is likely to occur, particularly in regions near the roof. While probabilities remain lower over most of the facade, non-negligible exceedance probabilities near the rooftop suggest a tangible risk of noise generation by the panel.

Figures 11 and 12 present the probability maps associated with the *Cullinan* and *Pressed RR* facade panels, respectively. As for the *Chloe* panel, two sound pressure level thresholds were considered: 40 dB, representative of a quiet environment, and 60 dB, corresponding to typical conversational noise levels. For the *Cullinan* panel, extensive regions of elevated noise probability are observed across all building floors for the 40 dB threshold: exceedance probability is greater than 20%. When the higher threshold of 60 dB is examined, probability remains above 15% over large portions

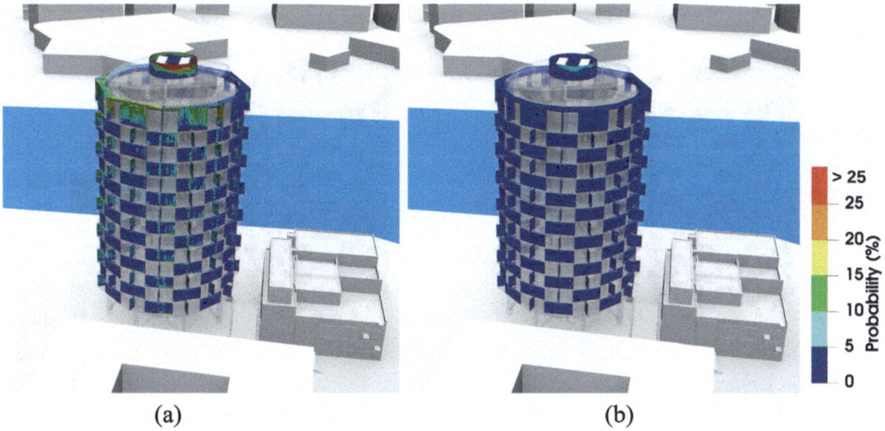

Fig. 10 Probability maps of wind-induced noise due to *Chloe* facade cladding for a threshold of $SPL > 40$ dB (**a**) and $SPL > 60$ dB (**b**). Figure created by the authors

of the upper two floors. These elevated probabilities are a direct consequence of the pronounced flow acceleration, enhanced turbulence levels, and significant vortex formation in the wake of the panel, as identified in the virtual wind tunnel simulations. In contrast, the *Pressed RR* panel is predicted to yield negligible wind-induced noise over almost the entire facade. This behaviour can be attributed to its planar perforation pattern and rounded geometric features, which limit flow acceleration and suppress the development of strong turbulent structures responsible for noise generation.

Among the three candidate cladding configurations considered, the *Cullinan* panel exhibits (Fig. 11) the highest risk of wind-induced noise generation. Consequently,

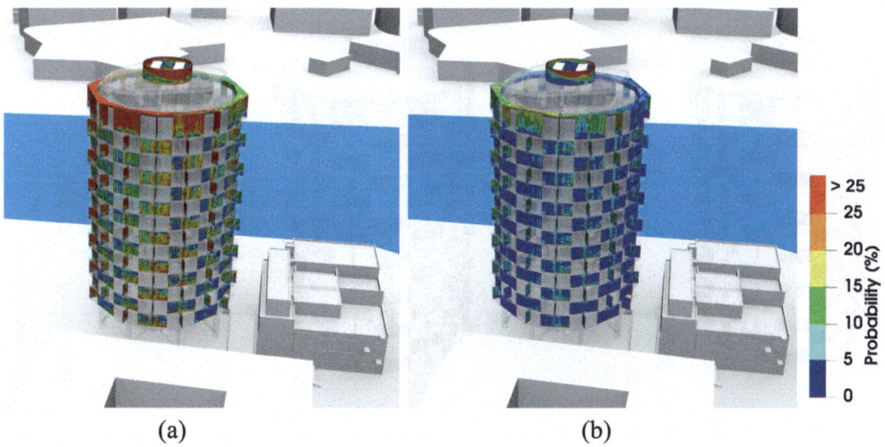

Fig. 11 Probability maps of wind-induced noise due to *Cullinan* facade cladding for a threshold of $SPL > 40$ dB (**a**) and $SPL > 60$ dB (**b**). Figure created by the authors

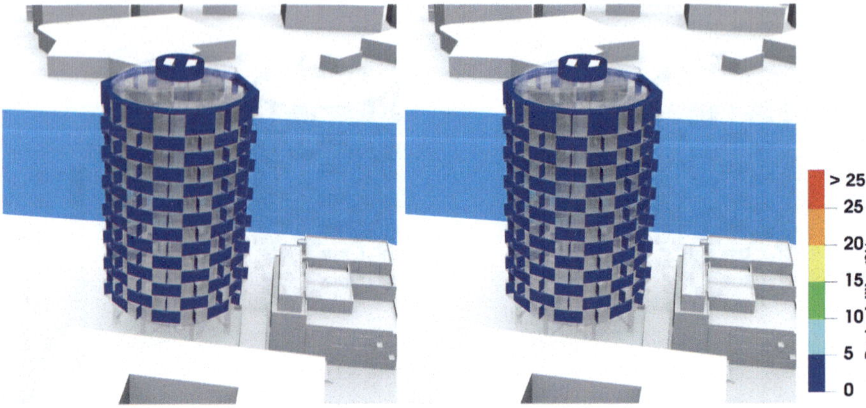

(a) (b)

Fig. 12 Probability maps of wind-induced noise due to *Pressed RR* facade cladding for a threshold of *SPL* > 40 dB (**a**) and *SPL* > 60 dB (**b**). Figure created by the authors

this configuration was excluded from further consideration. Figures 10 and 12 present the results for the *Chloe* and *Pressed RR* facade claddings, expressed in terms of probability maps for the 40 and 60 dB sound pressure level thresholds.

Despite this overall reduction, small localized regions near the rooftop exhibit probabilities of noise level higher than 40 dB on the order of 2% of the year (about 7 days) for the *Chloe* panel. Such regions are not observed for the *Pressed RR* panel. For the higher threshold of 60 dB, neither cladding configuration gives rise to extensive zones of potential acoustic discomfort across the facades (Figs. 13 and 14).

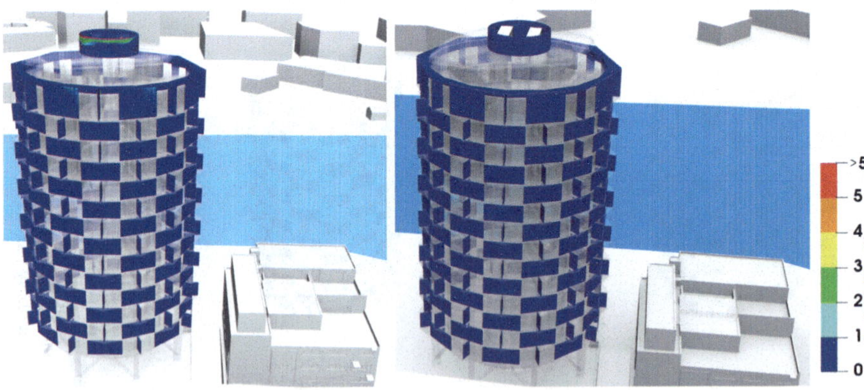

(a) (b)

Fig. 13 Probability maps of wind-induced noise due to *Chloe* facade (**a**) and *Pressed RR* facade cladding (**b**) for a threshold of *SPL* > 40 dB. Figure created by the authors

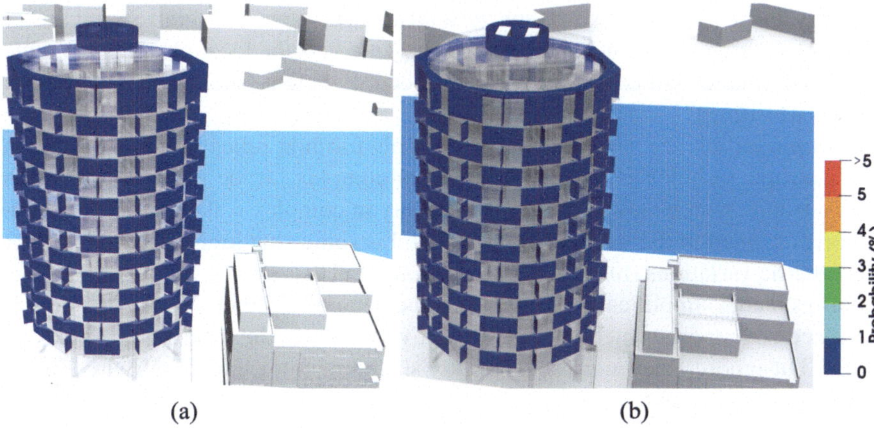

Fig. 14 Probability maps of wind-induced noise due to *Chloe* facade (**a**) and *Pressed RR* facade cladding (**b**) for a threshold of *SPL* > 60 dB. Figure created by the authors

The interpretation of the probabilistic risk maps offers a strong foundation for design decision-making: the concept of an exceedance probability is far more actionable for architects and urban planners than a single value and it allows for a discussion of risk tolerance: a 5% probability of exceeding 60 dB (conversational noise) for the Cullinan panel, concentrated on the upper floors, translates to a tangible problem for future penthouse occupants. In contrast, the 2% probability of exceeding 40 dB for the Chloe panel near the rooftop is a much lower concern, representing only about seven days of the year. The complete absence of noise risk for the Pressed RR panel at the 60 dB threshold is the clearest indicator of its superior acoustic performance. This method successfully translates a complex, multi-physics problem into a clear, visual, and probabilistic design metric.

5 Discussion

The primary advantage of this hybrid workflow is its speed and versatility. By decoupling the micro-scale element analysis from the macro-scale building simulation using the porous media approach, the computational burden is reduced by approximately 97%. This method successfully translates a complex, multi-physics problem into a clear, visual, and probabilistic design metric (e.g., visualizing a 2% exceedance probability versus a 20% probability).

To ensure proper interpretation of the results, the following assumptions and limitations inherent to the methodology must be stated:

- Steady-state modeling: the flow is modeled using steady-state RANS equations. Instead of resolving the time-dependent pressure fluctuations that constitute sound, this approach uses statistical turbulence metrics (turbulent kinetic energy and dissipation) to estimate the total acoustic power (Overall Sound Power Level). Consequently, the method effectively predicts the location and magnitude of noise sources but cannot resolve the spectral content (specific frequencies or tonal whistling).

- Isotropic turbulence assumption: Proudman's formula relies on the assumption of isotropic turbulence. In reality, flow near complex facade geometries is often anisotropic, which may introduce uncertainty in complex wake regions. Furthermore, as noted in the validation, the method is somewhat less sensitive to minor geometric variations (like hole size) than physical testing.
- Porous media simplification: in Phase 2, the detailed geometry of the cladding is replaced by a porous medium. This effectively models the macroscopic pressure drop but ignores the micro-scale vortex shedding from individual wires in the full-building context.
- Acoustic propagation: the current study focuses on the generation of noise at the source (the facade). It does not explicitly model the propagation of this sound to far-field receivers.

5.1 Technology Readiness Level (TRL) and Market Status

The methodology described herein corresponds to a Technology Readiness Level (TRL) of 8/9 (based on EU TRL scale 1–9). The fundamental physics (Navier-Stokes equations, Lighthill analogy) are well-established, and the digital workflow has been successfully validated and deployed in operational environments, such as the KORN 16 project. The workflow is software-agnostic and can be replicated using major CFD commercial packages (such as Ansys Fluent or Star-CCM+) or open-source (such as OpenFOAM), provided the solver supports custom source terms for porous media and access to turbulence statistics for post-processing via Proudman's formula. This ensures that the setup is reproducible by the broader engineering community.

6 Conclusions

This study demonstrates the efficacy of Computational Fluid Dynamics (CFD) as a design tool for building acoustics, even as an early-stage design optimization tool. By implementing a hybrid multi-scale workflow, the method presented herein successfully reduced the computational burden by approximately 97% compared to explicit meshing, enabling rapid iterative testing within a timeline compatible with agile architectural workflows.

The reliability of this methodology was confirmed through validation against the experimental datasets of Feng [4]. The numerical predictions fell within a $\pm$ 3 dB margin of error relative to physical measurements, a threshold of accuracy where variances are virtually imperceptible to the human ear. This validation provides the necessary confidence to utilize steady-state RANS simulations and Proudman's formula as a proxy for complex transient aeroacoustic phenomena in architectural applications.

Applied to the KORN 16 development in Hasselt, Belgium, the methodology successfully quantified the acoustic risks associated with mesh cladding panels. The analysis revealed that mesh cladding designs posed significant risks of noise generation due to flow acceleration, particularly near the rooftop, where local wind speeds are highest. However, other panels are able to maintain the desired visual transparency and shading performance while suppressing the turbulence structures responsible for noise.

By enabling the visualization of aeroacoustic risks before construction, this approach empowers architects and urban planners to prevent noise pollution, ensuring that the increasing geometrical complexity of modern facades does not come at the cost of environmental acoustic comfort.

Acknowledgments The authors would like to thank A-tract architecture for providing the design details of the KORN 16 project and for their valuable collaboration, which made the case study of this research possible.

References

1. A-tract architecture. KORN 16 Project. Available at: https://a-tract.be/projecten/korn-16/. Accessed 31 Dec 2025
2. Bellegoni M et al (2023) An extended SST k−ω framework for the RANS simulation of the neutral atmospheric boundary layer. Environ Model Softw 160:105583
3. Čehelová D et al (2017) Literature review on wind induced sound on buildings. Papabuild
4. Feng L (2012) Tone-like signal in the wind-induced noise of perforated plates. Acta Acust Acust 98(1):188–194
5. Franke J et al (2007) Best practice guideline for the CFD simulation of flows in the urban environment. COST Action 732
6. Franke J et al (2011) The COST 732 best practice guideline for CFD simulation of flows in the urban environment: a summary. Int J Environ Pollut 44(1–4):419–427
7. Joseph DD, Nield DA, Papanicolaou G (1982) Nonlinear equation governing flow in a saturated porous medium. Water Resour Res 18(4):1049–1052
8. Lighthill MJ (1952) On sound generated aerodynamically. I. General theory. Proc R Soc Lond Ser A 211(1107):564–587
9. Mahgoub AO, Ghani S (2021) Numerical and experimental investigation of utilizing the porous media model for windbreaks CFD simulation. Sustain Cities Soc 65:102648
10. Masson F et al (2024) Innovative design of facade elements to limit wind-induced noise in buildings. In: INTER-NOISE and NOISE-CON congress, vol 270, no 4, pp 7725–7735
11. Ploemen JCF et al (2011) Wind-induced sound on buildings and structures. In: Proceedings of the 13th international conference on wind engineering
12. Proudman I (1952) The generation of noise by isotropic turbulence. Proc R Soc Lond Ser A 214(1116):119–132
13. Srikumar SKR et al (2024) Application of a comprehensive atmospheric boundary layer model to a realistic urban-scale wind simulation. Build Environ:111330
14. Staco (2025a) Expanded metal Chloe. Available at: Staco.eu. Accessed 31 Dec 2025

15. Staco (2025b) Expanded metal Cullinan. Available at: Staco.eu. Accessed 31 Dec 2025
16. Staco (2025c) Pressed gratings, type RR. Available at: Staco.eu. Accessed 31 Dec 2025

Multi-objective Design and Optimization of Glass Façade Elements Using Machine-Learning-Based Computational Performance Engines

Michael A. Kraus, Ingo Stelzer, and Bediralp Çerme

Abstract The design of modern glass façades requires the simultaneous consideration of conflicting performance criteria, including acoustic insulation, structural integrity, and environmental sustainability. Traditional simulation workflows are often too computationally expensive for early-stage multi-objective optimization. This chapter presents a unified framework for the performance-based design of façade elements using machine-learning-based computational engines to provide instantaneous performance feedback. We introduce three specialized web-based applications: Kuraray's *SoundLabAI*, *StrengthLabAI*, and *CarbonReduceAI*. These tools with TRL 9 are integrated into a multi-objective optimization workflow that enables designers to navigate the trade-offs between technical requirements and carbon footprint. The results demonstrate that this ML-based approach achieves a high degree of accuracy compared to high-fidelity simulations while reducing computational time by several orders of magnitude, thereby facilitating a more holistic and iterative design process.

Keywords Glass façades · Performance-based design · Machine learning · Surrogate models · Multi-objective optimization · LCA · Performance-based design · Embodied carbon

Abbreviations

AI	Artificial Intelligence
ANN	Artificial Neural Networks
CO_2e	Carbon Dioxide Equivalent

M. A. Kraus (✉) · B. Çerme
ZM-I KI GmbH, Munich, Germany
e-mail: michael.kraus@zm-i.de

M. A. Kraus
Technical University of Darmstadt, ISM+D, Darmstadt, Germany

I. Stelzer
Kuraray Europe GmbH, Hattersheim am Main, Germany

A. Mejorin et al. (eds.), *Performance-Based FaÃ§ade Design*, Springer Tracts in Civil Engineering, https://doi.org/10.1007/978-3-032-28835-6_6

DL	Deep Learning
EPD	Environmental Product Declaration
FEA	Finite Element Analysis
GWP	Global Warming Potential
ML	Machine Learning
MOE	Mixture-of-Experts Architecture
MOO	Multi-Objective Optimization
TRL	Technology Readiness Level

1 Introduction

Glass façades are a hallmark of contemporary architectural expression, offering transparency, visual connectivity, and daylighting benefits. At the same time, they present complex engineering challenges: the envelope must satisfy stringent acoustic performance requirements, ensure structural safety under multiple load conditions, and increasingly adhere to sustainability benchmarks. Traditional methods for evaluating acoustic, structural, or environmental performance depend on specialized simulations or experimental campaigns, each with significant computational or logistical cost across a diverse set of software tools requiring expert specialization knowledge. In addition, the conventional separation between the architect's role (layout, geometry, aesthetics) and the civil engineer's role (structural soundness, building physics evaluations, etc.) leads to various technical as well as communication interfaces. Early design decisions often rely on conservative rules of thumb or isolated performance checks, offering little insight into trade-offs between competing objectives. Consequently, multi-objective design exploration remains fragmented, time-consuming, and technically demanding—particularly in early stages, when rapid performance feedback is essential.

The architectural glass industry is currently experiencing a paradigm shift driven by the dual requirements of increasingly complex aesthetic designs and stringent performance standards. **Performance-based façade design** has emerged as the leading methodology to address these challenges, shifting the focus from prescriptive code compliance to the quantitative evaluation of building envelope behavior across multiple domains. This chapter situates the use of Machine and Deep learning (ML/DL) as a core enabler of this transition, providing the speed necessary for truly integrated design exploration. ML/DL and data-driven modeling in combination with domain knowledge (here, civil and façade engineering) offer a transformative alternative. By training surrogate models on high-quality simulation or experimental data, complex prediction tasks can be approximated with high accuracy and dramatically reduced computational cost. In addition, integration of domain knowledge can significantly improve training speed and/or accuracy of the ML models. These ML surrogates unlock design-stage prediction at interactive speeds, enabling engineers and architects to rapidly explore large design spaces and integrate multi-criteria decision making into everyday workflows.

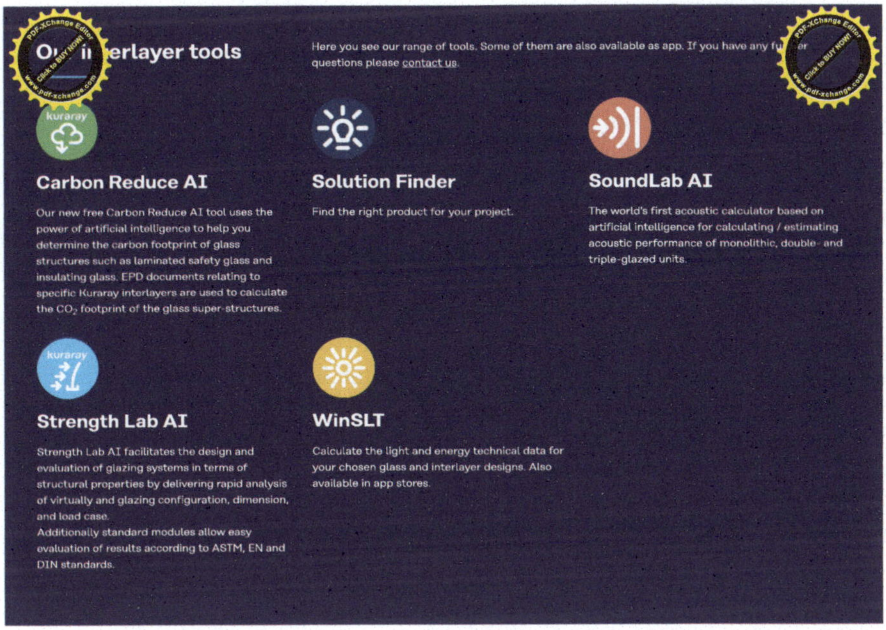

Fig. 1 Kuraray's computational engines for glass facade performance evaluation and optimization. Figure created by the authors

Within the context of glass façades, Kuraray's digital engineering initiative (cf. Fig. 1) has produced three specialized computational engines: *SoundLabAI*, *StrengthLabAI*, and *CarbonReduceAI*. SoundLabAI leverages machine learning to predict weighted sound insulation metrics such as R_w, STC, and OITC for arbitrary glazing assemblies, replacing complex experiments or numerical simulations with rapid, surrogate-based prediction. StrengthLabAI uses a customized Mixture-of-Experts deep learning architecture calibrated on parametrized finite-element datasets to approximate structural responses of monolithic and laminated glass systems for ultimate and serviceability limit states. CarbonReduceAI provides life cycle assessment (LCA) predictions based on design parameters, enabling sustainability evaluation without recourse to time-consuming LCA software.

Each surrogate engine builds on domain-specific data and tailored modeling strategies to ensure predictive accuracy and relevance for engineering practice. By integrating these engines into a unified multi-objective design and optimization framework, this chapter demonstrates how data-driven tools can democratize façade performance evaluation and support interactive, performance-oriented design workflows.

The optimizer developed in this framework performs discrete local search on glazing build-ups. Beginning from a user-defined baseline configuration, the method perturbs admissible glass thickness combinations and evaluates each candidate concurrently through asynchronous API calls to SoundLabAI, StrengthLabAI, and CarbonReduceAI. Normalized objective values are combined, filtered for feasibility, and ranked to identify superior designs under multi-objective criteria. This approach enables practitioners to identify trade-offs and optimal compromises among sound insulation, structural efficiency, and environmental performance without requiring deep expertise in numerical modeling.

The key innovation of this work lies not only in the development of accurate surrogate models but also in their integration within a unified optimizer tailored to façade design problems. By lowering computational and procedural barriers, the proposed framework fosters broader adoption of performance-based and sustainability-driven design strategies. The technology has been operationally validated over a three- to five-year horizon and is now deployed at a final maturity level of TLR 9.

The remainder of this chapter is structured as follows. Section 2 reviews the state of façade performance assessment and its limitations. The subsequent sections introduce the SoundLabAI (Sect. 3), StrengthLabAI (Sect. 4), and CarbonReduceAI (Sect. 5) tools with methodological detail. Section 6 presents the multi-objective optimization framework, followed by an application-oriented Sect. 7 with examples. Sections 8 and 9 consolidates discussion and conclusions, outlining practical insights and Sect. 10 avenues for future development.

2 Background

Performance-based design has become a central paradigm in contemporary façade engineering, driven by increasing demands on acoustic comfort, structural safety, and environmental sustainability. In contrast to prescriptive approaches, performance-based methods aim to quantitatively assess and balance multiple criteria already during early design stages. For glass façades, this paradigm is particularly relevant, as transparent building envelopes are sensitive to conflicting requirements such as sound insulation versus material efficiency or structural robustness versus embodied carbon. Despite its conceptual appeal, the practical implementation of performance-based façade design remains challenging due to fragmented analysis workflows and high computational effort.

This section reviews the state of the art in acoustic, structural, and environmental performance assessment as well as (multi-objective) optimization of glass façades, highlights limitations of conventional methods, and situates recent machine-learning-based approaches, particularly the author's prior work, within the broader research landscape.

2.1 Acoustic Performance Assessment of Glass Façades

The acoustic performance of glazed façades is commonly quantified using standardized metrics such as the weighted sound reduction index R_w and related spectrum adaptation terms according to ISO 717-1. These indicators describe the airborne sound insulation capacity of glazing systems and are typically obtained from full-scale laboratory measurements or advanced numerical simulations. While such approaches are well established and reliable, they are associated with considerable cost, long turnaround times, and limited flexibility with respect to parametric variation.

Existing tools for predicting the sound insulation performance of fenestration remain inadequate for reliable use in early-stage design [3, 4]. Although laboratory testing provides the highest accuracy, analytical models often diverge from experimental results—particularly for sealed or laminated glass—and fail to provide consistent "worst-case" estimates [3]. Numerical methods like the Finite Element Method (FEM) accurately capture fluid-structure interactions in the low-frequency range, while Statistical Energy Analysis (SEA) offers a robust framework for high frequencies; however, both require significant expertise and computational effort, limiting their practical efficiency [4].

As a consequence, acoustic considerations are often decoupled from early design exploration and addressed only once a glazing concept has largely been fixed. This limits the ability to systematically study the influence of glass thickness, interlayer configuration, cavity dimensions, or asymmetric build-ups on sound insulation performance.

To overcome these limitations, data-driven surrogate modeling approaches have recently been proposed. In particular, the development of *SoundLabAI* [7, 21] demonstrated that (ML) models trained on curated acoustic datasets can predict R_w, STC, and OITC values for a wide range of glazing assemblies with high accuracy and negligible computational cost. The work showed that ML-based predictors can effectively replace repeated laboratory testing or simulation in the design phase while remaining consistent with established acoustic standards and engineering expectations [7, 21]. These contributions laid the foundation for integrating acoustic performance directly into interactive design and optimization workflows.

2.2 Structural Performance of Glass Elements

Structural design of glass façades is governed by ultimate and serviceability limit states, typically relying on theories ranging from linear plate theory to non-linear Finite Element Analysis (FEA). For laminated glass, the interlayer's shear modulus G is critical for bending capacity [15, 17]. While standards such as DIN 18008 [5], EN 16612 [8], and ASTM E1300 [1] provide simplified analytical frameworks (e.g., effective thickness or ω-factors), these simplified approaches often lack accuracy for realistic façade systems involving complex geometries, layered build-ups, and geometrically nonlinear computational effects [10, 11, 18].

High-fidelity finite-element models can capture these complex effects but require expert knowledge, careful modeling choices, and significant computational resources. Consequently, structural assessments are rarely performed in large parametric studies or coupled with other performance criteria such as acoustics or sustainability.

To bridge these gaps, recent advances in data-driven structural mechanics have introduced surrogate-based evaluation. Building on earlier research on AI in glass engineering, the *StrengthLabAI* framework utilizes a customized Mixture-of-Experts (MoE) deep learning architecture to approximate stress distributions, deflections, and utilization ratios across a broad design space [18]. Trained on parametrically generated finite-element datasets, this surrogate preserves the predictive fidelity of numerical models while enabling rapid evaluation suitable for iterative design and integrated façade optimization workflows. This work represents a significant step toward embedding structural performance checks directly into integrated façade design workflows.

2.3 Environmental Performance and Life Cycle Assessment

Alongside acoustic and structural requirements, environmental performance has become a decisive criterion in façade design. Life cycle assessment (LCA) methods are commonly used to quantify indicators such as global warming potential (GWP) across production, use, and end-of-life phases. However, conventional LCA tools require extensive data input, specialized expertise, and are rarely compatible with fast design iterations.

As a result, sustainability considerations are often treated as a downstream verification step rather than an integral design driver. This disconnect hampers the realization of genuinely low-impact façade solutions, particularly when trade-offs with other performance objectives are not explicitly quantified.

Recent research and development efforts have therefore focused on simplifying and accelerating environmental assessments. In this context, surrogate-based LCA tools such as *CarbonReduceAI* aim to provide rapid, design-parameter-driven estimates of embodied environmental impacts, enabling sustainability metrics to be evaluated alongside acoustic and structural performance within a unified framework.

2.4 Limitations of Conventional Workflows

Despite progress in individual performance domains, current façade design workflows remain highly fragmented. Acoustic simulation tools, structural analysis software, and LCA platforms operate largely independently, each requiring dedicated models, inputs, and expert interpretation. Coupling these tools in iterative or optimization-driven workflows is computationally expensive and procedurally complex.

This fragmentation creates a high barrier for performance-based and sustainability-driven design, particularly in early project phases where rapid feedback is essential. Moreover, the lack of integration inhibits systematic multi-objective optimization and limits the ability to transparently explore trade-offs between competing criteria.

2.5 *Scientific Machine Learning in Glass and Façade Engineering*

The application of machine learning (ML) in glass and façade engineering has gained significant momentum, with early research highlighting the potential of deep learning for parameter identification and predictive modeling [15, 16]. These studies underscored that domain-specific datasets and tailored architectures are prerequisites for engineering-relevant accuracy and acceptance.

Building on this, the *SoundLabAI* and *StrengthLabAI* frameworks represent concrete realizations of ML-based performance-based surrogates. Notably, the *StrengthLabAI* project demonstrated that "classic" ML algorithms are outperformed by custom MoE neural networks [18], whose architecture and training are explicitly guided by structural engineering domain knowledge. By integrating these acoustic and structural simulators with environmental evaluation within a unified optimization framework, this chapter advances the state of the art toward integrated, multi-objective, and deployable façade design tools.

The following sections detail the individual computational engines, beginning with *SoundLabAI*, and demonstrate how their integration enables efficient, performance-based optimization of glass façade elements.

2.6 *Technology Readiness Level (TRL)*

The presented framework has reached Technology Readiness Level 9 (TRL 9) according to the European Commission's TRL definitions, indicating full system maturity and proven operation in relevant real-world environments. This assessment is based on cumulative testing and validation efforts conducted over the past three to five years, including prototype deployment, performance benchmarking, and iterative refinement during façade design projects in academic–industrial collaboration.

The integrated toolset, comprising the surrogate models *SoundLabAI*, *StrengthLabAI*, and *CarbonReduceAI*, as well as the unified optimization interface has been successfully applied in practical design studies, demonstrating functional reliability and user acceptance across multiple projects. The technology is operationally validated and available for project use within research and pilot industry settings; broader commercial release is planned but not yet publicly distributed. All tools are publicly available for use via the official homepages of Kuraray.

The authors were directly involved in the research, development, deployment, and validation of the underlying methods and tools but do not hold proprietary or commercial rights beyond their institutional frameworks.

3 *SoundLabAI*: Machine-Learning-Based Acoustic Performance Prediction of Glass Façade Elements

3.1 Motivation and Problem Context

In building acoustics, façade performance is quantified via single-number ratings: the weighted sound reduction index R_w (Europe), alongside STC and OITC (USA). These metrics condense complex frequency-dependent transmission loss into interpretable values, enabling engineers to specify requirements during procurement. Conventionally, R_w is derived by comparing certified laboratory spectra against standardized reference curves, such as DIN EN ISO 717-1 [6].

While laboratory testing remains the gold standard, its resource-intensive nature precludes exhaustive iteration. Conversely, high-fidelity vibro-acoustic simulations (e.g., FEM-based fluid–structure interaction) offer detailed insights but demand significant expertise and computational time. Consequently, acoustic validation is often delayed, limiting the exploration of critical trade-offs between mass, asymmetry, and damping strategies. *SoundLabAI* addresses this gap, providing a rapid ML-based engine that predicts R_w, STC, and OITC for diverse glazing assemblies by replacing physical testing with an accurate surrogate trained on curated experimental data, cf. Fig. 2.

3.2 Influential Design Parameters for Acoustic Performance

From façade acoustics and glazing technology, several design parameters are known to strongly influence sound insulation behavior:

- **Glass Thickness and Mass**: While increased mass generally improves insulation, critical frequency effects can compromise attenuation depending on pane geometry and boundary conditions.
- **Lamination**: Laminated panes mitigate coincidence effects through polymer interlayer damping, particularly in mid-to-high frequency regimes.
- **Cavity Configuration (IGU)**: For insulating units, cavity depth and gas filling (Argon/Krypton) dictate coupled resonance; wider separations improve performance but are limited by structural constraints.
- **Interlayer Characteristics**: Specialized acoustic interlayers significantly enhance transmission loss near resonant frequencies.

The complex, non-linear interaction of these variables motivates a data-driven approach capable of capturing quantitative relationships across realistic build-ups.

Fig. 2 SoundLabAI workflow: (left) online input mask of the setup, (middle) ML surrogate model for acoustic performance prediction, (right) online result output, from [7]

3.3 Surrogate Modeling Approach

SoundLabAI is framed as a supervised multi-target regression problem. Let $\mathbf{x} \in \mathbb{R}^d$ represent a mixed-type feature vector (thicknesses, interlayer types, gas filling) as detailed in Table 1. The model learns the mapping:

$$\hat{\mathbf{y}} = f_\theta(\mathbf{x}), \qquad \hat{\mathbf{y}} = \left[\widehat{R}_w,\ \widehat{\mathrm{STC}},\ \widehat{\mathrm{OITC}}\right]^\top. \tag{1}$$

Optimization minimizes the Mean Squared Error (MSE) over N samples:

$$\mathcal{L}(\theta) = \frac{1}{N}\sum_{n=1}^{N} \| f_\theta(\mathbf{x}_n) - \mathbf{y}_n \|_2^2 \,. \tag{2}$$

Following an evaluation of baseline regressors [7, 21], a **Random Forest** (cf. Fig. 3) was selected for its robustness with structured tabular data. Hyperparameter tuning via 5-fold cross-validation established an optimal tree depth of 9 to maximize generalization. More details on data, the training and evaluation are provided in the remainder of this subchapter.

Table 1 Feature space and data ranges for acoustic performance datasets

Feature	Feature type	Data range
Glass thickness	Continuous	[2–15] mm
Interlayer thickness	Continuous	[0.38–4.56] mm
Interlayer type	Categorical	Trosifol®Clear/Trosifol®UltraClear, Trosifol®SC Monolayer, Trosifol®SC Multilayer, Trosifol®ExtraStiff, SentryGlas®/SentryGlas®XtraTM
Cavity thickness	Continuous	[1–30] mm
Gas filling	Categorical	No gas, air, argon, krypton
Glass set-up	–	Mono, LG, IGU (double and triple)

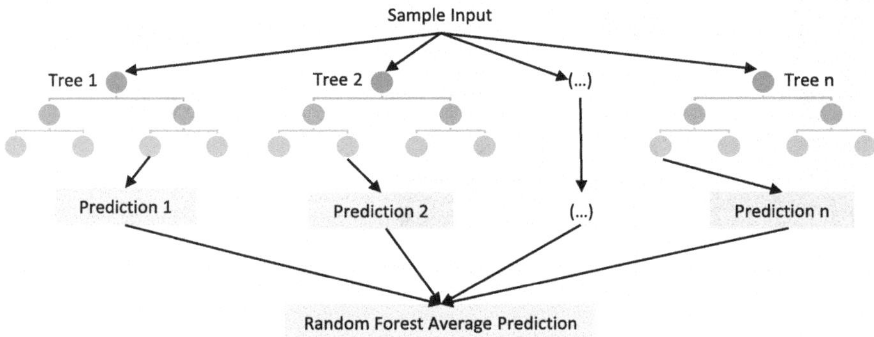

Fig. 3 Random Forest model architecture, from [7]

3.4 Dataset and Feature Engineering

The training dataset consists of structured measurement data compiled from an extensive experimental program on glazing systems with varying compositions. The feature set represents the glazing build-up as detailed in Table 1.

The data preparation follows a standard pipeline: (i) ingestion of structured tables, (ii) validation checks for missing values and inconsistent entries, (iii) separation into features and targets, and (iv) categorical encoding. For small-to-moderate dataset sizes typical of laboratory measurement campaigns, robust validation strategies such as cross-validation are essential to minimize overfitting and ensure generalization.

3.5 Training, Validation, and Model Quality

SoundLabAI was trained using 5-fold cross-validation for hyperparameter selection and evaluation. The ML surrogate achieves high fidelity on R_w prediction, with a

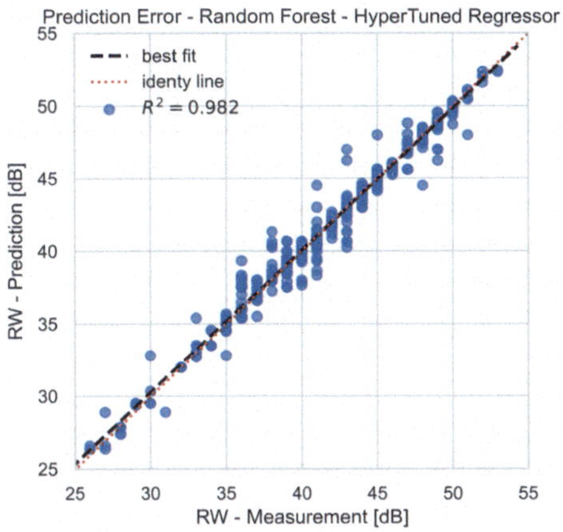

(a) Prediction error on the validation dataset, from [7].

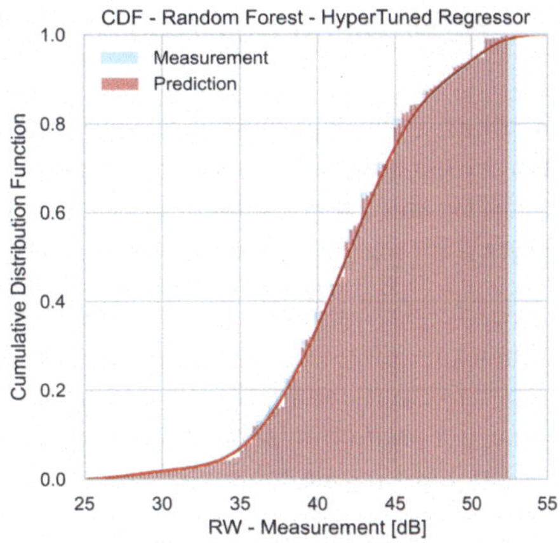

(b) Comparison of the cumulative distribution functions, from [7].

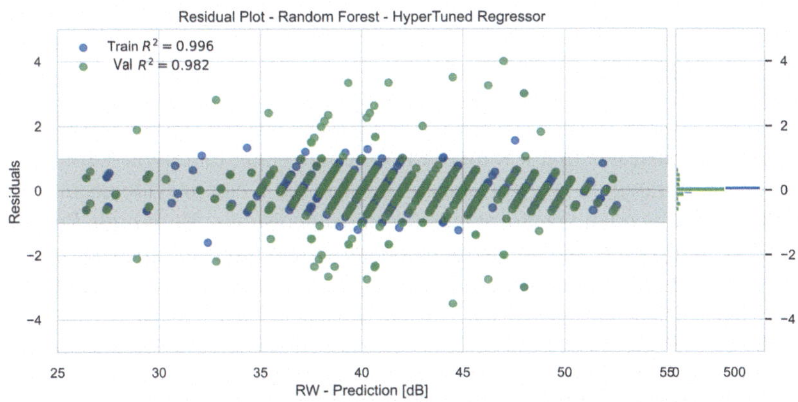

(c) Residual Plot for the Random Forest regressor, from [7].

Fig. 4 Comprehensive performance evaluation of the SoundLabAI surrogate

training R^2 of 0.996 and a validation R^2 of 0.982. These metrics, alongside the error plots in Fig. 4, confirm the model's reliability within the training distribution.

These results demonstrate that the Random Forest surrogate with depth 9 captures the relationship between glazing features and acoustic metrics with high fidelity, providing reliability for design-stage evaluation within the training data distribution.

In addition, the tool indicates when an experimental data point exists in the underlying database that closely matches the queried configuration. This supports transparent decision-making and mitigates over-interpretation of extrapolative predictions.

3.6 Deployment and User Workflow

SoundLabAI is deployed as a web-based application that allows users to define arbitrary glazing build-ups through an input mask and immediately obtain predicted R_w, STC, and OITC values, cf. Fig. 2. From a workflow perspective, the tool is designed for early-stage design exploration and routine engineering checks:

1. specify glazing type (monolithic/laminated; single/double/triple),
2. enter ply thicknesses and interlayer selections,
3. specify cavity and gas filling (if applicable),
4. execute prediction and interpret metrics together with uncertainty indicators.
5. (optional): iterate manually or automated (via the "optimizer" function).

Users can export the predicted results and, where available, the corresponding laboratory measurement reports for integration into architectural and engineering documentation.

3.7 Interim Summary

SoundLabAI demonstrates that a Random Forest surrogate, tuned via 5-fold cross-validation and calibrated on structured measurement data, can provide accurate and rapid predictions of acoustic performance for glazing systems. With high R^2 values on train and validation sets and an interactive deployment interface, the tool supports performance-based acoustic considerations in early and iterative design contexts.

4 *StrengthLabAI*: Machine-Learning-Based Structural Performance Prediction for Glass Façade Elements

4.1 Motivation and Problem Context

The structural assessment of glass façades is essential for ensuring safety across Ultimate (ULS) and Serviceability (SLS) Limit States, as mandated by international design standards and norms. Current practice typically relies on either simplified analytical models, which often lack the necessary precision for complex assemblies, or high-fidelity FEA. While accurate, FEA entails significant modeling overhead and prohibitive computational costs, particularly when accounting for the geometrically non-linear behavior of laminated glass systems. These bottlenecks often relegate rigorous structural analysis to the final stages of design, forcing early-stage explorations to rely on empirical estimations rather than data-driven validation. Consequently, the integration of structural performance into rapid, multi-objective design workflows remains a significant challenge.

To bridge this gap, we introduce *StrengthLabAI*, a rapid ML engine designed to approximate both linear and non-linear structural responses—specifically stresses and deformations—with high fidelity and negligible computational cost (cf. Fig. 5). By employing a customized MoE deep learning architecture trained on extensive synthetic FEA datasets, *StrengthLabAI* effectively maps complex structural design features to key performance indicators. The integration of this surrogate model into an accessible web-based platform enables the instantaneous evaluation of structural performance during early design phases, thereby facilitating robust multi-criteria decision-making within the architectural and engineering industry.

4.2 Influential Design Parameters for Strength Performance

The structural behavior of glass facade elements is governed by a combination of geometric and material (non-)linearities:

- **Aspect Ratio and Thickness**: The ratio of pane dimensions to thickness dictates the transition from plate bending to membrane action, significantly affecting stress distributions.
- **Shear Coupling (Lamination)**: In laminated glass, the interlayer's shear modulus G is temperature- and time-dependent, determining the degree of composite action between glass plies.
- **Boundary Conditions**: Support types (cantilever, two-side line-supported, Navier-supported (four-side line-supported), point-fixed, or cantilevered) and load types (wind pressure, line loads) fundamentally alter the stress concentrations.
- **Geometric Nonlinearity**: Large deflections relative to thickness necessitate non-linear considerations to prevent underestimating pane stiffness.

These complex interactions motivate a neural network architecture capable of capturing non-linear manifolds across diverse structural cases.

4.3 Surrogate Modeling Approach

The development of *StrengthLabAI* is framed as a supervised regression task to predict maximum principal stress σ_{max} and deflection w_{max}. Let $\mathbf{x} \in \mathbb{R}^d$ represent the structural feature vector, encompassing geometry, material properties, and loading conditions. To account for the wide range of response magnitudes and the long-tailed distribution characteristic of structural data, we employ a **MoE** architecture, cf. Fig. 6:

$$\hat{\mathbf{y}}(\mathbf{x}) = \sum_{i=1}^{n} p_i(\mathbf{x}) \cdot e_i(\mathbf{x}; \theta_i), \quad \text{where} \quad \hat{\mathbf{y}} = \begin{bmatrix} \hat{\sigma}_{max} \\ \hat{w}_{max} \end{bmatrix} \tag{3}$$

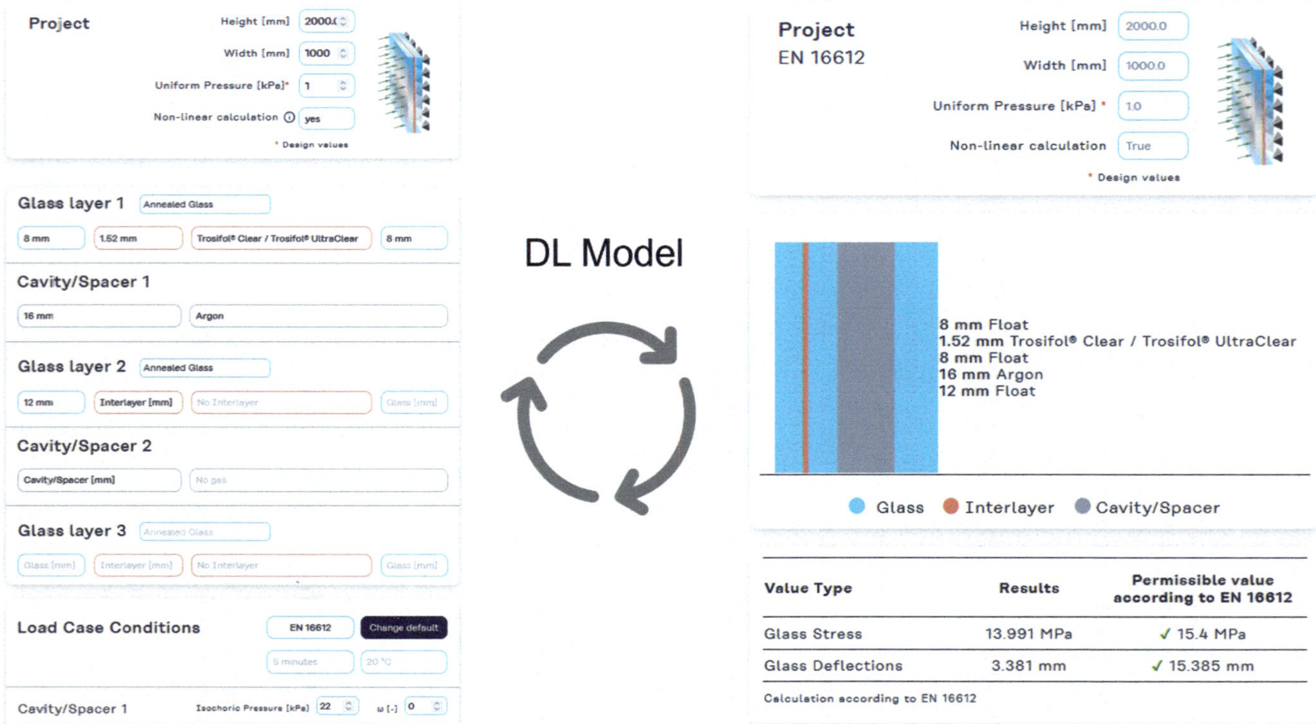

Fig. 5 StrengthLabAI workflow: (left) online input mask for structural systems and resp. parameters for load, geometry, (middle) MoE surrogate model for stress and deflection prediction, (right) online result output, from [18]

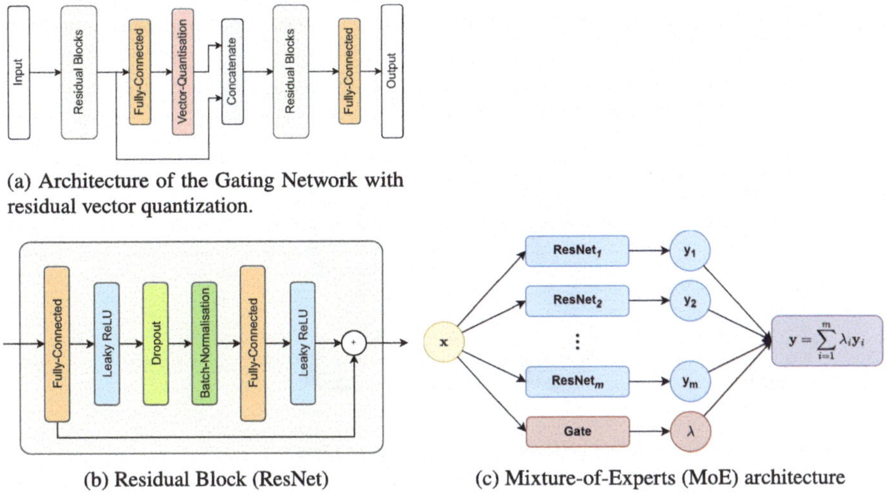

(a) Architecture of the Gating Network with residual vector quantization.

(b) Residual Block (ResNet)

(c) Mixture-of-Experts (MoE) architecture

Fig. 6 Overview of the deep learning architectures: **a**, **b** show the internal components and modular units, while **c** illustrates the integrated Mixture-of-Experts framework, from [18]

Here, p_i denotes a gating network that dynamically routes inputs to specialized experts e_i via softmax activation. To enhance decision confidence, the gating mechanism incorporates a vector quantization layer and a trainable temperature coefficient τ. Optimization is driven by a composite loss function—incorporating Mean Squared Error (MSE), Huber, and Mean Squared Logarithmic Error (MSLE)—to ensure robustness across varying output scales.

4.3.1 Synthetic Data Generation Pipeline

The surrogate is developed for nine distinct Design Cases covering monolithic, laminated, and insulating glass elements under a range of line and area loads, multiple support conditions (e.g., cantilever, simply supported panels), and both ULS and SLS limit states, cf. Fig. 2. This broad coverage ensures that the surrogate model can be confidently applied across typical façade design scenarios encountered by practitioners.

Training high-capacity surrogate models requires an extensive and representative dataset. We implemented a parametric pipeline where each *Design Case* is defined by a vector $\mathbf{x}_j$ spanning engineering-relevant intervals (see Table 3).

A MATLAB-based routine utilizes full factorial sampling to programmatically execute simulations in ANSYS via APDL. The FEA models employ `SOLID186` elements to capture three translational degrees of freedom per node, accommodating geometric non-linearities and the near-incompressible behavior of interlayers. The resulting datasets, ranging from 10^5 to 1.1×10^7 samples, are stored in Parquet format for high-performance access.

Table 2 Design Cases covered by *StrengthLabAI*: structural system and loading specification, from [18]

Design Cases covered by *Strength Lab AI*		
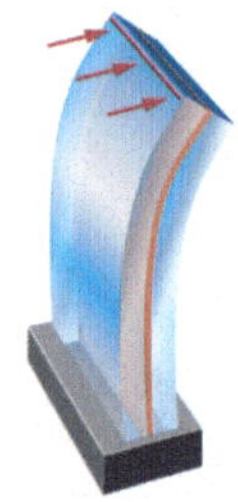	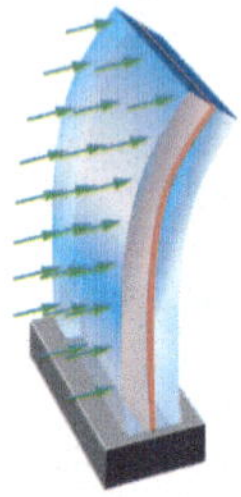	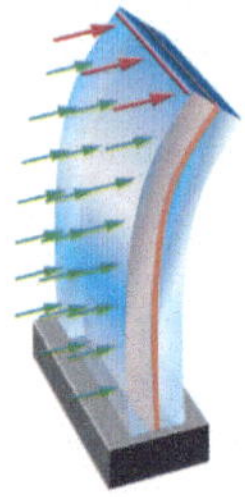
Design Case 1	Design Case 2	Design Case 3
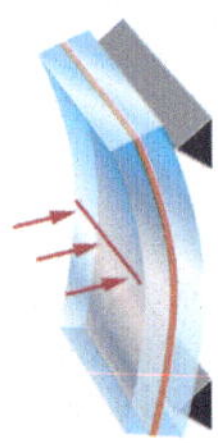	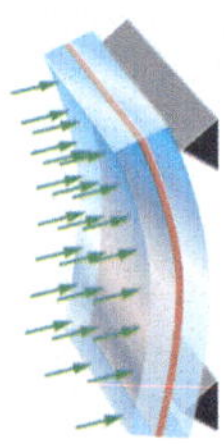	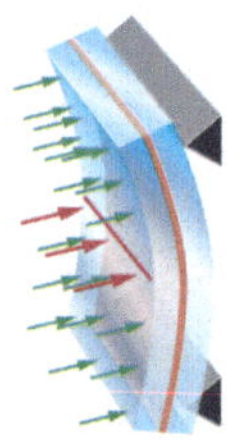
Design Case 4	Design Case 5	Design Case 6
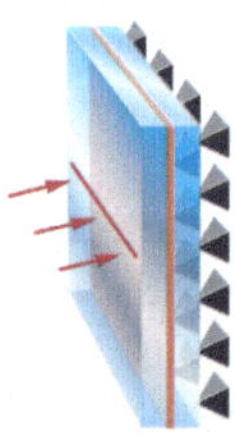	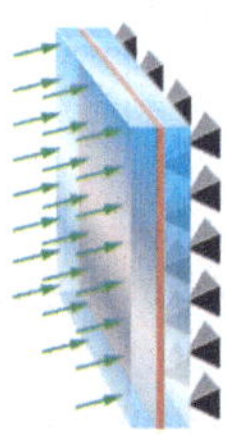	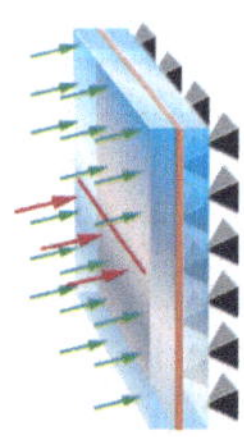
Design Case 7	Design Case 8	Design Case 9

Table 3 Design space spanned by the Design Cases together with parameter sampling intervals

Design Case	1	2	3	4	5	6	7	8	9
E_{glass} (MPa)	70,000; 71,000								
$E_{interlayer}$ (MPa)	3e−3: 6e3 (500 logarithmic steps)								
Height h (mm)	300: 50: 3250								
Width b (mm)	500: 50: 6000								
Line load q_d (kN/m)	0.25: 0.25: 10	–	0.25: 0.25: 10		–	0.25: 0.25: 10		–	0.25: 0.25: 10
Line load position h_q (mm)	500: 50: 3250	–	500: 50: 3250		–	500: 50: 3250		–	500: 50: 3250
Area load q_d (kN/m^2)	–	0.25: 0.25: 10		–	0.25: 0.25: 10		–	0.25: 0.25: 10	
Thickness t_{glass} (mm)	[2 3 4 5 6 8 10 12 15 20 25]								
Thickness $t_{interlayer}$ (mm)	[0.38; 0.76; 1.14; 1.52; 2.28]								

Prior to training, the data undergo rigorous preprocessing:

- **Filtering**: Outlier responses exceeding realistic engineering limits are excluded.
- **Feature Scaling**: Input variables are standardized (zero mean, unit variance) to ensure numerical stability.
- **Partitioning**: Data are split into training, validation, and test sets (e.g., 70/10/20), with early stopping and learning rate scheduling (`ReduceLROnPlateau`) to prevent overfitting.

4.3.2 Comparative Evaluation and Validation

Before finalizing the MoE architecture, baseline models including XGBoost and residual feedforward neural networks (ResNets) were evaluated. While XGBoost utilized gradient boosting with extensive hyperparameter tuning and the ResNet employed dropout and batch normalization, neither could consistently generalize across the full distribution of structural behaviors. The MoE surrogate overcomes these limitations by partitioning the task; experts are targeted toward specific distributional segments (e.g., lower vs. upper ranges) using customized loss functions.

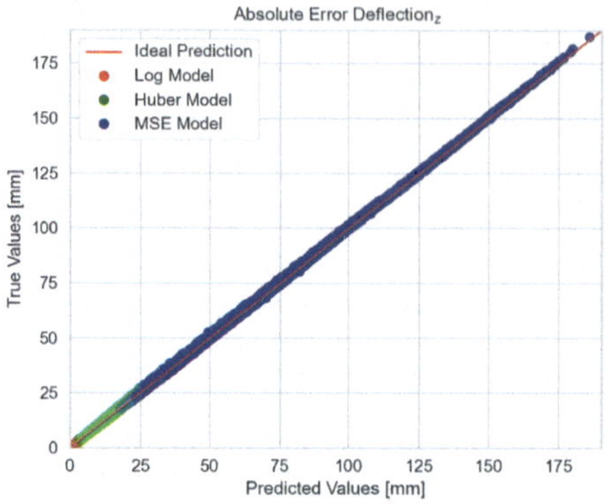

(a) Out-of-plane deflection w_z.

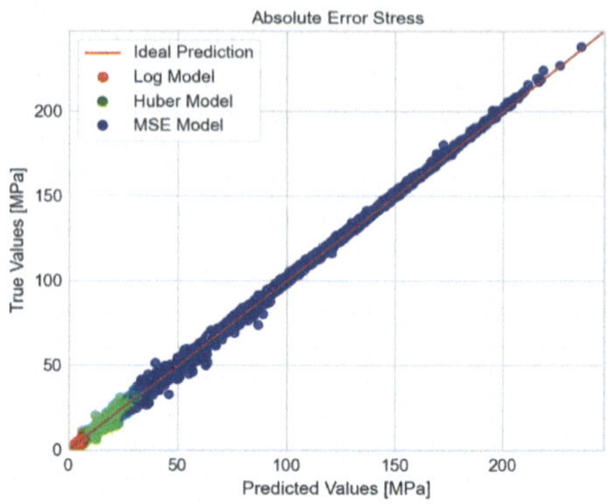

(b) Maximum principal stress σ.

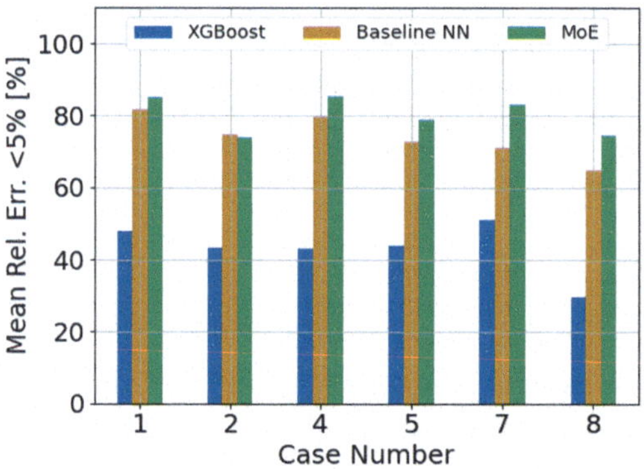

(c) Comparative performance metrics across all design cases.

Fig. 7 StrengthLabAI performance results: **a**, **b** surrogate predictions for Case 8 and **c** comparative evaluation showing MoE superiority over baseline models, from [18]

Model performance is evaluated using multiple metrics:

- **Mean Absolute Error (MAE)**: average magnitude of prediction errors,
- **Coefficient of Determination** (R^2): goodness of fit across the dataset,
- **Thresholded Relative Error** ($\mathcal{L}_{\mathrm{rel}}$): fraction of predictions within specific relative error bounds (e.g., 1%, 5%, 10%).

Relative error metrics are particularly relevant due to the exponential and long-tailed distribution of structural responses, ensuring balanced evaluation across low and high output magnitudes.

The predictive performance of the surrogate models was evaluated across multiple structural Design Cases. As illustrated in Fig. 7, the MoE architecture consistently outperforms both XGBoost and baseline neural network models. Specifically, the MoE model achieves a significantly higher fraction of predictions with a mean relative error below 5% across all tested scenarios, demonstrating superior robustness in capturing complex, non-linear structural responses.

To ensure engineering validity, the surrogate was verified against industry-standard software (Dlubal RFEM 5, MEPLA). Approximately 20 external validation cases per Design Case confirmed that *StrengthLabAI* maintains high fidelity relative to established FEA benchmarks while providing near-instantaneous inference.

4.4 Deployment and User Workflow

The tool is deployed as a web application that facilitates rapid iteration. The user workflow follows a five-step process:

1. select structural system (Design Case),
2. define geometry and laminate build-up,
3. input load magnitudes and support conditions,
4. execute the MoE prediction engine,
5. verify utilization against code-defined limits (e.g., DIN 18008, EN16612/EN16613, ASTM E1300).

The system enables users to export a summary report and detailed prediction data for professional documentation.

4.5 Interim Summary

StrengthLabAI demonstrates how a tailored Mixture-of-Experts surrogate, calibrated on extensive synthetic FEA data and integrated into a user-oriented web tool, can efficiently replace traditional structural simulations for glass façade performance evaluation. High predictive accuracy across a range of structural cases, combined with verification against industry tools, supports its application in daily engineering practice and its inclusion within broader multi-objective design workflows.

5 *CarbonReduceAI*: Environmental Impact Performance Prediction of Glass Façade Elements

5.1 Motivation and Problem Context

Life Cycle Assessment (LCA) is a standardized methodology (ISO 14040 [12] resp. ISO 14044 [13]) for quantifying the environmental burdens of product systems across their entire life cycle. In the context of high-performance façades, the **Global Warm-**

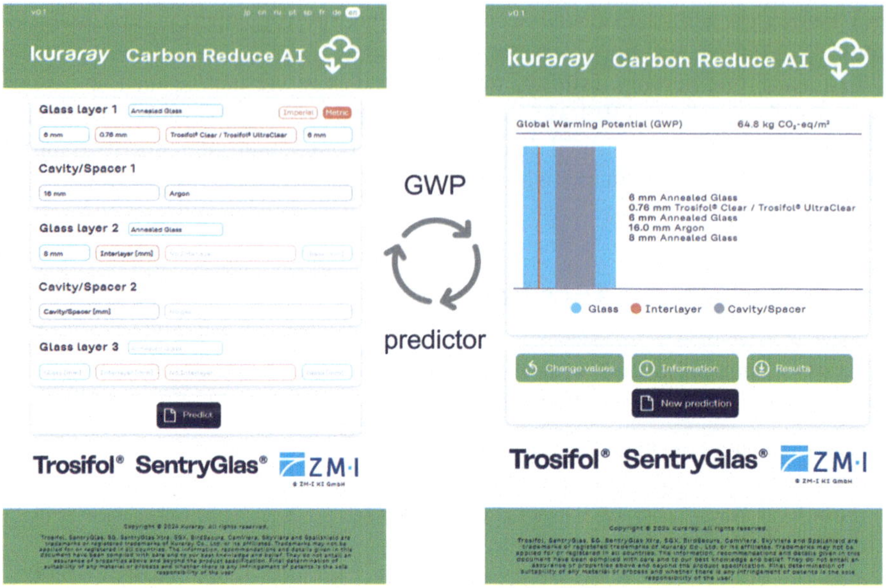

Fig. 8 CarbonReduceAI workflow: (left) online input mask of the setup, (middle) performance prediction, (right) online result output. Figure created by the authors

ing Potential (GWP), measured in kg CO_2-eq, serves as a primary metric for evaluating embodied carbon. To facilitate rapid environmental feedback during early-stage design, we introduce *CarbonReduceAI*—a deterministic analytical engine that approximates GWP for arbitrary glass assemblies within a multi-objective optimization framework, cf. Fig. 8.

5.2 *System Boundaries and Functional Unit*

The tool adopts a **cradle-to-gate** boundary, focusing on raw material extraction and manufacturing phases. The functional unit is defined as the environmental impact per unit area of the façade (1 m^2). This scope excludes operational energy and end-of-life recovery (Module D), prioritizing the immediate embodied carbon trade-offs associated with material selection and geometric configuration.

5.3 Influential Design Parameters for Environmental Performance

The embodied carbon footprint of a façade system is primarily dictated by the material intensities and manufacturing processes of its components:

- **Glass Volume and Type**: The thickness and number of glass plies (monolithic vs. IGU) directly scale the GWP due to the energy-intensive nature of primary glass production.
- **Interlayer Material**: The specific mass and chemical composition of polymer interlayers (e.g., PVB, ionoplast) contribute varying GWP values per unit area.
- **Gas Fillings and Spacers**: While smaller in volume, the secondary components in insulating glass units add to the total embodied carbon of the assembly.
- **Framing and Integration**: The structural depth and material choice of the surrounding façade profile significantly influence the overall life cycle impact.

Categorical variables (e.g., glass type or interlayer type identifiers) are mapped to corresponding physical properties (such as density and specific emission factors) using lookup tables. The Python implementation then derives numerical features suitable for quantitative impact calculation.

5.4 Global Warming Potential Calculation Logic

The surrogate implements embodied GWP estimation through a *bottom-up calculation* of material volumes and corresponding emission factors. For each layer in the glazing assembly, the following steps are executed:

1. **Volume Computation**: Layer thickness is converted to volumetric representation (in m^3) using a unit conversion factor.
2. **Material Emission Factor Lookup**: Each material is assigned a GWP factor (kg CO_2-eq per m^3), derived from databases or internal material specifications.
3. **Contribution Aggregation**: The total embodied carbon is computed as the summation of the contributions from each constituent layer.

Mathematically, for a set of n material layers with volumes V_i and associated emission factors f_i, embodied GWP is given by:

$$\mathrm{GWP}_{\mathrm{total}} = \sum_{i=1}^{n} V_i \cdot f_i. \tag{4}$$

The implementation currently predicts a single environmental impact indicator 'GWP', although the structure allows for extension to additional impact categories in future developments.

5.5 Deployment and User Workflow

The deployment of CarbonReduceAI is fully integrated into the optimizer module of the unified framework. The workflow follows these steps:

1. **Input**: Define the material set-up within the design mask.
2. **Perturbation**: The optimizer generates candidate façade configurations.
3. **Prediction**: The LCA API calculates GWP values in real-time.

The system enables users to export a summary report and detailed prediction data for professional documentation.

5.6 Interim Summary

CarbonReduceAI implements rapid embodied carbon calculation for glass façade systems. By providing volume-weighted GWP estimates, the tool supports environmental performance evaluation stand-alone or within broader multi-objective contexts via the design optimizer integration, ensuring that sustainability is a primary driver from the earliest stages of design.

6 Multi-objective Optimization Framework

6.1 Problem Formulation, Design Variables and Design Space

The design of glass façade systems involves multiple performance criteria that must be simultaneously addressed, including acoustic performance, structural adequacy, and environmental impact. In mathematical terms, this constitutes a **multi-objective optimization problem**, where a vector of objective functions $\mathbf{F}(\mathbf{x}) = [f_1(\mathbf{x}), \ldots, f_k(\mathbf{x})]$ is defined on a feasible domain of design variables $\mathbf{x}$. A solution $\mathbf{x}^*$ is considered Pareto optimal if no objective can be improved without simultaneously degrading at least one other objective.

In the present optimizer implementation, the decision variables are the **discrete thickness values** of glass layers (and optionally interlayers) for a glazing system. These variables take values from predefined finite sets (e.g., valid glass thicknesses such as 2 mm, 3 mm, …, 15 mm) to reflect practical manufacturing and design constraints.

6.2 Perturbation Strategy for Candidate Generation

To ensure computational efficiency during early-stage design, the optimizer employs a **local perturbation strategy** rather than an exhaustive global search. For a design variable with a current index i in the thickness set, the set of candidate indices for a perturbation degree d is defined as:

$$\mathcal{I}(i, d) = \{j \in \mathbb{Z} \mid \max(0, i - d) \leq j \leq \min(i + d, N)\} \tag{5}$$

where N is the maximum index of the valid thickness array. This local search limits the combinatorial explosion of variants while ensuring all candidates remain physically plausible.

Each candidate configuration is serialized into a JSON payload and evaluated via asynchronous HTTP requests. This parallelization maximizes throughput across the following surrogate endpoints:

- **Acoustics**: Predicted weighted sound reduction index R_w (via *SoundLabAI*).
- **Strength**: Utilization metrics for stress σ and deflection w (via *StrengthLabAI*).
- **LCA**: Embodied Global Warming Potential GWP (via *CarbonReduceAI*).

6.3 Normalization and Weighted Aggregation

To achieve fair comparison across disparate performance metrics, the optimizer employs min-max normalization to map all objectives to a common [0,1] scale.

For each objective function f_i, the normalized value f_i^{norm} is computed by first identifying the minimum and maximum values across all candidate configurations:

$$f_i^{min} = \min_{\mathbf{x} \in \mathcal{C}} f_i(\mathbf{x}), \quad f_i^{max} = \max_{\mathbf{x} \in \mathcal{C}} f_i(\mathbf{x}) \tag{6}$$

where $\mathcal{C}$ represents the set of candidate configurations.

The normalization formula depends on whether the objective should be maximized or minimized: For objectives to be maximized (e.g., acoustic insulation parameters):

$$f_i^{norm}(\mathbf{x}) = \frac{f_i(\mathbf{x}) - f_i^{min}}{f_i^{max} - f_i^{min}} \tag{7}$$

And for objectives to be minimized (e.g., GWP, maximum stress):

$$f_i^{norm}(\mathbf{x}) = \frac{f_i^{max} - f_i(\mathbf{x})}{f_i^{max} - f_i^{min}} \tag{8}$$

Those normalization formulas ensure that higher normalized values consistently represent better performance across all objectives. When $f_i^{max} = f_i^{min}$ (all candidates have identical performance), the normalized value is set to 1.0 as default.

Subsequently, a composite utility score $U(\mathbf{x})$ is derived through weighted linear aggregation:

$$U(\mathbf{x}) = \sum_{i=1}^{k} w_i \cdot f_i^{norm}(\mathbf{x}), \quad \text{subject to} \quad \sum_{i=1}^{k} w_i = 1, \quad w_i \geq 0 \tag{9}$$

where k denotes the number of objectives (2 or 3 depending on optimization mode) and w_i stand for the user-defined weights representing design priorities.

6.4 Ranking and Selection Workflow

The optimization process follows a systematic workflow to identify and rank optimal glass configurations:

1. Compute performance metrics for the user's initial design using surrogate models.
2. Create candidate configurations through systematic thickness variations ($\pm$ 1 discrete step per glass pane), filtering duplicates and constraint violations.
3. Predict performance metrics for all candidates simultaneously using asynchronous HTTP requests to surrogate model APIs, reducing evaluation time from minutes to seconds.
4. Transform all objectives to [0, 1] scale by collecting values across candidates, computing global min/max, and applying normalization formulas.
5. Compute normalized ratios ρ_i and aggregate utility $U(\mathbf{x})$.
6. Compute composite scores for each candidate, and rank them in descending order by utility score.
7. Return top-performing configurations for multi-criteria decision-making.

6.5 Summary

The presented optimization framework provides a robust methodology that enables performance-based design exploration of glass façades. By implementing a local search strategy through discrete parameter perturbation, the system avoids combinatorial complexity while ensuring architectural feasibility. The core of the framework lies in its ability to couple rapid, asynchronous surrogate evaluations for acoustic, structural, and environmental criteria with a normalized weighted-sum aggregation. This approach enables the instantaneous ranking of design candidates according to multi-objective utility, allowing practitioners and researchers to navigate the complex trade-offs between mechanical safety, sound insulation, and embodied carbon during the decisive early stages of the design process.

7 Application Example: Surrogate-Based Multi-objective Optimization of a Vertical Double Insulating Glazing

This section demonstrates a realistic application of the surrogate-based modeling and optimization framework introduced in this book by revisiting a double insulating glazing system from [14]. The example involves a 1500 mm × 2500 mm insulating glass unit with an outer 8 mm fully tempered glass (FTG) pane and an inner laminated safety glass (LSG) composed of two 5 mm float glass (FG) panes with a 0.76 mm PVB interlayer. The assembly is supported on all edges, and environmental and service loads are defined in accordance with standard façade design practice acc. to DIN 18008.

Figure 9 illustrates the system layout and layer configuration. Load Cases (LC) considered include climatic effects (summer and winter pressure differences due to geodetic and atmospheric changes), wind suction, and horizontal live load (e.g., impact at 1.0 m height) are summarized in Table 4. Stresses and deformations are evaluated using the AI tools of the previous chapters for both, serviceability and ultimate limit states.

The baseline example (without interlayer bond) demonstrates that both outer and inner panes satisfy structural criteria: maximum design stresses remain below material capacities, and deformations meet serviceability limits. These reference results serve as a benchmark against which surrogate predictions and optimization outcomes are compared.

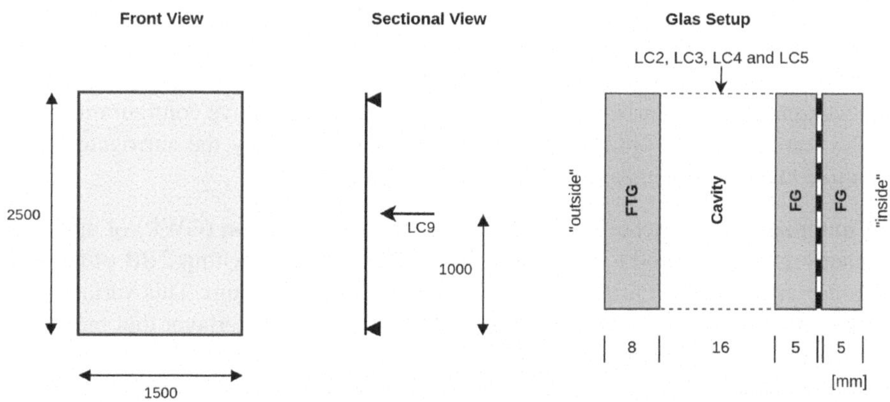

Fig. 9 Geometry and layer build-up of the Example B12 vertical double insulating glazing (FTG outerside, LSG from FG with PVB interlayer inside). Figure created by the authors

Table 4 Summary of load cases and resulting pressures for Example B12 (kN/m^2)

LC No.	Description	Design magnitude
LC2	Climatic load (summer, geodetic)	+ 7.20 (kPa)
LC3	Climatic load (summer, met. + temp.)	+ 8.80 (kPa)
LC4	Climatic load (winter, geodetic)	− 3.60 (kPa)
LC5	Climatic load (winter, met. + temp.)	− 12.50 (kPa)
LC8	Wind suction (characteristic)	− 1.12 (kPa)
LC9	Horizontal live load (impact, 1.0 m)	− 1.00 (kN/m)

To embed this example in the multi-objective framework, we define three performance criteria:

- **Structural utilization** (f_{strength}): peak stress or utilization ratio predicted by the strength surrogate,
- **Acoustic performance** (f_{sound}): weighted sound reduction index R_w predicted by the sound surrogate,
- **Environmental impact** (f_{LCA}): embodied carbon expressed as global warming potential (GWP) per unit area predicted by the LCA surrogate.

Each criterion is normalized relative to the baseline and aggregated with equal weighting (1/3 each) into a single objective for ranking. Surrogate predictions are obtained via API calls for each candidate design during optimization.

The optimization begins with the B12 baseline configuration and generates candidate variants by perturbing discrete design variables (e.g., glass thickness combinations and interlayer choices) within practical bounds. Each candidate is evaluated with the surrogate models, and objective ratios are computed relative to the baseline. The weighted sum scalarization guides the ranking of alternative configurations.

Two representative optimized configurations identified by the surrogate-based optimizer demonstrate meaningful design improvements:

- **Configuration A**: Achieves a reduction in embodied carbon (GWP) of approximately 15% compared to baseline, while maintaining R_w within 2 dB of the reference and structural utilization below the baseline maximum. This variant uses a slightly thinner outer pane combined with an alternate interlayer that preserves stiffness and acoustic damping.
- **Configuration B**: Improves acoustic performance by 3 dB relative to baseline while preserving structural integrity and keeping embodied carbon within 5% of the reference value. This design emphasizes thickness adjustments and material selection that favor mass and damping effects.

Figure 10 compares predicted performance metrics for the baseline and two optimized designs, illustrating how surrogate models capture relative changes across key objectives.

These results demonstrate that surrogate-based optimization enables rapid exploration of trade-offs among acoustics, structure, and sustainability—without requiring detailed simulations for every candidate—making it well suited to early-stage design workflows.

8 Discussion

This work has presented an integrated framework for performance-based design and multi-objective optimization of glass façade elements, combining fast surrogate models for acoustics (SoundLabAI), structural performance (StrengthLabAI), and environmental impact (CarbonReduceAI) with a pragmatic and computationally efficient optimization strategy. By replacing repeated high-fidelity simulations with highly-accurately trained surrogate models and lightweight evaluation pipelines, the framework enables rapid and interactive exploration of optimal façade design alternatives, which is particularly valuable in early design stages where design freedom is high and decision-making is most influential. The tools provided in this chapter cut computational cost in terms of typical evaluation times by at least 50% compared to conventional FEM-based workflows.

The case studies presented in Sect. 7 demonstrate that façade design problems are inherently multi-disciplinary and governed by competing performance objectives. Improvements in one domain, such as enhanced airborne sound insulation through increased glass thickness, often come at the expense of other objectives, notably increased material consumption and embodied carbon. Similarly, reductions in structural utilization do not necessarily correlate with improved acoustic or environmental performance. These observations are consistent with findings in the broader façade optimization literature, where complex and often non-intuitive interactions between design variables and performance criteria necessitate systematic multi-objective decision support approaches [2, 20]. Within this context, the presented framework provides actionable insights by quantifying relative performance changes and exposing trade-offs in a transparent and interpretable manner.

The case studies presented in Sect. 7 illustrate that façade design problems are inherently multi-disciplinary and governed by competing performance objectives. Enhancing airborne sound insulation through increased glass thickness, for example, may lead to higher embodied carbon, while reduced structural utilization may not improve acoustic performance. These findings confirm patterns observed in façade optimization literature, emphasizing the need for systematic, multi-objective decision support approaches [2, 20]. Within this context, the presented framework provides actionable insights by quantifying performance deltas (typically 5–15% variation among competing objectives) and transparently exposing trade-offs between criteria that are otherwise difficult to assess.

Fig. 10 Comparison of surrogate predictions for baseline and optimized configurations across structural utilization, R_w, and GWP. Figure created by the authors

A key central contribution of this work lies in the effective deployment of surrogate models as enablers of such multi-objective exploration. *SoundLabAI* leverages data-driven learning on high-fidelity acoustic simulation datasets to provide accurate predictions of weighted sound reduction indices for arbitrary glazing configurations, substantially reducing the need for computationally expensive numerical acoustics or physical testing [7]. *StrengthLabAI* employs a customized mixture-of-experts architecture to robustly predict stresses and deflections across a wide range of monolithic and laminated glass systems, covering both serviceability and ultimate limit state indicators with high predictive accuracy [18]. *CarbonReduceAI* complements these models by providing embodied carbon estimates derived from material volumes and emission factors, enabling sustainability considerations to be directly embedded into façade design decision processes. Together, these tools form a coherent web-based surrogate ecosystem that supports rapid feedback and iterative refinement, which would be infeasible using conventional simulation-based workflows.

The optimization strategy adopted in this work follows a heuristic perturbation-based approach tailored to the discrete nature of typical façade design variables, such as standardized glass thicknesses and laminate configurations. Rather than aiming to exhaustively sample the full design space or compute complete Pareto fronts, the optimizer focuses on generating locally perturbed alternatives around a user-defined baseline and evaluating them efficiently using the surrogate models. This strategy balances search breadth with computational efficiency and has proven effective in identifying locally optimal or near-optimal configurations that provide balanced improvements across multiple objectives. The use of min-max normalization and weighted aggregation allows designers to express preferences between competing criteria while maintaining interpretability and responsiveness in the optimization process.

Despite its demonstrated effectiveness, several limitations of the current framework should be acknowledged. First, the weighted-sum aggregation of objectives, while simple and computationally efficient, does not explicitly capture the full set of non-dominated solutions typically associated with Pareto-optimal analysis, where Pareto-based algorithms such as NSGA-II could be employed to reveal a broader spectrum of trade-offs and design alternatives [20]. Second, the current implementation of CarbonReduceAI focuses on global warming potential (GWP) as a primary indicator of embodied environmental impact. While GWP is widely used and highly relevant, comprehensive life cycle assessment practice encompasses additional impact categories, including resource depletion, acidification, eutrophication, and toxicity, as well as different life cycle stages and temporal considerations [9]. Extending the surrogate to include additional impact categories and more detailed life cycle modeling would broaden its applicability, especially in regulatory or certification-driven design contexts. Finally, uncertainty quantification is not yet explicitly integrated into the surrogate predictions or optimization outcomes. As surrogate models inevitably introduce approximation errors—particularly in extrapolation regions—future work could incorporate uncertainty-aware modeling and robust optimization techniques to provide confidence bounds and risk-sensitive design recommendations. Such approaches are gaining traction in structural and environmental optimization research and would further enhance trust and reliability in surrogate-assisted design workflows [19].

9 Conclusions

This work delivers a practical and extensible framework for surrogate-based multi-objective optimization of glass façades, integrating acoustic, structural, and environmental considerations into a cohesive web-based workflow at TRL 9. The approach bridges high-fidelity simulation and real-world design practice by enabling rapid, interpretable, and performance-driven decision support. Preliminary benchmarks indicate substantial computational savings (at least 50% speed up), improved design iteration rates (at least up to three-fold), and clearer trade-off visualization, enhancing adoption potential in industrial façade design contexts. By aligning with current research in façade optimization and sustainability analytics, and by outlining pathways for future enhancement, such as uncertainty integration and extended life-cycle modeling, the framework provides a robust foundation for continued academic and industry application in advanced façade engineering.

10 Future Work and Outlook

Ongoing and future developments aim to enhance the applicability and accessibility of the proposed framework. Planned extensions include the integration of an expanded database encompassing both acoustic and sustainability data to support broader generalization across various façade design domains. Methodologically, the incorporation of uncertainty quantification and Pareto-front-based optimization (e.g., NSGA-II) will further strengthen model robustness and interpretability. Finally, the web-based toolset will be continuously refined based on user feedback and project-scale applications.

Acknowledgements We gratefully acknowledge the valuable contributions of Dr.-Ing. Michael Drass, Rafael Bischof, Dr.-Ing. Henrik Riedel, Leon Schmeiser, and Dr.-Ing. Alexander Pauli to the design, training, and implementation of the tools presented in this chapter. Furthermore we acknowledge the financial support of Kuraray Europe GmbH and ZM-I.

References

1. ASTM E1300-16 (2016) Standard practice for determining load resistance of glass in buildings. American Society for Testing and Materials, West Conshohocken, PA
2. Bianchi F, Loonen R, Favoino F (2024) Multi-criteria performance-based design of adaptive building façades. Build Environ 248:110944. https://doi.org/10.1016/j.buildenv.2023.110944

3. Chen F, Huang Y, DeGanyar T (2019) Sound insulation of monolithic or laminated single- and double-glazing panels. In: Prediction methods for sound insulations, Aachen
4. DeGanyar T, Chen F, Huang Y, Grotemeyer R (2019) Sound insulation of fenestration systems: a comprehensive web-based simulation program and validation
5. DIN 18008-1:2020-05 (2020) Glass in building—design and construction rules—part 1: terms and general bases. Deutsches Institut für Normung, Berlin, DE
6. DIN EN ISO 717-1:2021-05 (2021) Acoustics—rating of sound insulation in buildings and of building elements—part 1: airborne sound insulation (ISO 717-1:2020). Deutsches Institut für Normung [DIN]/International Organization for Standardization, Berlin, DE
7. Drass M, Kraus MA, Riedel H, Stelzer I (2022) SoundLabAI—machine learning for sound insulation value predictions of various glass assemblies. Glass Struct Eng 7(1):101–118. https://doi.org/10.1007/s40940-022-00167-z
8. EN 16612:2019 (2019) Glass in building—determination of the lateral load resistance of glass panes by calculation. European Committee for Standardization, Brussels, BE
9. Fnais A, Nguyen K, Rezaei F (2022) Life cycle assessment of building façades: a review of methodological developments and challenges. J Build Eng 57:104879. https://doi.org/10.1016/j.jobe.2022.104879
10. Galuppi L, Royer-Carfagni G (2013) The effective thickness of laminated glass: inconsistency of the formulation in a proposal of en-standards. Compos B Eng 55:109–118
11. Galuppi L, Royer-Carfagni G (2020) Green's functions for the load sharing in multiple insulating glazing units. Int J Solids Struct 206:412–425
12. ISO 14040:2006 (2006a) Environmental management—life cycle assessment—principles and framework. International Organization for Standardization, Geneva, Switzerland. https://www.iso.org/standard/37456.html
13. ISO 14044:2006 (2006b) Environmental management—life cycle assessment—requirements and guidelines. International Organization for Standardization, Geneva, Switzerland. https://www.iso.org/standard/38498.html
14. Kasper R, Pieplow K, Feldmann M (2016) Beispiele zur Bemessung von Glasbauteilen nach DIN 18008, 2. auflage edn. Ernst & Sohn, Berlin. ISBN 978-3-433-03153-7
15. Kraus MA (2019) Machine learning techniques for the material parameter identification of laminated glass in the intact and post-fracture state. Doctoral thesis, Universität der Bundeswehr München. Supervisors: Siebert G, Schneider J, Müller G
16. Kraus MA, Drass M (2020) Artificial intelligence and machine learning for structural glass engineering applications—overview, case studies, and future potentials. Glass Struct Eng 5(4):567–592. https://doi.org/10.1007/s40940-020-00130-0. Overview of AI/ML applications in structural glass engineering
17. Kraus MA, Schuster M, Kuntsche J, Siebert G, Schneider J (2017) Parameter identification methods for visco- and hyperelastic material models. Glass Struct Eng 2(2):147–167
18. Kraus MA, Bischof R, Riedel H, Schmeiser L, Pauli A, Stelzer I, Drass M (2024) StrengthLabAI: a mixture-of-experts deep learning approach for limit state analysis and design of monolithic and laminate structures made of glass. Glass Struct Eng 9(4):607–655. https://doi.org/10.1007/s40940-024-00277-w
19. Moustapha M, Sudret B (2022) Surrogate-based multi-objective optimization with uncertainty quantification: a review. Struct Multidiscip Optim 65(6):1–28. https://doi.org/10.1007/s00158-022-03214-5
20. Shan M, Hammad A (2023) Multi-objective optimization of building façade design: a review of methods and applications. Autom Constr 150:104838. https://doi.org/10.1016/j.autcon.2023.104838
21. Stelzer I, Kraus MA, Drass M, Riedel H (2023) SoundLabAI tool—machine learning for sound insulation value predictions. In: Proceedings of glass performance days (GPD) 2023, 2023. Conference contribution on the SOUND LAB AI tool presented at GPD Finland

R-IOT: Concept for Predictive Cyber-Physical Facade Maintenance

Jens Böke, Natchai Suwannapruk, Puttakhun Vongsingha, Lars Anders, and Paul-Rouven Denz

Abstract The facade mediates environmental boundary conditions and user requirements. It has a significant impact on the interior comfort and energy efficiency of a building. In view of today's ambitious sustainability goals, there is an increasing requirement for energy and material reduction in the construction sector. Building stock is becoming more important as the facade construction market is shifting towards renovations and modernizations. Facade renovation measures often reveal further undetected wear and damage, exposing performance losses and leading to unplanned project expansions with additional construction time and costs. In building automation, a wide range of sensor and monitoring technologies exists. Cyber-physical systems refer in this context to the tight integration of physical components with digital information processing and control. The presented study addresses the question of how such systems can be used in building operation for predictive facade maintenance in order to continuously identify possible failures and ensure their performance. The main finding of the study is the holistic nature of cyber-physical maintenance concepts, which require integrated measures in both the physical and cyber domains. As a result of the study, the R-IOT concept addresses comprehensive facade sensor integration with a Design for Disassembly construction approach to facilitate maintenance. The development offers new potential for extending the service life of facades with additional opportunities by the future use of artificial intelligence.

Keywords Building automation · Facade renovation · Energy performance · Design for Disassembly · Sensor network · Health monitoring · Maintenance strategy

Abbreviations

BAS Building automation system

J. Böke (✉) · N. Suwannapruk · P. Vongsingha · L. Anders · P.-R. Denz
Priedemann Fassadenberatung GmbH, Grossbeeren/Berlin, Deutschland, Germany
e-mail: jens.boeke@priedemann.net

A. Mejorin et al. (eds.), *Performance-Based Façade Design*, Springer Tracts in Civil Engineering, https://doi.org/10.1007/978-3-032-28835-6_7

CPS	Cyber-physical system
DfD	Design for disassembly
SHM	Structural health monitoring
TRL	Technology readiness level
UI	User interface

1 Introduction

1.1 Background on Facade Service Life and Renovation

Facades mediate between the external environment and interior of buildings with a significant impact on the energy performance, user comfort and appearance [1]. They are complex systems of various functional layers, distinct performance requirements, and different risk failure types and lifespans [2]. Their overall service life largely depends on the individual integrated components and materials. Structural and infill components have a longer lifespan, compared to other interface components such as gaskets, sealants and thermal breaks. Those less durable parts need consideration already after approximately 25–30 years [3]. Depending on the facade typology, construction methods and proper maintenance, the service life can reach up to 50 years [4]. Figure 1 illustrates the typical life span of a facade broken down into the durability of the individual materials and components used.

A high demand for facade renovations exists today in view of the aging building stock. Therefore, the European Renovation Wave initiative was introduced to accelerate the building modernization rate by 2030 [5]. Starting from the main reason of counteracting wear and tear, several further factors determine individual decisions for building renovations [6]. They are usually linked to economic considerations, such as reducing energy costs by improving the energy performance, accessing incentives, change of building use, or maintaining the monetary value of the building [7]. Continuous revisions of energy standards and certifications over the course of a building's service life create additional need and accelerate the renovation cycle.

The scope of renovations varies by project, ranging from minor repairs to complete replacements of the facade [8]. In many cases, repairs are only carried out on the basis of observed or reported symptoms, leaving the underlying causes unresolved and allowing the defects to persist and recure [9]. The actual extent of necessary interventions often only becomes clear during the renovation work on site. When the facade is opened up, hidden signs of wear and tear and problems that had previously gone undetected come to light [10]. Consequences are unplanned tasks, delays and additional costs. Renovations of the building envelope often lead to the relocation of occupants with direct economic losses due to business interruption and lowered level of productivity [11]. Compared to demolition and reconstructing of a building with significant waste production and extensive utilization of grey energy during the

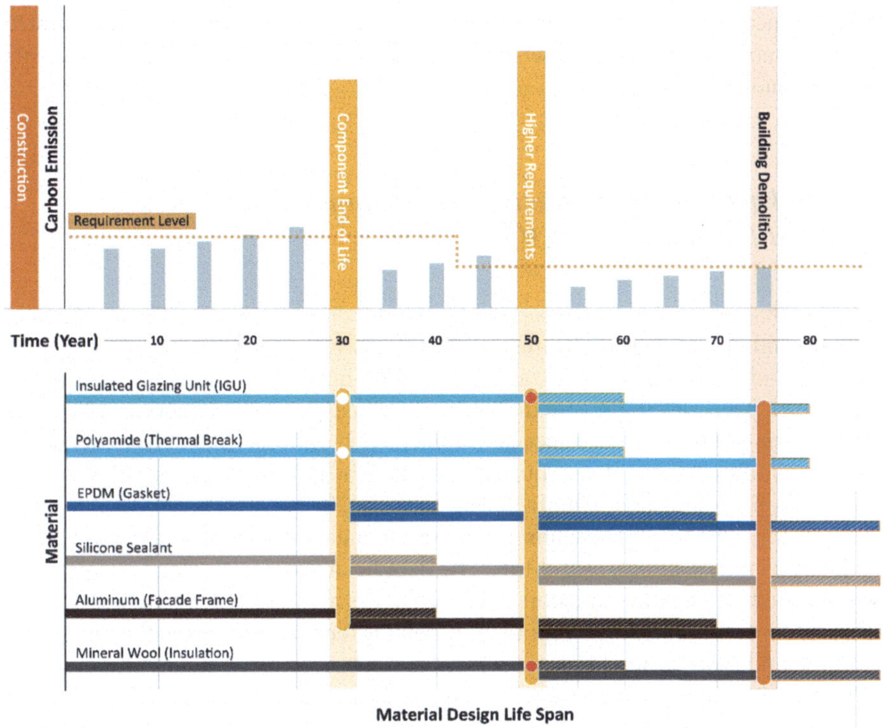

Fig. 1 The typical service life of facades, taking CO_2 emissions and material lifespans into account. *Source* Authors

process, renovations are still considered as a lower-threshold and more sustainable strategy [12].

While the construction sector is still largely based on linear business models with a "take, make, dispose" mindset, building circularity is becoming increasingly important in today's sustainability debate [13]. In addition to operational energy consumption, this raises awareness of the material and energy required for construction. Various circular economic concepts such as Cradle-to-Cradle or the R-Strategies have been introduced by different stakeholders like the Ellen MacArthur Foundation in recent years. Building circularity aims to maximize the utility and value of material resources throughout their lifespan [14]. The application of circular principles in the construction industry requires design consideration of adaptability, flexibility and disassembly to enable reversibility and recyclability of the building products [15].

The resulting awareness of components and materials reuse after their end of life leads to increasing relevance of strategies such as Design for Disassembly (DfD) [15, 16]. DfD as a strategy originates from manufacturing industry with the aim to minimize waste [17]. In the construction sector, its potential already becomes identified for recovering material resources for reuse, repair, remanufacturing and recycling [18]. By anticipating the disassembly process at the end-of-life through the utilization mechanical connections that are accessible and selecting components size that are suitable for handling, DfD provides ease of access of specific components during maintenance [19].

1.2 Background on Predictive Maintenance and Structural Health Monitoring (SHM)

Predictive maintenance is characterized by the direct monitoring of the mechanical condition and operational performance of systems to ensure efficient maintenance schedules and operational optimizations based on data [20]. It facilitates informed maintenance decisions to prevent disruption in processes and productions [21]. The strategy of predictive maintenance is established in various industries, especially in sectors with high reliability requirements, such as power plants, public services, transportation systems and manufacturing sectors [22, 23]. Machinery health prognosis typically consists of four subsequent stages: data acquisition, health indication, health stage and remaining useful life. The data from acquisition stage is processed and translated into health indicators. The indicators are interpreted to a health stage of the machine. The identified degradation trend is then used to assess the remaining useful life and pre-specified failure thresholds [24].

Predictive maintenance concepts involve automation technologies with sensors, data acquisition- and transmission, a database with data analysis and modeling, condition assessment, performance prediction, alarm devices, a visualization user interface, and software for error detection [25]. They can be categorized into three main techniques based on the source of the data: Existing sensor techniques (Process sensor) utilize operational output data to detect anomalies in the system. Test sensor techniques apply supplementary sensors to monitor the behavior of the operation, for example vibration and acoustic sensors to detect leaks. Test signal-based techniques sends signals into the system, for example measuring resistance and capacity to identify cracks, corrosion or wears of the equipment [26].

In the construction sector Predictive maintenance is realized by means of structural health monitoring (SHM), especially in civil engineering, infrastructure and large-scale structures. Here it serves as a tool for disaster mitigation and monitoring of structures throughout their service life [27]. In addition to structural health, functionality, integrity and safety, a crucial aspect of SHM is to detect the occurrence of damage and propagation behavior [28]. A wide range of sensors have been employed in SHM, including accelerometers, temperature sensors and strain gauges [29]. Existing research demonstrated the application of SHM in the monitoring of aging highways [30], or the vibration behavior of long span bridges [31].

1.3 Background on Building Automation Systems (BAS) and State of the Art in Facade Health Monitoring

In the domain of facades, automation technologies are mainly applied as hierarchically organized sensors, controllers, and actuators of Building Automation Systems (BAS). The implementation of BAS is increasing due to its potential to optimize the building operation, enhancing energy efficiency and improving occupant's comfort and satisfaction [32]. According to ISO 16484–3, the functionality of BAS can be categorized into six functionalities including: Grouping and Zoning, Event Notification, Alarm Notification, Historical Data Access, Scheduling and Scenarios. Through these functions, BAS supports comprehensive building control and management, enabling applications such as automated HVAC control, optimized lighting operation, and real-time equipment monitoring [33].

Various existing research studies relate to the implementation of monitoring systems and sensor networks in facades. Luna-Navarro et al. [34], for example, investigate sensor networks in glazing units to measure the impact of the facade on the comfort and behavior of occupants. Arnesano et al. [35] propose a prefabricated facade module with integrated sensors and shading control to maximize user comfort and energy efficiency. Existing research also addresses cyber-physical automation concepts for facades, which aim in particular to increase operational performance through coordinated automations of adaptive facades [36]. A key feature is the decentralized structure of such systems, which enables a high degree of flexibility and resilience.

Cyber-physical systems (CPS) are based on the close integration of physical components with their digital control. In cyber-physical facades, the physical domain refers to devices and components with integrated automation technology. The digital control includes data, the software system and the communication network. The state of the physical system is digitally represented on the cyber level to providing data to support user decision-making and to regulate the physical system through automated or user informed control [37]. As a cyber aspect of CPS, digital twins serve as virtual representation of the physical aspects with the functions of a repository: storing, updating and sharing information and conditions of the physical asset throughout its lifetime [38]. The dynamic nature of the digital twins allows for real-time data from sensors to be utilized for advance modeling and simulation, enabling continuous monitoring, analysis and optimization of the physical system [39].

1.4 Problem Statement and Objectives

According to the current state of the art, renovations are the method of choice for renewing and modernizing worn and inefficient facade constructions. Unrecognized and previously hidden damage often comes to light during renovation measures and indicates functional limitations and performance losses in the previous operation of the building, leading to unplanned extensive renovations with additional costs, time, and material consumption. In many disciplines outside the construction sector, the strategy of predictive maintenance is employed to facilitate the reliability and longevity of physical assets. Continuous monitoring can contribute to smooth operation of technical systems and extends their service life through appropriate maintenance measures. Although the technological prerequisites, such as mechanical devices and available automation technologies, exist in the field of facades, this approach has not yet been transferred.

Against this background, the aim of the study is to demonstrate the feasibility and potential of predictive maintenance strategies for facades and to provide a design reference as a guideline for the cyber-physical implementation of such systems. The objective is to develop an approach to maintain the performance of the facade throughout the entire life cycle of a building. The long-term objective is to minimize the resource expenditure and downtime in building operation due to renovations by implementing a continuous, demand-oriented maintenance concept. The study considers predefined performance levels, the integration of a monitoring system that records the behavior and condition of facade components, and a disassembly strategy for easy access and maintenance.

2 Methodology

The study follows a Research by Design approach, utilizing the design process itself as a method of inquiry [40]. Starting point of the investigation was the Metals in Construction Design Challenge 2024, calling for the development of a sustainable and at the same time resilient curtain wall system for an existing high-rise building in New York. Central aspect of the competition task was a desired facade service life of at least 75 years in order to match the lifespan of the overall building [41]. The competition task served as a case study for the development of the R-IOT facade concept. In the design process, decisions are systematically made based on the desired criteria for integrated cyber-physical predictive maintenance. An important followed principle is the potential for generalization of decisions and technology applications. This is to produce new transferable knowledge and value beyond the specifically proposed design application in the focused use case scenario [42].

In the design process, both the digital data evaluation at the cyber level and the physical are initially examined separately and then combined into an overall concept. At the cyber level, the study investigates the data processing workflow in different stages and the application of User interface (UI) as a visualization tool for performance monitoring and maintenance planning. On the physical level, the focus lies on the development of a DfD design with consideration of maintenance procedures and potential disassembly after the facade's service life. A literature review is conducted to determine the fundamentals and current state of the art in the field of facade renovations, predictive maintenance concepts, sensor-based monitoring systems and design for disassembly. The design is developed using the Rhinoceros 3D planning environment due to its extensive capabilities for digital negotiation of various design parameters. At different stages of the design process, the planning status is discussed in workshops and reviewed by stakeholders and experts from the facade industry.

3 Results

Result of the study is a holistic facade concept, developed under the title R-IOT. Key aspect is the cyber-physical monitoring system to record the building physical characteristics and status information of the facade in combination with a DfD maintenance approach. On the one hand, it consists of sensors and hardware, which are integrated into the physical functional levels of the facade. During ongoing building operation, the integrated sensor technology provides comprehensive system health information and performance data. The data is evaluated at the cyber level by means of a digital twin, comprising a visualized dashboard representation of the facade system and a database. The digital twin also serves as a repository of monitored data which is used as a basis for component health assessment and maintenance schedule. Additionally, the repository logs the details of the past maintenance and performed retrofits. Figure 2 shows the overall concept of the system at the interface of physical facade components and evaluation on the cyber level in the following detailed steps: data acquisition, diagnosis, evaluation, and reporting.

3.1 Data Acquisition

Environmental conditions with direct impacts on the building, such as temperature, relative humidity, air quality, solar irradiance, windspeed and precipitation level are monitored by a weather station. This gathered data serves as the baseline to assess the facade's performance and for deriving mitigation plans. Sensors are installed in facade components such as profiles, infills, and claddings to monitor the impact of the environment. For instance, in the glazing layer, surface temperature and heat flux state are monitored to evaluate thermal behavior. These raw values are then calculated towards the U-value, which quantifies the rate of heat transfer through the element. Also the G-value is monitored as the total solar energy transmittance, defining the effectiveness of the glazing as a thermal barrier. Additional parameters such as air infiltration are also monitored at the gasket element to identify possible air ingress and gasket defects. As part of the acquisition, the raw monitoring data is cleaned and processed into a structured data set. The structured data becomes visualized in a dashboard by numerical values and graphs as shown in Fig. 3. The dashboard also tracks and summarizes the maintenance schedule and lifespan of the facade components. It provides alert when its performance drops below the defined

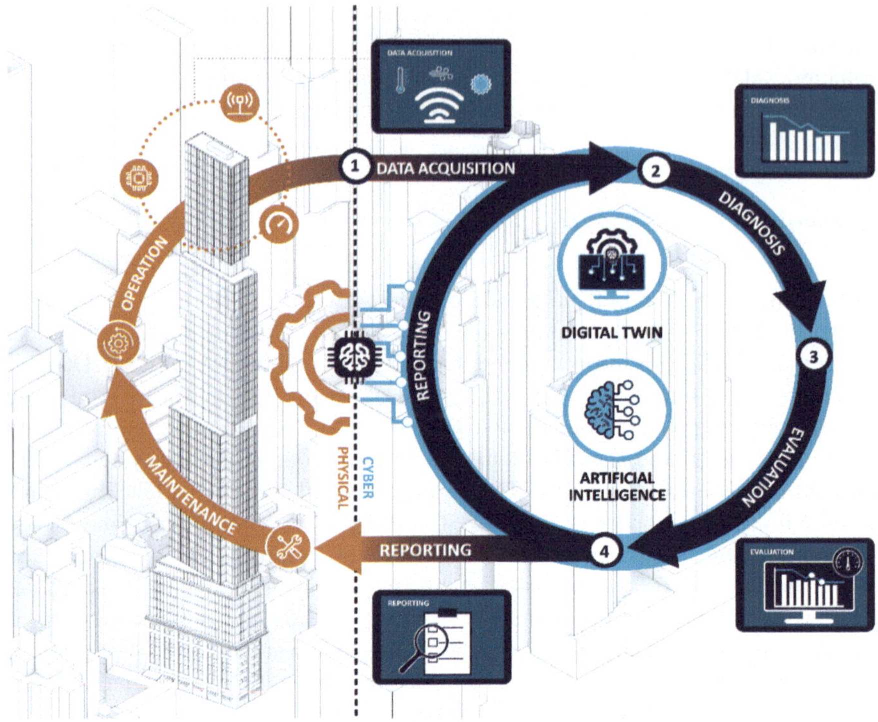

Fig. 2 Principle of the cyber-physical monitoring system. *Source* Authors

threshold or when maintenance action is required. This approach enables a strategic maintenance schedule to be planned by the facility manager.

Indoor conditions such as temperature, relative humidity, illumination and Co_2 level are monitored. The energy consumption of the building power consumption is also monitored, this includes a detailed documentation of the energy usage from lighting, appliances and HVAC system. By understanding the exterior conditions and the impact it has on the interior space, the performance of the facade as an interface can be further analyzed.

3.2 Diagnosis

During diagnosis, the structured data set is interpreted by cross-referencing and establishing correlations in order to identify potential problems. Damage patterns and signs of wear are detected by observing and cross refencing the stimuli factors with the condition monitoring of the facade components. For instance, the performance of the facade's insulation is evaluated by referencing the external temperature to the indoor temperature and the facade infill surface temperature as well as humidity

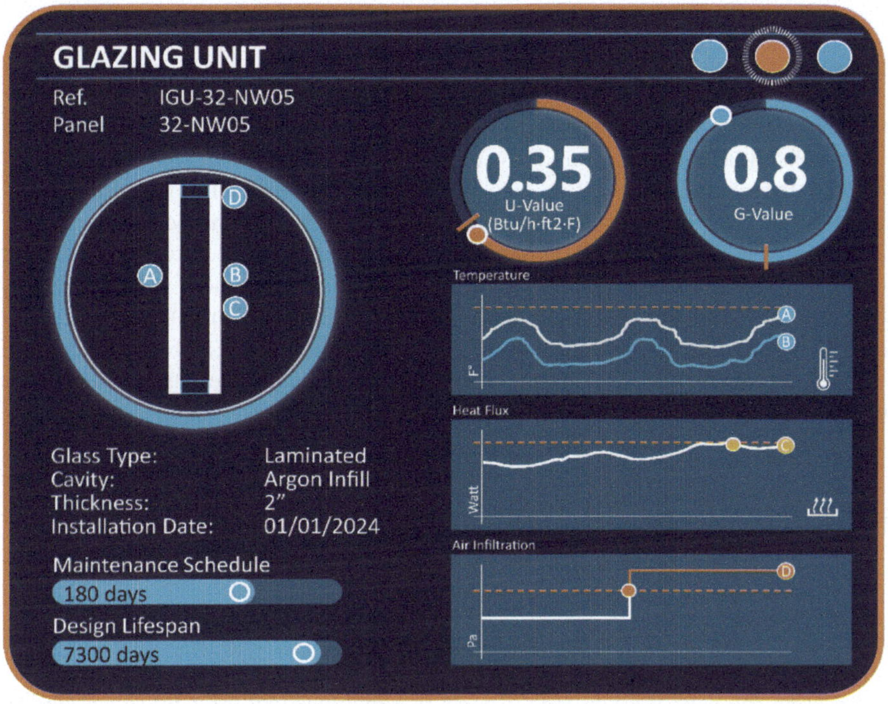

Fig. 3 Example for data acquisition by the sensor system at the glazing unit. *Source* Authors

conditions. Another example is the assessment of strain levels in facade profiles under high wind load conditions to detect anomalies that indicate early signs of structural deterioration.

3.3 Evaluation

The evaluation process identifies action requirements and necessary maintenance measures in the operation of the facade. This involves comparing the diagnosed condition data of the facade with predefined minimum requirements such as the U-value or strain level, as well as with project-specific predefined target criteria. Facade component conditions, occupant comfort, and building energy efficiency are taken into account. The automated detection of deviations between the actual condition of the facade with the specified target criteria is transferred to the localization of problem areas with associated maintenance requirements.

3.4 Reporting and Documentation

In reporting, status data and identified recommendations for action are provided to the building's facility management via the visualized dashboard in Fig. 4. The evaluation report consolidates the structured raw data from the system with recommendations on maintenance and intervention measures. The cyber-physical monitoring system supports the decision with its assessment of the need for action, localization of the problem, and comparison with recorded past errors and measures taken. The decision on the current measure, such as cleaning, repair, or replacement of components, remains at the disposal of facility management. To allow the monitoring system to learn from previous data and decisions, ongoing documentation and archiving are of crucial importance. Recorded data, diagnosis, and renovation measures are therefore documented in the digital twin.

3.5 Design for Disassembly (DfD) to Facilitate Physical Actions of Maintenance

In the thought line of the R-IOT concept, maintenance work must be conducted during building operation in accordance with the specified intervention requirements. This underscores the need for a holistic concept that includes low-threshold accessibility on physical level to enable efficient and rapid cleaning, repair, and replacement of components. An important foundation is the acknowledged differentiated understanding of the facade as a layered system of products, components and materials with individual ease of accessibility and different service-life.

R-IOT proposes a dismountable, modular unitized facade system that employs the DfD principle to provide ease of maintenance. Its high degree of standardization enables the reusability of components and interchangeability between panels. Plug-and-play connections allow the elements to be assembled or dismounted with ease. By minimizing the use of chemical bonding, the design provides reversibility and enables the components to be accessible and removable for maintenance. According to Fig. 5, the developed facade system distinguishes focal layers including: the infill, the spandrel element and the terminal.

The infill serves as the main mediator between the interior and exterior. It reflects the architectural intention by design options of either a transparent, opaque or partially opaque element. It additionally incorporates additional auxiliary functions such as sun shading or operable window system. The spandrel element serves as a structural interface between the facade system and the building structure. It is engineered to support the load of the infill elements which sit on top and transfer the load to the building structure through the brackets. The element is designed to accommodate a wide range of cladding elements, depending on the design intent of the architect. An important objective of the development is low-threshold maintenance access at minimized disruptions without downtimes for building operation and without

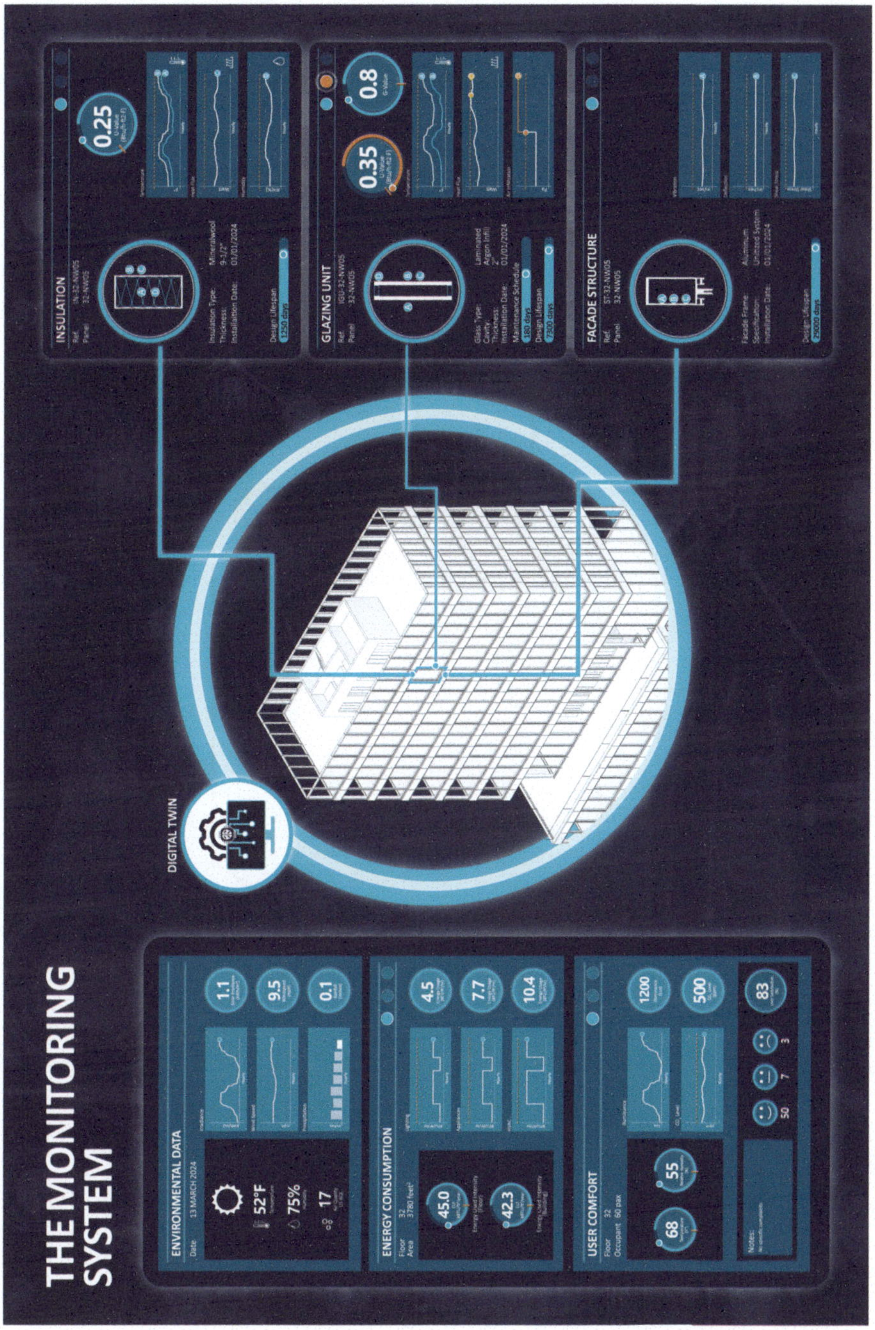

Fig. 4 Visualized dashboard as interface between the cyber-physical monitoring system and facility management. *Source* Authors

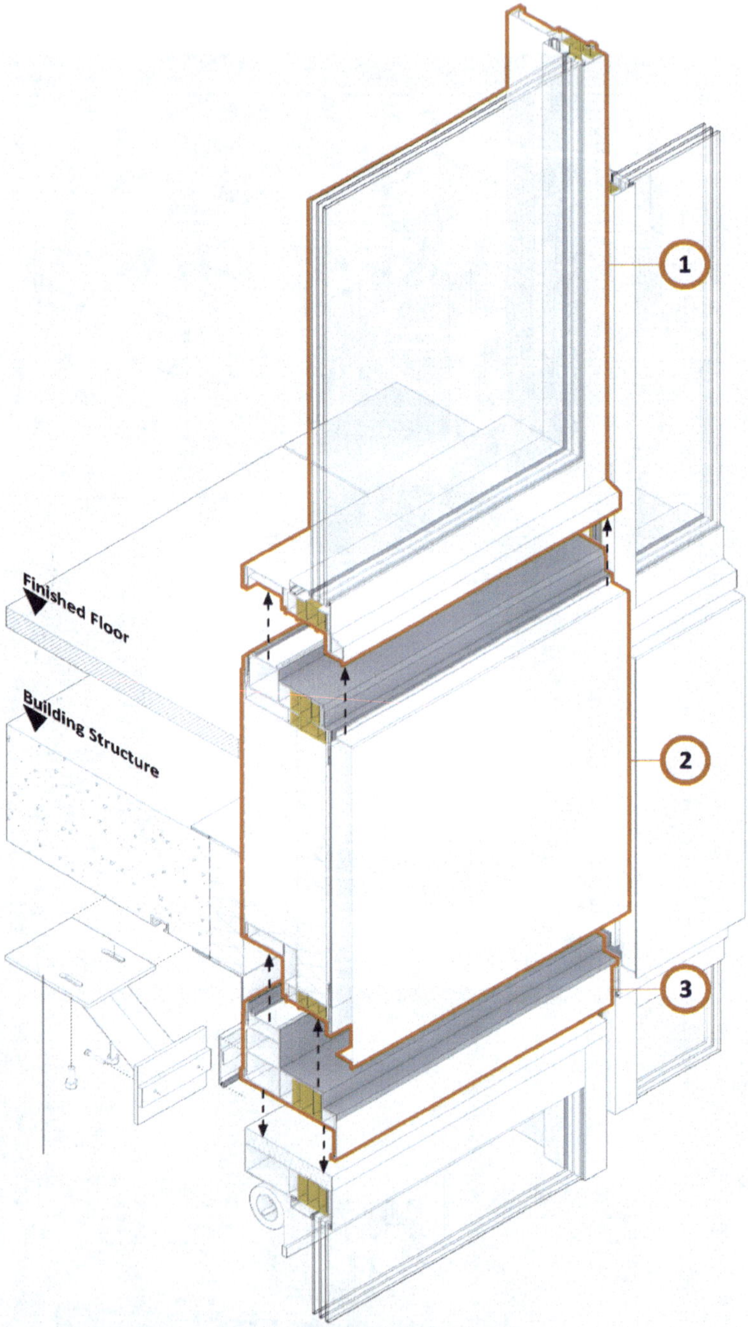

Fig. 5 Reversible R-IOT construction principle with the main aspects: infill (1), the spandrel element (2) and the terminal (3). *Source* Authors

on-site preparations such as scaffolding and heavy machinery. Unlike conventional facades, R-IOT therefore includes a terminal element that allows accessing of the entire construction from the inside. During the disassembly process of the spandrel element, the terminal element is the anchor point for the temporary bracket to support the facade module on the lower floor. Once dismounted, the elements can be further dismantled as shown in Fig. 6 to perform the necessary operations such as gasket repairs, glazing replacement or insulation upgrades.

The implemented cyber-physical predictive monitoring enables an approach to demand-oriented maintenance, which mediates between the wear and tear with performance losses of the facade and the investment costs of the necessary maintenance. The appropriate maintenance measure is decided on a case-by-case basis and includes cleaning work, repairs, or the replacement of components and assemblies. Figure 7 illustrates the fundamental shift of thinking in the R-IOT strategy compared to conventional renovation practices, now focusing on continuous measures to maintain the facade above required performance level.

The R-IOT concept is based on the idea of circular construction in the sense of applied R-Strategies. A conscious decision to "refuse" replacing a still functional

Fig. 6 Design concept for dismantling the R-IOT facade system from the inside without the need for scaffolding. *Source* Authors

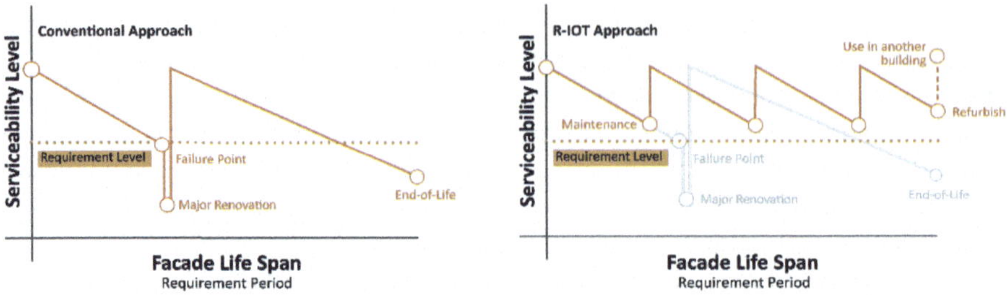

Fig. 7 Conventional Renovation strategy versus R-IOT approach. *Source* Authors

component is considered before repairing it or replacing it by a reused component, is considered. In this way of thinking, replaced components of the facade system do not necessarily become waste, but flow into a recycling cycle according to their condition, as shown in Fig. 8.

Beyond the individual facade, data-based health monitoring of individual constructions is intended to enable further synergies, for example through the neighborhood coordination of maintenance work on facades of different buildings. By combining comparable maintenance work, knowledge about required and otherwise available material resources, or the consolidation of orders for spare parts that

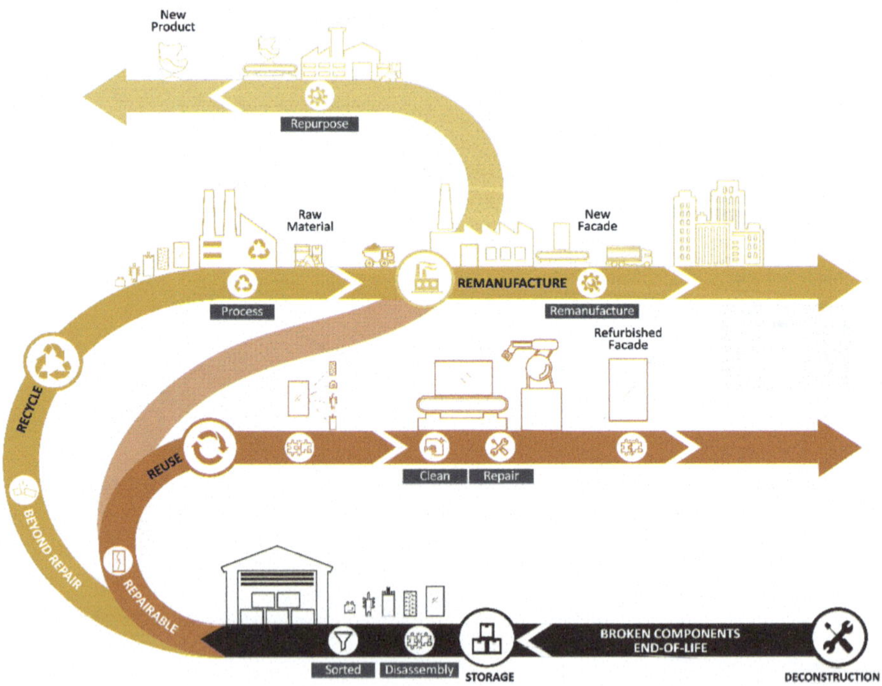

Fig. 8 Building circularity in the maintenance strategy of R-IOT. *Source* Authors

are needed multiple times, additional efficiency gains in facade maintenance are untapped. Within the scope of the study, R-IOT was initially developed only as a single exemplary facade construction. However, the long-term vision of the project is to create ecosystems of cyber-physically monitored facades in smart cities as illustrated in Fig. 9.

4 Discussion

On the physical level, the main achievement of the project lies in the combination of sensor integration with the DfD design approach. A finding of the study is that damage patterns and problems on the facade cannot be derived from a single source of information in a specific component or material. Accordingly, the focus is more on comparing different integrated sensors with each other and taking past system status data into account. Against this background, the interpretation of different data constellations with regard to problems that can be narrowed down and localized is a decisive factor. Further research is needed here to establish correlations between different sensor data in order to make meaningful diagnoses about the condition of the facade and potential problems. Within the scope of the concept, only initial examples such as the recording of glazing unit performance via differently positioned temperature sensors could be evaluated. The systematic design of sensor technology for various issues relating to facade performance should be the subject of future research.

The evaluation of sensor data at the cyber level was also implemented on an exemplary basis. It is based on individual algorithms for comparing the current situation with performance expectations. From the perspective of the research project, there is a particular potential for investigating the possible application of artificial intelligence, especially machine learning methods to further automate decision-making and increase accuracy through detailed pattern recognition in complex data situations. In such concept the sensor data is used for training the system to identify recurring damage scenarios and align maintenance schedules.

The potential of R-IOT lies in its consideration as an integrated facade system based on both equally relevant cyber- and physical sublayers. This finding of a holistic system understanding is consistent with the literature identified on general cyber-physical system definitions in the background study. Efficiency results from the high degree of integration of both levels within the facade concept and is determined at the interfaces between sensor-based data collection and the resulting predictive maintenance recommendations via reporting.

Fig. 9 R-IOT concept as an ecosystem for smart cities. *Source* Authors

5 Conclusion and Outlook

R-IOT was successfully developed as a concept and demonstrates the potential of integrating cyber-physical maintenance strategies into the facade. At the current stage, the concept corresponds to an early Technology Readiness Level (TRL) 1–3 in the research phase. The exemplary design process can serve as a reference for both further research and the application of strategies of predictive maintenance in the field of facades. Examples of exploitable aspects include the consideration of reversible connections and the avoidance of adhesive connections in detailed planning, or the positioning of specific sensor technology at the functional levels of the facade. In this sense, the study serves as a guideline for further research and actual application of corresponding systems to the facade. In this context, interfaces to additional currently relevant topics in the facade industry, such as BAS or the interaction with strategies of building circularity, were established. An important remaining future research task is to enable the interpretation of specific sensor data with regard to common damage patterns.

Acknowledgements The R-IOT project was developed as competition entry in the working environment of the Priedemann Facade-Lab. We thank all colleagues for the exchange of ideas and insights that made the concept possible. We also acknowledge the Metals in Construction Design Challenge 2024 and its jury for their critical and appreciative review of this competition entry.

References

1. Herzog T, Krippner R, Lang W (2004) Facade construction manual. Birkhauser-Publishers for Architecture, Basel, Boston
2. Chew MYL, Silva ND, Tan PP, Das S (2006) Grading of Risk Parameters for Facade Maintainability
3. Yalaz ET, Tavil AU, Celik OC (2018) Lifetime performance evaluation of stick and panel curtain wall systems by full-scale testing. Constr Build Mater 170:254–271. https://doi.org/10.1016/j.conbuildmat.2018.03.061
4. Centre for Window and Cladding Technology (CWCT) (2005) Robustness, durability, tolerances and workmanship. In: Standard for systemised building envelopes. CWCT, Bath
5. European Environment Agency (2024) Addressing the environmental and climate footprint of buildings. Publications Office, LU
6. International Energy Agency (2017) Deep energy retrofit—Case studies: IEA EBC Annex 61 Subtask A (26 case studies)
7. Wilson C, Crane L, Chryssochoidis G (2014) Why do people decide to renovate their homes to improve energy efficiency?
8. Martinez A, Patterson M, Carlson A, Noble D (2015) Fundamentals in Façade retrofit practice. Procedia Eng 118:934–941. https://doi.org/10.1016/j.proeng.2015.08.534
9. Farahmandpour K (2015) Conventional and nonconventional repair of Curtain Wall systems. Tennessee, United States of America
10. Kim JJ, Miller JA, Kim S (2020) Cost impacts of change orders due to unforeseen existing conditions in building renovation projects. J Constr Eng Manag 146:04020094. https://doi.org/10.1061/(ASCE)CO.1943-7862.0001888

11. Hassanain MA, Al-Mudhei A (2006) Business continuity during facility renovations. J Corp R Estate 8:62–72. https://doi.org/10.1108/14630010610679871
12. Menegaki M, Damigos D (2018) A review on current situation and challenges of construction and demolition waste management. Curr Opin Green Sustain Chem 13:8–15. https://doi.org/10.1016/j.cogsc.2018.02.010
13. Benachio GLF, Freitas M do CD, Tavares SF (2020) Circular economy in the construction industry: a systematic literature review. J Clean Prod 260:121046. https://doi.org/10.1016/j.jclepro.2020.121046
14. Ellen MacArthur Foundation (2015) Towards a circular economy—business rationale for an accelerated transition
15. Rahla KM, Mateus R, Bragança L et al (2021) Implementing circular economy strategies in buildings—from theory to practice. Appl Syst Innov 4(2):26. https://doi.org/10.3390/asi4020026
16. Ostapska K, Rüther P, Loli A, Gradeci K (2024) Design for disassembly: a systematic scoping review and analysis of built structures designed for disassembly. Sustainable Prod Consumption 48:377–395. https://doi.org/10.1016/j.spc.2024.05.014
17. Boothroyd G, Alting L (1992) Design for assembly and disassembly. CIRP Ann 41:625–636. https://doi.org/10.1016/S0007-8506(07)63249-1
18. Guy B, Ciarimboli N (2008) DfD: design for disassembly in the built environment: a guide to closed-loop design and building. Hamer Center
19. Crowther P (1999) Design for disassembly to recover embodied energy. In: Sustaining the future: energy, ecology, architecture: PLEA '99, Brisbane, Australia
20. Mobley RK (2002) An introduction to predictive maintenance, 2nd edn. Butterworth-Heinemann, Amsterdam
21. Sakib N, Wuest T (2018) Challenges and opportunities of condition-based predictive maintenance: a review. Procedia CIRP 78:267–272. https://doi.org/10.1016/j.procir.2018.08.318
22. Pech M, Vrchota J, Bednář J (2021) Predictive maintenance and intelligent sensors in smart factory: review. Sensors 21:1470. https://doi.org/10.3390/s21041470
23. Nangia S, Makkar S, Hassan R (2020) IoT based predictive maintenance in manufacturing sector
24. Lei Y, Li N, Guo L et al (2018) Machinery health prognostics: a systematic review from data acquisition to RUL prediction. Mech Syst Signal Process 104:799–834. https://doi.org/10.1016/j.ymssp.2017.11.016
25. Bao Y, Chen Z, Wei S et al (2019) The state of the art of data science and engineering in structural health monitoring. Engineering 5:234–242. https://doi.org/10.1016/j.eng.2018.11.027
26. Hashemian HM (2011) State-of-the-art predictive maintenance techniques. IEEE Trans Instrum Meas 60:226–236. https://doi.org/10.1109/TIM.2010.2047662
27. Jiazeng S, Xi Z, Ning LC et al (2024) Predictive maintenance and its applications in civil engineering structures: a review. J Southeast Univ (Engl Ed) 40:245–256. https://doi.org/10.3969/j.issn.1003-7985.2024.03.004
28. De Simone MC, Lorusso A, Santaniello D (2022) Predictive maintenance and structural health monitoring via IoT system. In: 2022 IEEE workshop on complexity in engineering (COMPENG). IEEE, Florence, Italy, pp 1–4
29. Sonbul OS, Rashid M, Sonbul OS, Rashid M (2023) Towards the structural health monitoring of bridges using wireless sensor networks: a systematic study. Sensors 23(20):8468. https://doi.org/10.3390/s23208468
30. Sazonov E, Haodong L, Curry D, Pillay P (2009) Self-powered sensors for monitoring of highway bridges. IEEE Sensors J 9(11):1422–1429. https://doi.org/10.1109/JSEN.2009.2019333
31. Tronci EM, Nagabuko S, Hieda H et al (2022) Long-range low-power multi-hop wireless sensor network for monitoring the vibration response of long-span bridges. Sensors 22(10):3916. https://doi.org/10.3390/s22103916
32. Domingues P, Carreira P, Vieira R, Kastner W (2016) Building automation systems: concepts and technology review. Comput Stand Interfaces 45:1–12. https://doi.org/10.1016/j.csi.2015.11.005

33. International Organization for Standardization (2007) Building automation and control systems (ISO 16484–3:2005)
34. Luna-Navarro A, Fidler P, Law A et al (2021) Building impulse toolkit (BIT): a novel IoT system for capturing the influence of façades on occupant perception and occupant-façade interaction. Build Environ 193:107656. https://doi.org/10.1016/j.buildenv.2021.107656
35. Arnesano M, Bueno B, Pracucci A et al (2019) Sensors and control solutions for Smart-IoT façade modules. In: 2019 IEEE international symposium on measurements & networking (M&N). IEEE, Catania, Italy, pp 1–6
36. Böke J, Knaack U, Hemmerling M (2020) Prototype of a cyber-physical façade system. J Build Eng 31:101397. https://doi.org/10.1016/j.jobe.2020.101397
37. Pishdad-Bozorgi P, Gao R, Shelden D (2020) Introduction to cyber-physical systems in the built environment, pp 23–41
38. Pan Y, Hu Z, Brilakis I (2023) Digital twins and their roles in building deep renovation life cycle. In: Lynn T, Rosati P, Kassem M et al (eds) Disrupting buildings. Springer, Cham, pp 83–96
39. Li H, Hong T (2025) A digital twin platform for building performance monitoring and optimization: performance simulation and case studies. Build Simul 18:1561–1579. https://doi.org/10.1007/s12273-025-1290-2
40. Roggema R (2016) Research by design: proposition for a methodological approach. Urban Sci 1:2. https://doi.org/10.3390/urbansci1010002
41. Ornamental Metal Institute of New York (2024) Metals in construction design challengel $15,000 prize. In: Metals in construction magazine 2026 design challenge. https://metalsinconstruction.org. Accessed 12 Dec 2025
42. Hauberg J (2011) Research by design—a research strategy. Lusofona J Arch Educ 5:46–56

Façade Access for Tall and Complex Buildings: A Performance-Based Design Framework

Randall Knox and Martin Wunder

Abstract Façade access systems are an integral component of performance-based façade design, supporting the safe inspection, maintenance, and long-term life-cycle performance of building envelopes. As buildings increase in height and architectural complexity, façade access must be considered as a fundamental design parameter rather than a secondary operational requirement. This chapter presents a performance-based decision-making framework for the selection of façade access systems in tall and complex buildings. The framework defines key input parameters (including building height, façade geometry, surface area, maintenance requirements, and regulatory constraints) and establishes a structured decision logic linking these inputs to appropriate access strategies. A taxonomy of façade access systems is developed, and their application is evaluated in relation to performance criteria such as safety, operational efficiency, and maintainability. The framework is applied through case studies to demonstrate how different building conditions lead to specific access solutions, including Building Maintenance Units, monorail systems, and hybrid approaches. The results show that façade access system selection can be generalized as a performance-based process that is transferable across building types and geographic contexts. The chapter contributes to the formalization of façade access engineering as a distinct component of façade design and provides a structured methodology for integrating access systems into the overall building design process.

Keywords Façade access · Building maintenance unit · Suspended platforms · Tall buildings · Façade engineering · BIM · Maintenance access · Performance-based design

R. Knox (✉) · M. Wunder
Global BMU, New York, USA
e-mail: randall@globalbmu.com

M. Wunder
e-mail: martin@globalbmu.com

A. Mejorin et al. (eds.), *Performance-Based Façade Design*, Springer Tracts in Civil Engineering, https://doi.org/10.1007/978-3-032-28835-6_8

Abbreviations

AWP	Aerial work platform
BIM	Building information modeling
BMU	Building maintenance unit
ISA	Intermittent stabilization anchor
SPP	Self-powered suspended platform
TRL	Technology readiness level

1 Introduction

Maintaining building façades is a critical component of lifecycle management, public safety, and long-term asset preservation. Façade access systems provide the means by which cleaning, inspection, repair, replacement, and maintenance operations are performed on exterior envelopes and selected interior surfaces such as atria and skylights. As buildings have grown taller and architectural forms have become increasingly complex, façade access has evolved from a relatively straightforward logistical issue into a specialized engineering discipline closely intertwined with façade design, structural planning, and operations strategy.

In performance-based façade design, building envelopes are evaluated not only for structural and environmental performance but also for maintainability and operational safety. Façade access systems therefore represent a critical component of façade performance, supporting inspection, cleaning, repair, and replacement activities throughout the building lifecycle [1, 2].

Historically, many buildings could be serviced through simple suspended platforms, portable davits, or ground-based equipment. However, the rapid development of high-rise and supertall buildings, especially in dense urban centers, has changed the problem fundamentally. Contemporary towers frequently incorporate setbacks, sculpted massing, sloping façades, recessed curtain walls, terraces, deep overhangs, large podiums, double-skin façades, and integrated sustainability features such as photovoltaic panels and green roofs. These design characteristics create maintenance challenges that cannot be fully addressed by conventional access solutions alone [3, 4].

Façade access must therefore be considered as an integral part of the architectural and engineering process. A building that cannot be safely and efficiently maintained over its life cycle is functionally incomplete, regardless of its aesthetic or structural achievements. The integration of access systems at an early design stage supports safety, code compliance, operational efficiency, and preservation of architectural intent. Poorly planned access can lead to costly retrofits, incomplete façade coverage, unsafe maintenance procedures, and conflicts with other building systems.

This chapter presents a global perspective on façade access for tall and complex buildings. It defines the scope of façade access engineering, categorizes the principal system types, discusses regulatory and operational constraints, and introduces a performance-based decision framework for selecting appropriate solutions based on building height, façade geometry, and operational performance requirements. It also examines stabilization methods, BIM-based coordination, emerging technologies, and case studies that illustrate real-world implementation. The goal is to provide a structured reference for professionals engaged in façade access planning and design in diverse urban and architectural contexts.

This chapter is structured as a performance-based decision-making framework for façade access system selection. It defines key input parameters, decision logic, and resulting access strategies applicable to tall and complex buildings.

Performance-based façade design evaluates building envelopes in terms of safety, durability, maintainability, and lifecycle performance [1]. Within this framework, façade access systems are essential because they enable inspection, cleaning, and repair operations required to maintain façade performance over time.

Façade access performance can be evaluated across:

- Safety performance
- Maintainability
- Operational efficiency
- Lifecycle cost
- Regulatory compliance

In this chapter, supertall buildings refer to structures exceeding 300 m, consistent with CTBUH classifications [4].

2 Façade Access System Taxonomy

Façade access systems are classified into three primary categories based on their level of integration with the building envelope, deployment configuration, and operational performance characteristics:

1. **Building-Integrated Systems**: Building Maintenance Units (BMUs), monorails, and gantry systems.
2. **Self-Powered Suspended Access Systems**: Davit systems and self-powered platforms (SPPs) used at roof level, terraces, and setbacks.
3. **Ground-Based Access Systems**: Aerial Work Platforms (AWPs), including telescopic and articulated boom lifts.

Self-powered suspended access systems are permanently installed or dedicated building systems, despite their modular configuration, and should not be confused with temporary rental equipment.

In practice, multiple systems are often combined to provide complete façade coverage across towers, podiums, atriums, and specialized architectural features.

2.1 Building-Integrated Systems

Building-integrated systems include Building Maintenance Units (BMUs), monorail systems, and gantry systems. BMUs are typically installed on the roof of tall buildings and provide controlled suspended access to curtain wall façades. Monorail systems provide continuous suspended paths along façades or interior spaces where conventional vertical access methods are not feasible. Gantry systems are commonly used to service large skylights, atriums, and recessed glazing systems in complex buildings. They are typically justified where building height, complexity, façade area, or maintenance intensity make other types of suspended platform systems inefficient or unsafe.

BMUs are the most common permanent access solution for tall and supertall buildings. They generally comprise a roof-mounted machine with hoists, a jib or boom, suspended wire ropes, and a working platform. They may be fixed in one location or traverse along tracks or runways. Depending on the building geometry, the BMU may incorporate slewing (rotation), telescoping (extension of the jib), luffing (vertical angle adjustment), articulating, or mast-lowering functions. Because BMUs are widely deployed and codified, they represent mature technology and are generally assessed at TRL 9.

Monorail systems provide a continuous suspended path along façades or interior spaces. They are particularly valuable for podium roofs, soffits, atria, skylights, and areas where conventional vertical drops are interrupted by projections or deep recesses. Monorails may be recessed into soffits, suspended from structural framing, or mounted on brackets. They can use manual or powered trolleys and may support platforms or rope-based access rigs. Like BMUs, monorails are mature systems with broad international use.

Gantry systems consist of structural frames or bridge-like assemblies that move along rails or tracks to provide access over large glazed roofs, atria, domes, and recessed spaces. Their geometry is highly project-specific. Although gantries are well established, they are more specialized than BMUs or monorails and are usually custom engineered to suit a building's form.

2.2 Self-Powered Suspended Access Systems

Self-powered suspended access systems include portable davits and self-powered platforms (SPPs). These systems provide flexible access solutions for a wide range of building heights and configurations. Davit systems support suspended platforms from bases located at roof or terrace levels. Self-powered platforms utilize traction hoists mounted on the platform and are capable of being configured in modular sections to accommodate different façade conditions. These systems are widely used in mid-rise buildings and in areas where building-integrated systems such as BMUs are not required.

Davit systems support suspended platforms from portable arms inserted into permanently installed roof sockets or bases. These systems may be conventional aluminum or aluminum and steel units or ultra-lightweight variants such as carbon fiber davits. Portable davits are especially useful on setbacks and terraces, although each operational level must be carefully planned with respect to storage, rigging space, and structural support.

Self-powered suspended platforms use traction hoists mounted directly on the platform. These systems are flexible, modular, and highly effective for many mid-rise applications. However, height limitations are imposed by regulatory requirements, rope weight, emergency rescue considerations, and operational safety. As a result, roof-powered systems become necessary above a certain elevation threshold.

2.3 Ground-Based Access Systems

Ground-based systems include aerial work platforms (AWPs), telescopic booms, articulated boom lifts, and scissor lifts. These systems offer flexibility and rapid deployment but are limited by reach, site logistics, wind exposure, and ground bearing constraints. They are most effective for podiums, low-rise façades, interior spaces, and select exterior conditions where vehicle positioning is feasible.

Ground-based methods may also include water-fed poles for lower glass surfaces. These have practical benefits in selected conditions but are limited by reach and by the requirement for sufficient stand-off distance.

2.4 Special Access Contexts

Façade access also extends beyond conventional vertical façades. Interior atria, skylights, domes, underhangs, canopies, photovoltaic installations, signage, and green walls may each require dedicated access strategies. In many projects, the most effective solution is not a single piece of equipment but a coordinated system combining BMUs, monorails, davits, AWPs, and specialized maintenance rigs.

3 Methodology: Performance-Based Access System Selection Framework

3.1 Framework Structure

The proposed façade access framework consists of three components:

Inputs:

- Building height
- Façade geometry
- Façade surface area
- Maintenance requirements
- Regulatory constraints

Decision Logic:

Evaluation of access feasibility based on reach, safety, stabilization requirements, and operational efficiency.

Outputs:

Selection of façade access systems, including BMUs, davits, monorails, gantries, or hybrid solutions.

Façade access includes window cleaning, façade inspection, repair, sealant replacement, localized façade maintenance, glass replacement, inspection of building-integrated lighting and signage, and access to selected interior glazed spaces. It also encompasses the systems and procedures required to perform these tasks safely, repeatedly, and efficiently over the building's service life.

As building envelopes have become more technologically and architecturally complex, special access needs have expanded. Green roofs and green walls require maintenance access that may conflict with traditional equipment storage. Photovoltaic panels require safe means of cleaning and inspection. Interior atria and skylights, often highly visible architectural features, require discreet access systems compatible with the design intent. Historic structures may require reversible or minimally invasive solutions. These conditions reinforce the importance of treating façade access as part of the larger building performance strategy.

Façade surface area may exceed 40,000–60,000 m^2 in tall buildings, making maintenance efficiency critical [4]. Lifecycle maintenance costs must therefore be considered alongside initial installation decisions [5].

4 Regulatory Context and Operational Safety

Façade access systems operate within a tightly regulated framework intended to protect workers, building occupants, and the public [6]. The governing standards vary by jurisdiction, but most address load capacity, structural anchorage, wire rope systems, fall protection, emergency descent, maintenance procedures, and rescue planning.

4.1 Global Standards

In the United States, principal references include OSHA 1910.66, ASME A120, and IWCA I-14.1, along with local rules that may be particularly stringent in cities such as New York and jurisdictions such as California. Internationally, EN 1808 in Europe, CSA Z271 in Canada, GB 19154 in China, and AS 1418.13 in Australia are among the key applicable standards. These frameworks differ in detail but share common concerns regarding structural integrity, safe operation, training, and emergency procedures.

4.2 Height and Rescue Constraints

A major operational distinction exists between self-powered suspended platforms and roof-powered systems. As suspension height increases, the cumulative weight of wire ropes, the characteristics of traction hoists, and emergency rescue limitations become more significant. This is one reason why self-powered platforms are generally restricted to lower heights, while BMUs become necessary for taller buildings.

4.3 Environmental and Urban Factors

Wind, thermal exposure, street conditions, surrounding structures, and roof congestion all affect façade access design. Dense urban environments can restrict ground-based access, require careful control of platform swing, and complicate staging, rescue, and storage. Supertall buildings may experience significant wind effects and building movement, increasing the importance of stabilization systems and controlled operating procedures.

4.4 Inspection and Maintenance Obligations

Many jurisdictions require periodic façade inspections, particularly for tall buildings and urban occupancies. These programs reinforce the need for dependable access systems that allow close-up examination of curtain walls, masonry, sealants, and specialized envelope elements. Rope access may be appropriate as a supplemental method in some jurisdictions, but for repeated maintenance and comprehensive lifecycle service, permanent or semi-permanent access systems remain essential.

5 Stabilization Systems

Suspended work platforms are subject to movement caused by wind, building geometry, operator actions, and building sway. Stabilization is therefore fundamental to worker safety, protection of façade materials, and the precision required for maintenance and replacement operations [7].

5.1 Mullion Guides

Mullion guide stabilization uses guide shoes on the platform engaging with dedicated vertical tracks integrated into the façade. This is especially important at greater heights and in projects where continuous stabilization is required. Continuous stabilization systems are typically required on tall buildings to meet safety and operational requirements defined in façade access standards such as ASME A120. Guide shoe designs vary, including split assemblies that can be inserted at intermediate levels and actuated systems with retractable components. Track dimensions must be coordinated carefully with façade tolerances to prevent binding or jamming. The façade contractor typically designs the guide track to meet criteria established by the façade access consultant, including resistance to wind loads, operational forces, and loads associated with glass replacement or sloping conditions.

5.2 Intermittent Stabilization Anchors

Intermittent stabilization anchors (ISAs) are discrete façade-mounted anchor points used to stabilize suspended platforms. ISAs are used for stabilization at lower elevations. These are typically installed at intervals such as every third floor and may be external button types or recessed detent-pin sockets depending on local code requirements. The platform connects progressively to these anchors using lanyards as it descends. ISAs provide an effective solution where continuous guide tracks are not required, but they must be carefully integrated with façade aesthetics and maintenance procedures.

5.3 Guide Rails and Guide Rope Systems

Guide rails may be integrated into façades or special structures where direct platform control is required along complex routes. Guide rope systems, also known as soft rope systems, are secondary control lines that allow the platform to be positioned relative to complex façade geometries. Guide rope systems are especially useful where the platform must traverse horizontally or diagonally, or where the façade includes curved or angled surfaces. These systems allow the platform to be drawn into position relative to complex planes, enabling access beyond simple vertical drops.

5.4 Stabilization as a Design Issue

The selection of stabilization strategy should occur early in design because it affects the façade interface, support loads, visual appearance, and operational procedures. These systems must satisfy performance requirements including wind stability and load capacity. Stabilization is not a secondary accessory to access equipment; it is a primary determinant of whether the system can safely serve the intended envelope.

6 Methodology/Performance-Based Framework

Selecting a façade access system is best understood as a performance-based design problem. The goal is not merely to choose equipment that can theoretically reach the façade, but to select a system that delivers safe, complete, efficient, and maintainable coverage over the building's full service life. Key performance criteria include:

- wind stability limits
- platform load capacity
- safety redundancy requirements
- glass replacement load allowances.

6.1 Primary Selection Variables

Six core variables govern façade access strategy:

Building height. Height strongly influences the transition from self-powered suspended platform access to roof-powered BMU systems. Taller buildings generally require BMUs, particularly where façade area is extensive or repeated maintenance tasks are anticipated.

Façade geometry. Recesses, setbacks, underhangs, negative slopes, curved façades, and articulated massing may require specialized cradles, monorails, guide rope systems, or articulated BMUs.

Façade area. Large envelope areas increase the importance of operational efficiency. Even where self-powered platform systems are technically possible, the cumulative labor and repositioning burden may justify roof-powered BMU systems.

Operational efficiency. Access systems should minimize setup time, repositioning, partial coverage gaps, and disruptions to building operations. Efficiency becomes particularly important in towers with large curtain wall zones and frequent maintenance cycles.

Maintenance load requirements. Cleaning may require relatively light equipment, but glass replacement, lighting maintenance, or façade repairs may require auxiliary hoists, glass handling units, or higher stabilization loads.

Architectural integration. Visual impact, preservation of amenity spaces, storage, parking, and compatibility with roof structures and screen walls all influence the final selection.

6.2 Indicative Selection Matrix

A simplified selection matrix can be expressed as follows:

- **Low-rise, simple façades**: portable davits and self-powered suspended platforms, or AWPs.
- **Mid-rise, moderate complexity**: davits and self-powered platforms, plus monorails where appropriate. Platform stabilization by intermittent stabilization anchors.
- **Tall buildings**: BMUs with integrated platform stabilization systems such mullion guides.
- **Supertall or highly complex buildings**: custom BMUs, multiple BMUs, gantries, guide rope systems, and specialized cradles.

This matrix is only a starting point. Real projects often combine systems, especially where towers sit on podiums or contain multiple façade typologies.

6.3 Hybrid Strategies

A single building may require one strategy for podium façades, another for the main tower, and another for atria or recessed skylights. For example, a tower may use a central BMU for the primary envelope, davits and self-powered platforms for podium terraces, monorails beneath underhangs, and a gantry for a large atrium skylight. This layered approach reflects the reality that façade access must respond to the building's geometry rather than impose a uniform equipment concept on all zones.

7 Access Systems for Low- and Mid-Rise Zones

Even on tall projects, lower zones such as podiums, terraces, and setbacks often require solutions distinct from those serving the main tower.

7.1 *Davits and Powered Platforms*

Portable davits and powered suspended platforms remain highly effective for moderate building heights. Permanent sockets or bases are integrated with the structure, and removable davit arms support the suspension points. The resulting loads must be coordinated with the building structure. Self-powered platforms are often modular and can be assembled in different lengths, with powered end sections capable of functioning independently as single-person units.

These systems provide flexibility and are valuable where permanent roof-powered equipment is unnecessary. However, they require adequate rigging space, storage, and carefully planned operational procedures, particularly at setbacks where davits cannot typically move between levels during use.

7.2 *Ultra-Lightweight Davits*

With the growth of exterior terraces and tenant amenity spaces, conventional davits can create architectural and operational conflicts. Ultra-lightweight carbon fiber davits offer a solution by allowing one set of equipment to be transported between levels and stored remotely. This reduces the visual and spatial burden on terraces and improves compatibility with contemporary high-end amenity design.

7.3 *Under-Paver Davit Bases*

Where podium roofs or terraces are intended as occupied amenity areas, projecting davit sockets are often undesirable. Recessed under-paver bases with matching removable covers enable access systems to be integrated without permanent visual intrusion. Such details demonstrate how façade access planning can preserve architectural intent while maintaining operational capability.

7.4 Ground-Based Access at the Podium

Aerial work platforms and articulated boom lifts are useful at lower elevations where access paths, bearing capacity, and setup locations permit. Because these machines impose significant localized loads, podium slabs, pavers, vaults, and landscaped surfaces must be evaluated carefully. Ground-based access may be preferred where roof rigging is impractical, but in dense urban conditions it is often constrained by street occupation limits, overhead obstructions, and restricted maneuvering space.

7.5 Water-Fed Poles

For selected lower glazed areas, water-fed poles can provide economical access for cleaning. Their use depends on having sufficient ground space and modest reach requirements. They are a limited but sometimes useful supplementary tool.

8 Access to Skylights, Atria, and Underhangs

Complex buildings frequently include large interior and semi-exterior glazed elements that require dedicated maintenance systems.

8.1 Skylights and Domes

Skylights may be accessed through gantries, monorails, or fall-restraint systems depending on geometry. Gantry systems are especially effective for barrel vaults, domes, and broad glazed roofs, where a bridge-like frame can traverse along rails and allow workers to access the glazing safely. Gantries may also carry supplementary platforms or extendable sections to reach recessed or end-wall zones.

Monorails provide a more visually integrated alternative in some projects. Recessed c-channel monorails can be incorporated into soffits, while extruded monorails offer a more refined appearance and can follow curved or sloped paths. Powered climbing trolleys are used where the monorail itself is inclined.

8.2 Underhangs and Soffits

Underhangs are difficult because conventional vertical descent may not provide access to the soffit or the façade below the projection. Solutions include monorails under the overhang, diverter systems that redirect the suspension ropes, forward-approach cradles, and articulated BMUs. In some cases, a BMU platform is used to rig a secondary system that services the underhang zone.

8.3 Interior Atria

Interior atria require access systems that are both technically capable and architecturally discreet. Monorails and gantries are common solutions, especially where large glazed walls or overhead roofs require regular cleaning. Because such spaces are often publicly visible, equipment integration and storage are particularly sensitive design issues.

9 Façade Access Above 490 ft. (149 m) and Beyond

At greater heights, the operational and regulatory limitations of self-powered platforms make roof-powered systems necessary. This is the domain in which the BMU becomes the defining access technology.

9.1 Building Maintenance Units

BMUs are permanent, roof-powered machines with drum hoists located at roof level, suspending a working platform by wire ropes. They are used on tall and supertall buildings because they overcome the height limitations of traction-hoist platforms and permit more complex movements and heavier operations, including glass replacement.

A BMU may be a fixed central crane-type machine mounted on plinths, or a track-mounted unit traversing a runway to reach multiple drops. Crane-type units often include slewing and telescoping jibs, luffing features, mast extension or retraction, and platform spreader bars (structural members used to maintain platform alignment and load distribution) that help align the cradle parallel to the façade. Track-mounted BMUs may run on concrete runways or steel beam tracks, requiring coordinated supports, electrical feeds, tie-downs, and sometimes catwalk access for maintenance.

9.2 Design Considerations

BMU design must consider maximum and minimum outreach, roof obstructions, screen walls, emergency generator flues, aircraft warning lights, lightning protection, and roof circulation. The machine must not only reach the furthest façade point but also retract sufficiently to service the nearest surfaces. It must be able to maneuver past architectural and mechanical obstructions without compromising maintenance coverage.

Large or highly articulated supertall buildings may require multiple BMUs serving distinct zones. In certain cases, a BMU may be parked within an intermediate mechanical level or behind an operable façade panel, allowing the system to be concealed when not in use.

9.3 Storage and Parking

Architectural integration of BMUs often depends on successful concealment strategies. These may include retractable masts, luffing jibs that can be lowered behind screen walls, lift tables, storage garages, or roof enclosures integrated into the architectural composition. Effective parking design preserves the visual integrity of the skyline profile while ensuring the equipment remains serviceable and accessible.

10 Accessing Complex Forms

Tall buildings increasingly depart from rectilinear forms, creating access problems that require specialized equipment strategies.

10.1 Forward Approach

Forward-approach platforms are specialized suspended platforms that extend outward from the suspension line to reach recessed façade areas or underhangs. Rather than hanging directly beneath the suspension points, the cradle projects outward toward the work surface through articulated, telescopic, or scissor-type geometry. This enables close access to surfaces that would otherwise be unreachable from a conventional suspended platform.

10.2 Articulated and Variable-Geometry Platforms

Some platforms include extendable end sections or variable geometry specifically designed to access recesses, corners, and non-planar conditions. Such systems are particularly valuable in tower crowns, sculpted setbacks, and heavily articulated façades where normal cradle geometry would be insufficient.

10.3 Negative Slopes

Inward-sloping façades (negative slope conditions) create pull-in forces that require dedicated stabilization and rope management. Access may involve removable diverters at the slope transition, multiple sets of guide shoes, and guide tracks designed to handle the resulting forces. Careful coordination is required to ensure the platform can transition safely from vertical to sloped zones and return without damaging the façade or destabilizing the system.

10.4 Guide Rope for Curved and Angled Movement

Guide rope systems allow a platform to move horizontally, diagonally, or inward across a façade. Using reelers, pins, and controlled tensioning, the platform can be drawn into successive positions along curved or irregular planes. This is an important strategy for buildings with complex setbacks and compound geometry where a simple vertical drop would leave areas inaccessible.

11 Special Operations

Façade access systems must often support activities beyond routine cleaning.

11.1 Glass Replacement

Glass replacement is one of the most demanding façade access operations. BMUs frequently incorporate auxiliary hoists and glass handling units equipped with suction systems or vacuum lifters. The auxiliary hoist must be sized for the heaviest anticipated panel plus rigging and handling equipment. Depending on the façade geometry, pulleys, guides, or pantograph cradles may be required to maneuver the glass into recessed or sloped positions. These operations generate additional loads on the façade, stabilization systems, and the building structure and therefore must be considered in the early building design phases and in the original façade access design criteria and basis of design [8].

11.2 Historic and Preservation Contexts

Historic buildings often require highly tailored access solutions that avoid damage to protected surfaces and maintain reversibility. Temporary custom work platforms, internal access routes, and specialized rolling rigs may be required where permanent interventions are constrained by landmark regulations or fragile finishes.

11.3 Special Architectural Features

Signage, lighting, green walls, rooftop façades, crown structures, and photovoltaic arrays all create access requirements that may exceed the assumptions of routine window cleaning. A comprehensive façade access strategy must identify these maintenance tasks explicitly rather than leaving them to ad hoc future solutions.

12 Digital Coordination and BIM Integration

Digital coordination has become essential to façade access planning, especially in large or geometrically complex projects.

12.1 Early Integration

Façade access should ideally enter the project during schematic design and continue through design development and construction documentation. Early involvement enables the design team to reserve space for machine paths, platform storage, rigging zones, stabilization interfaces, and maintenance circulation before these areas are consumed by architectural, structural, or MEP requirements.

12.2 Parametric Equipment Modeling

Parametric BIM models allow BMU and platform movements to be simulated realistically. Functions such as jib slewing, luffing, mast extension, track travel, and platform positioning can be represented digitally. These simulations show whether the equipment can reach every intended drop and whether movements conflict with screen walls, roof structures, mechanical equipment, lighting elements, or neighboring façade features. This simulation enables designers to identify potential safety risks and operational conflicts before construction, reducing the likelihood of incomplete façade coverage and improving maintenance efficiency.

12.3 Multidisciplinary Collaboration

Façade access sits at the intersection of architecture, structure, MEP, façade engineering, and building operations. BIM facilitates coordination among these disciplines and makes the access strategy visible to the broader design team. This reduces the likelihood that access needs will be deferred or inadvertently compromised [9].

12.4 Lifecycle Planning

Beyond initial coordination, digital models can support maintenance simulation, inspection planning, and long-term facility management. As digital twin practices evolve, extending BIM models into data-linked representations of the operational building, façade access equipment may increasingly be linked with maintenance schedules and building performance data, strengthening the role of access engineering within lifecycle management.

13 Emerging Technologies

New technologies are beginning to influence façade maintenance, although most remain supplementary rather than substitutive.

Robotic cleaning systems have been developed for selected curtain wall conditions, especially on rectilinear façades. These systems are still limited in their ability to address complex geometry, variable dirt conditions, and the nuanced manual work often required for high-quality cleaning.

Drones are increasingly useful for inspection and defect detection, particularly for preliminary surveys and post-event assessments. However, drones do not replace the need for close-access maintenance systems capable of cleaning, repair, and replacement work.

Sensor-integrated façades may support predictive maintenance by identifying leakage, movement, or surface degradation before visible failure occurs. Automated BMU control systems may also improve repeatability and positioning precision.

Although these developments are promising, conventional access systems remain the backbone of façade operations for tall and complex buildings. Emerging technologies should therefore be understood as augmenting, not replacing, BMUs, monorails, gantries, davits, and suspended platforms.

14 Technology Readiness and Market Deployment

Technology readiness levels provide a framework for evaluating engineering systems [10]. The maturity of façade access technologies varies considerably. Established systems such as BMUs represent mature technologies, while robotic systems remain at lower readiness levels.

- **BMUs**: TRL 9; mature and globally deployed
- **Davits**: TRL 9; codified and widely used
- **Swing stages and self-powered platforms**: TRL 9; standardized and regulated
- **Monorails**: TRL 9; proven and versatile
- **Gantries**: TRL 8–9; specialized but mature
- **AWPs**: TRL 9; globally standardized
- **Guide rope systems**: TRL 9; established in specialized applications
- **Robotic façade cleaning**: TRL 2–5; limited deployment, early development
- **Drone inspection**: generally intermediate maturity as an inspection supplement

The TRL perspective is useful because it distinguishes between technologies that are operationally dependable today and those that remain experimental or project-specific. This is particularly important when the building owner requires guaranteed maintainability over decades [9].

15 Results and Application of the Framework

15.1 Case Study: 989 Sixth Avenue, New York City

989 Sixth Avenue in Midtown Manhattan, also known as Casoni, illustrates the integration of multiple façade access strategies within a dense urban tower. At approximately 785 ft., the building requires a fixed crane-type BMU for the main tower. The BMU incorporates an extendable mast allowing the machine to be lowered into a concealed parking position behind roof screen walls, preserving the architectural profile when not in use. Stabilization is provided through a combination of mullion guides, intermittent stabilization anchors, and guide rope systems that allow the platform to maneuver at setbacks. An auxiliary hoist and glass handling unit support glass replacement operations. The project demonstrates how tall-building façade access depends not only on machine selection but also on parking strategy, stabilization design, and geometric problem solving.

Framework Application.

Inputs

- Height: 785 ft.
- Geometry: curtain wall with setbacks

- Large façade area

Decision Logic.

Height exceeds SPP capability → requires BMU.
Setbacks require stabilization + guide rope.

Output.

BMU with integrated stabilization and auxiliary hoist.

Lessons

- Early BMU integration improves design outcomes
- Hybrid stabilization is required for complex geometry
- Framework is transferable to similar tall towers

15.2 Case Study: New York Life Insurance Company Building

The New York Life Insurance Building presents a very different access challenge. Following storm damage to its gold-leaf clay tile roof in 2012, restoration required a specialized access strategy respectful of the landmark character of the building. The project used temporary rigging, internal access routes, and a custom work platform equipped with oversized soft wheels designed to minimize pressure on delicate finishes. This case illustrates the importance of reversible access strategies (systems that can be installed and removed without permanent alteration or damage to the historic fabric) along with surface protection and custom engineering in preservation contexts. It also reinforces a broader lesson: façade access engineering extends beyond modern curtain wall towers and must be adaptable to exceptional building typologies and heritage requirements.

Framework Application.

Inputs

- Historic building
- Fragile roof
- Limited intervention allowed

Decision Logic.
Conventional systems unsuitable due to loading.

Output.

Custom lightweight access systems.

Lessons

- Access design must respond to preservation constraints
- Custom solutions may be required
- Framework applies beyond high-rise buildings

These findings demonstrate that façade access strategies must be selected based on height thresholds, façade articulation, specific conditions, and operational requirements. This section demonstrates application of the proposed framework to real-world projects, illustrating how input parameters and decision logic lead to specific façade access strategies. These case studies demonstrate the applicability of the framework across different building types and constraints, forming the basis for the generalized observations discussed in the following section.

16 Discussion

Façade access practice varies internationally according to regulation, labor conditions, construction culture, and market expectations. Some jurisdictions rely more heavily on rope access or temporary systems for low- and mid-rise buildings, while others emphasize permanent integrated infrastructure at earlier height thresholds. Dense global cities tend to encourage more carefully planned access because street occupation, public safety exposure, and façade inspection regimes create strong incentives for predictable and repeatable maintenance systems.

Despite regional variation, a consistent global best practice is emerging: façade access should be integrated early in the design process, aligned with maintenance needs, and coordinated across disciplines. Buildings designed without an adequate access strategy often experience operational inefficiencies, unsafe improvisation, visual compromise through late-added equipment, or incomplete maintenance coverage. These risks increase with building height and geometric complexity, making performance-based selection and early coordination globally relevant.

Façade access system selection is a performance-based process driven by building height, geometry, and operational requirements. The framework presented here is transferable across building types and geographic contexts [3, 4].

17 Conclusion and Future Directions

Façade access is an essential component of building performance, not a secondary maintenance concern. As buildings become taller, more sculptural, and more technologically sophisticated, access systems must respond with equal sophistication

[11]. The most successful façade access strategies are those developed early, coordinated thoroughly, and tailored to the building's geometry, maintenance demands, regulatory context, and architectural intent.

A global performance-based approach demonstrates that no single system suits every condition. BMUs dominate tall and supertall applications because of their reach, power, and ability to support complex operations such as glass replacement. Davits, powered platforms, monorails, gantries, and AWPs remain indispensable in lower zones, specialized spaces, and supplementary roles. Stabilization systems, especially mullion guides, intermittent anchors, and guide rope methods, are critical to safe and effective operation. Digital tools such as BIM and parametric simulation now make it possible to resolve conflicts and verify coverage well before construction, significantly improving outcomes.

Emerging technologies including drones, robotics, and automated controls will continue to improve inspection efficiency and selected maintenance functions. However, for the foreseeable future, they will complement rather than replace conventional façade access systems. The enduring priorities remain safety, maintainability, operational efficiency, and integration with the building's architecture and lifecycle strategy.

Façade access engineering is a critical subdiscipline within façade engineering and building lifecycle planning. Integrating access design with architectural, structural, and maintenance workflows improves safety, preserves design intent, and ensures that complex envelopes remain serviceable over time. As global cities continue to build taller and more complex structures, the professionalization of façade access engineering will become increasingly important to sustainable and maintainable building practice. The framework presented in this chapter provides a structured method for façade access system selection that can be applied across a wide range of building typologies.

Author Disclosure The authors are affiliated with Global BMU, a company involved in façade access consulting and equipment development. The chapter presents general engineering principles and does not promote specific proprietary products.

References

1. Knaack U, Klein T, Bilow M, Auer T (2014) Façades: principles of construction. Birkhäuser, Basel
2. Straube J, Burnett E (2005) Building science for building enclosures. Building Science Press, Westford
3. Al-Kodmany K (2018) The Vertical City: a sustainable development model. WIT Press, Chicago
4. Wood A (ed) (2014) Best tall buildings: a global overview. Council on Tall Buildings and Urban Habitat, Chicago
5. Mobley RK (2008) Maintenance engineering handbook, 7th edn. McGraw-Hill, New York
6. Harris F, McCaffer R (2013) Modern construction management, 7th edn. Wiley-Blackwell, Chichester

7. Ali MM, Moon KS (2007) Structural developments in tall buildings: current trends and future prospects. Archit Sci Rev 50(3):205–223
8. Wurm J (2007) Glass structures: design and construction of self-supporting skins. Birkhäuser, Basel
9. Greeno R, Osbourn D (2012) Mitchell's introduction to building. Routledge, London
10. Blockley D (2012) Engineering: a very short introduction. Oxford University Press, Oxford
11. Schittich C (2001) Building skins: concepts, layers, materials. Birkhäuser, Basel

Curtain Wall Systems: Typologies, Performance Requirements, and Installation Criteria Within the European Regulatory Framework

Paolo Rigone and Paolo Giussani

Abstract Curtain wall systems represent a key component of contemporary building envelopes, combining architectural expression with advanced performance requirements. As non-load-bearing external elements, curtain walls function as complex environmental filters, integrating thermal, acoustic, lighting, and aerodynamic functions while responding to increasingly stringent energy and sustainability regulations. This paper provides a systematic overview of the principal curtain wall typologies adopted in current practice, stick systems, unitized systems, point-fixed glass façades, structural sealant glazing systems, and double-skin façades, within the European regulatory framework. Beyond a descriptive classification, the study critically examines the adequacy and limitations of harmonized standards, particularly EN 13,830, in addressing emerging façade technologies, structural glass applications, customized systems, and increasing architectural complexity. Special attention is given to performance requirements related to wind resistance, air permeability, watertightness, thermal transmittance, acoustic insulation, durability, and fire performance, as well as to the role of CE marking and European Technical Assessments. The paper also analyses installation tolerances, structural interfaces, and movement accommodation as decisive factors in long-term performance. Regulatory gaps are identified concerning post-breakage behavior of structural glass, extreme-event resilience, digital simulation-based verification, and adaptive façade systems. The chapter concludes by highlighting the need for future regulatory evolution to align certification procedures with ongoing technological innovation in façade engineering.

Keywords Curtain walls · Facade application · Tolerances · Façade performances · Building Façade system

---

P. Rigone (✉)
Politecnico di Milano and Principal Studio di Ingegneria Rigone, Milan, Italy
e-mail: paolo.rigone@polimi.it

P. Giussani
Politecnico di Milano and Facade Consultant at Studio di Ingegneria Rigone, Milan, Italy

A. Mejorin et al. (eds.), *Performance-Based Façade Design*, Springer Tracts in Civil Engineering, https://doi.org/10.1007/978-3-032-28835-6_9

1 Introduction

The building envelope, and curtain wall systems in particular, act as a complex filter between the indoor and outdoor environments, integrating multiple scientific disciplines, including thermal performance, acoustics, wind engineering, and lighting engineering. These aspects combine both direct and indirect solar control with the increasingly advanced construction technologies currently available on the market.

The evolution of the building envelope from traditional solutions toward more complex systems such as curtain walls reflects a progressively refined design approach focused on indoor comfort, energy efficiency, and environmental performance, especially for buildings whose performance must be certified within an urban context.

The reference product standard for curtain wall systems is EN 13,830:2020 "Curtain walling—Product standard", which defines a curtain wall as [1]: "part of the building envelope made of a framework usually consisting of horizontal and vertical profiles, connected together and anchored to the supporting structure of the building, and containing fixed and/or openable infills, which provides all the required functions of an internal or external wall or part thereof, but does not contribute to the load bearing or the stability of the structure of the building. Curtain walling is designed as a self-supporting construction which transmits dead loads, imposed loads, environmental load (wind, snow, etc.) and seismic load to the main building structure".

Curtain wall performance requirements must be defined in relation to both the building typology and the environmental context, including morphological characteristics. According to the standard UNI 11,173:2015 and its annexes it gives selection criteria of performance characteristics [2] of watertightness [3], air permeability [4], and resistance to wind [5, 6] load (environmental performance). During the design phase, additional aspects must also be considered, such as thermal insulation, acoustic insulation, safety in use, fire safety, burglary resistance, and accessibility, all of which are extensively addressed in the relevant regulatory framework [7, 8]. The stringent regulatory framework governing the building envelope and the high-performance requirements outlined above must be accompanied by equally important considerations related to installation phases and tolerance criteria. These aspects are directly linked to the ability to maintain the required performance not only at the time of installation, but throughout the entire service life of the building envelope. While the European regulatory framework provides a harmonized basis for the design, testing, and circulation of curtain wall systems within the European Economic Area, the rapid evolution of façade technologies and their main components raise critical questions regarding the adequacy and adaptability of current standards. Emerging materials, structural glass applications, complex smart and adaptive façade systems, and digital tools increasingly challenging the conventional assumptions embedded in harmonized product standards such as mentioned EN 13,830. Moreover, climate change and the growing frequency of extreme environmental events call into question whether laboratory-based verification procedures fully reflect real in-service performance and long-term durability of the façade systems. In this context, the present chapter does not aim merely to describe the existing regulatory framework, but to critically examine its scope, limitations, and capacity to accommodate innovation in façade engineering. By integrating technological, regulatory, and installation perspectives, the chapter identifies key gaps and outlines potential directions for future regulatory evolution.

The chapter is organized into six sections. Following this introduction, Section two presents the main terminology and classification of curtain wall systems, describing their principal elements as well as the key components that are fundamental for assessing the overall performance of the selected system. Section three outlines the classification of curtain wall systems according to the adopted construction technologies and the geometric and functional characteristics of their components. Section four describes the regulatory framework related to the assessment criteria recognized by all European Member States, with reference to CE marking for the circulation of products within the entire European Economic Area (EEA). Section five discusses the installation phase, with particular emphasis on aspects related to structural tolerances, alignment, movement joints, fixing constraints, and aesthetic requirements, all of which are directly related to the achievement and long-term preservation of building envelope performance. The last section six highlights the important aspects related to the installation framework, focusing on the type of façade and their specific tolerance criteria.

2 Terminology and Classification

The principal technical terms (see Fig. 1) used in the field of curtain wall systems and their design are defined according to the standard EN 13,119 Curtain walling terminology [9] as follows:

1. Unitised Construction

Pre-assembled, interlinking, storey height or multi-storey height facade modules, complete with infill panels.

2. Double Skin Curtain Walling

A type of curtain walling kit comprising inner and outer skins and an air cavity, the whole designed and supported as an integrated system fulfilling the functions of the curtain walling kit.

3. Stick Construction

Carrier framework of site assembled components supporting glass, insulating glass units and infill panels.

4. Transom

Horizontal structural framing member of a curtain walling.

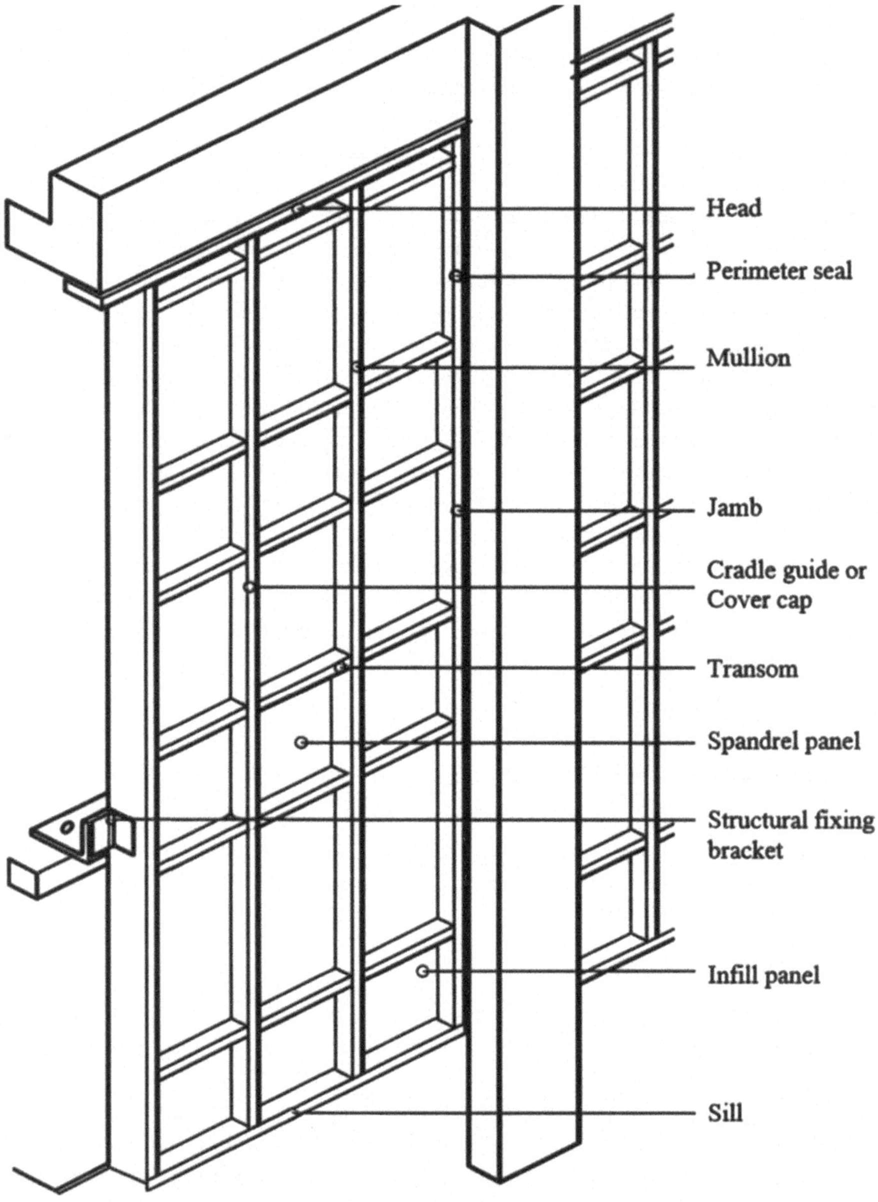

Fig. 1 Façade scheme components. Credits: UNICMI

5. Head

Horizontal structural framing member positioned at the top of the curtain walling or at the top of an area of glass, windows, panels or doors.

6. Sill

Horizontal framing member positioned at the base of an area of glass, windows, panels or doors.

7. Mullion

Vertical structural framing member of a curtain walling.

8. Cover Cap (Mullion/transom)

Profiled external cover applied, usually snap fitted, over the face of the profile pressure plate to provide an architectural finish.

9. Pressure Plate

Length of profiled material attached to mullions and/or transoms around the perimeter of a pane of glass, insulating glass unit or infill panel to provide restraint and usually compress the glazing gasket.

10. Drainage Hole

Opening through which water drains to the building exterior.

11. External Glazing Gasket

Preformed resilient profiled length of sealing material installed between the external face of a glass pane, insulating glass unit or infill panel and the surrounding frame, glazing bead or pressure plate.

12. Movement Joint

Joint to accommodate thermal or other movements arising from the curtain walling or the supporting structure.

13. Infill Panel

Translucent or opaque filler or facing material, either of one piece or an assembly, installed within a surrounding frame.

14. Spandrel Area

Area of a curtain walling between two horizontal zones, normally between glazing and concealing the edge of the floor slab.

15. Insulating Glass Unit

Composite unit of two or more panes of glazing hermetically sealed to provide an improved performance of thermal and acoustic insulation.

16. Coping

Component which protects the junction between the curtain walling and the roof edge from weather.

17. Cradle Guide

Specially profiled continuous recess or projection designed into or separately attached to the mullion to provide a location guide for an access cradle.

18. Pressure Equalisation

Method of sealing and compartmenting the wall that enables the rapid minimization of differential air pressure between cold facade cavities or glazing rebates and the external air.

19. Setting Block

Small block of suitable material, placed under the lower edge of a pane of glass, insulating glass unit or infill panel when setting it in a frame.

20. Glazing Bead

Length of profiled material used around the periphery of a pane of glass, insulating glass unit or infill panel to secure it in its frame.

21. Spigot

Profiled structural connecting piece shaped to provide continuity between two length of hollow profile of a framing member.

22. Structural Fixing Bracket

Assembly of structural components designed to transmit all actions on the curtain walling back to the building structure while allowing any required movement.

23. Thermal Break

Element of low thermal conductivity incorporated into an assembly to reduce the flow of heat between more conductive materials.

24. Vapor Control Layer

Layer comprising a material or coating with greater resistance to vapor transmission than the other layers of the wall and designed to control vapor movement through the wall.

25. Structural Sealant Glazing SSGS

Type of curtain walling where the glass is retained by a perimeter structural sealant and potentially combined with supplementary mechanical restraints.

3 Classification of Curtain Wall Systems

Curtain wall systems (excluding inclined glazing and ventilated façades) may be classified according to the adopted construction technology and the geometric-functional characteristics of their components. The most used systems include:

1. Stick and frame construction systems
2. Unitized construction systems
3. Point-fixed façades systems
4. Structural sealant glazing systems
5. Double-skin façades systems

3.1 Stick and Framed Construction Systems

The façade is composed of a framework, generally made of steel or aluminum, consisting of vertical (mullions) and horizontal (transoms) members connected together to provide regular support for the infill panels (see Fig. 2). The mullions are responsible for transferring to the load-bearing structure the actions resulting from the self-weight of the façade and from environmental loads, primarily wind actions [10]. Façade mullions transmit these loads to the building structure through dedicated fixing brackets, anchored either at the top, head, or at the bottom to the floor slab by means of mechanical anchors or steel inserts previously cast into the concrete. The fixing solution may alternatively be realized using point-fixed elements, connected through drilling operations typically performed on the upper surface of the slab.

In addition to acting as stiffening elements, transoms allow the correct fixing of the infill components. The infill panels may be glazed, either fixed or operable, or opaque, typically located at slab-edge zones. These opaque panels perform thermal and acoustic insulation functions and conceal the slab edge as well as the elements located at the upper or lower side of the structural slab. The system is completed by sealing gaskets, which ensure adequate watertightness and control of air permeability (see Fig. 3). This façade typology requires on-site assembly of all the components described above.

A similar façade system of the stick construction system is represented by framed façades, in which the infill panels are pre-framed and fixed directly to the mullions without intermediate transoms (see Fig. 4). These framed infill panels are prefabricated in the workshop and mechanically connected to the façade grid previously installed on site. This system modification allows a reduction in installation time and the transfer of part of the assembly operations to the factory, resulting in improved execution accuracy. The main advantages of this construction system include adaptability and scalability (flexibility) in response of changing design requirements; generally lower cost (cost efficiency) compared to other curtain wall systems; ease of storage and limited space requirements for on-site assembly. The main drawbacks are primarily related to on-site installation, which requires the use of fixed

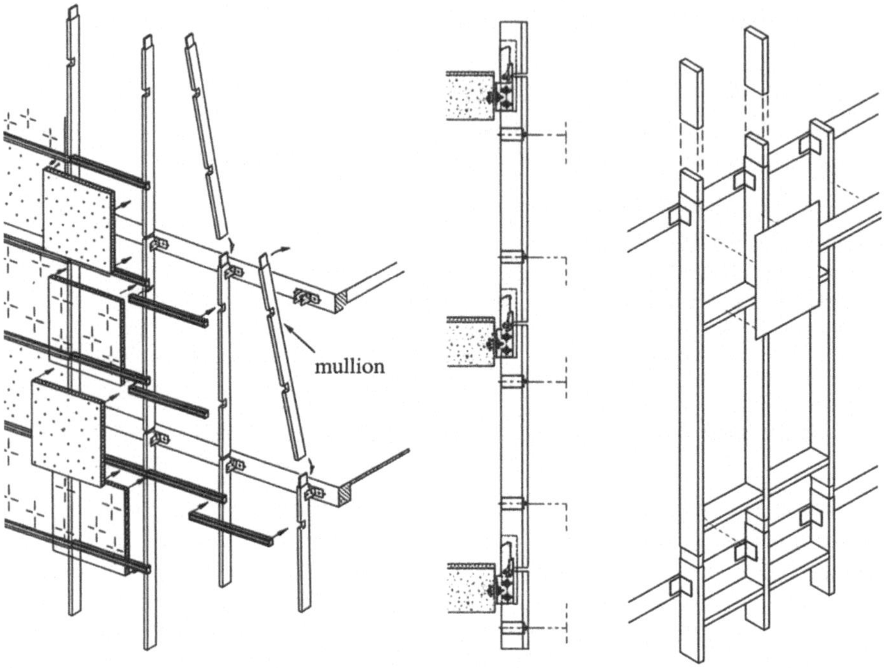

Fig. 2 Stick construction scheme. Credits: UNICMI

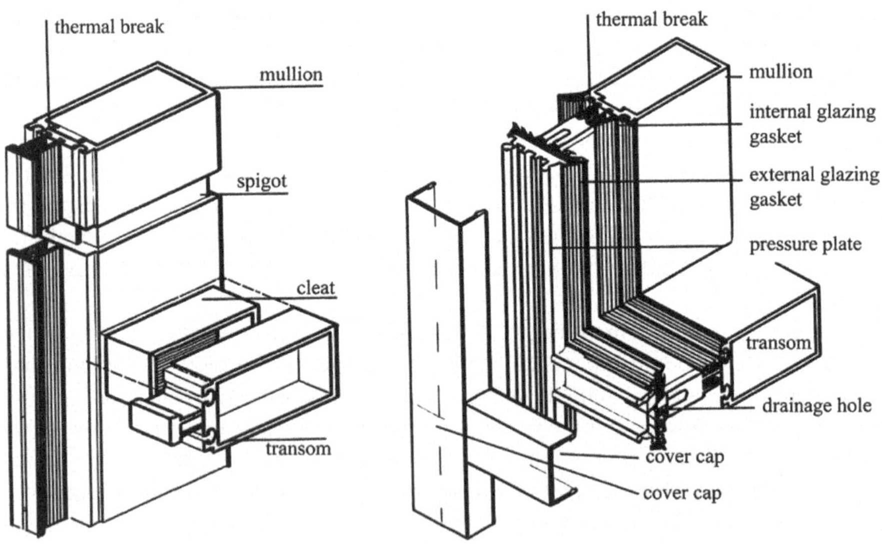

Fig. 3 Detailed façade components. Credits: UNICMI

or mobile external scaffolding. Depending on the project, this may represent a significant limitation, particularly for high-rise buildings. As regulatory considerations and limitations, under standard EN 13,830, stick systems are assessed as complete façade assemblies through laboratory testing of representative specimens. However, performance verification assumes standard geometries, controlled tolerances, and predefined load scenarios that are strictly related to the specific project and design specifications. In complex projects involving irregular floor heights, large cantilevers, or customized profiles, the applicability of existing test evidence may become limited. In such cases, additional testing or project-specific verification may be required. Furthermore, the product standard does not explicitly address long-term performance drift due to ageing of gaskets, sealants, and thermal breaks, which may influence durability beyond initial certification. Conversely, this system is well suited to projects lacking repetition of a standard module, although recent advances in factory production technologies increasingly facilitate the assembly of geometrically complex modules. Concentrating all assembly phases on site requires stringent quality control protocols and careful supervision of in-situ installation operations.

3.2 Unitized Construction Systems

Unitized construction systems are composed of a single modular, box-like element that integrates both the infill component and the metal substructure [11]. Each module is installed on site by means of dedicated fixings that connect the unit directly to the building's main load-bearing structure (see Fig. 5). To form a continuous façade, the modules are positioned adjacent to one another, creating a homogeneous external plane. Unitized façades may incorporate different glazing fixing technologies, including retained glazing with glazing beads, external pressure plates, structural sealant glazing, and point-fixed systems. Installation follows a predefined sequence based on the coupling of profiles, typically male/female connections, and the installation of gaskets at the joints (vertical and stack/horizontal joints). Assembly generally proceeds from the bottom upwards and in horizontal bands, although alternative procedures may be adopted depending on site logistics or building shape and complexity. Compared to stick systems, the installation of unitized modules is generally faster and requires fewer finishing operations and the initial costs are significantly higher.

Unitized modules enable rapid installation while eliminating the need for external scaffolding or temporary access structures. During installation, individual units are lifted (see Fig. 6) using mechanical hoists (with also the support of fixed or adjustable spreader bar). Once positioned, the façade modules are unloaded and distributed at the relevant floor level. The installation process typically begins with the positioning of anchorage brackets, usually installed in advance of module delivery, over a height of theoretically three to four stories approximately, followed by the installation of the modules themselves.

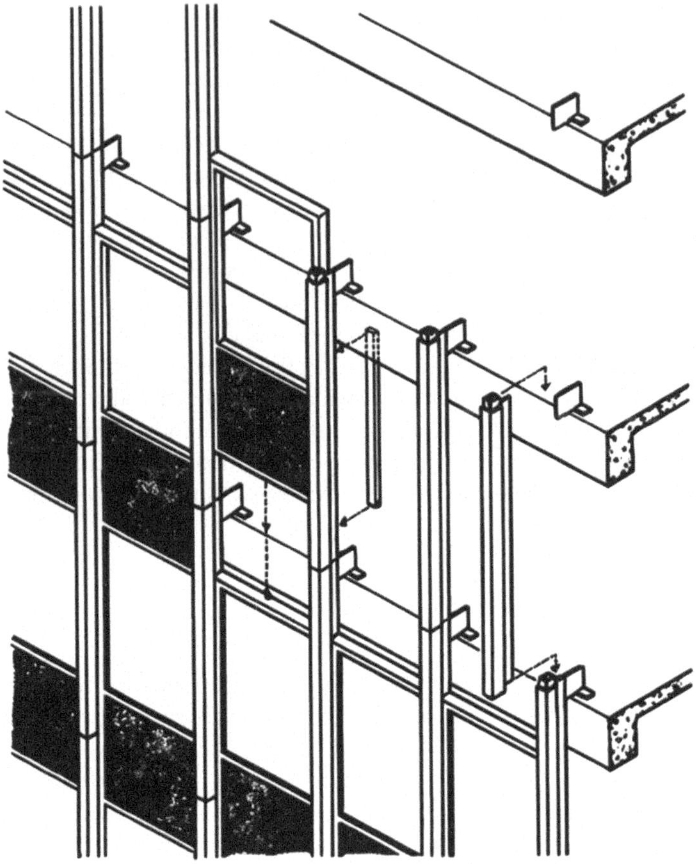

Fig. 4 Framed construction façades scheme. Credits: UNICMI

Installation may be carried out using either the site crane (direct installation) or hoisting equipment located several stories above the installation level. The main disadvantages of unitized systems include the increased handling and potential storage times when modules are of large dimensions; the need for careful control of transport, storage, and on-site handling conditions, as unsuitable hygrothermal exposure, accidental impacts, or improper lifting operations may compromise dimensional tolerances and cause localized cracking or damage to metal surfaces, glazing, and sealing gaskets; logistical constraints related to delivery scheduling and the verification of maximum module dimensions with respect to road transport limits and potential obstacles, such as bridges or narrow passages, particularly in view of recent architectural and building trends that increasingly favor larger unitized modules. Unitized systems are typically certified through full-scale laboratory testing in accordance with the testing protocol given by EN 13,830. While factory-controlled production enhances quality assurance, certification remains based on representative modules

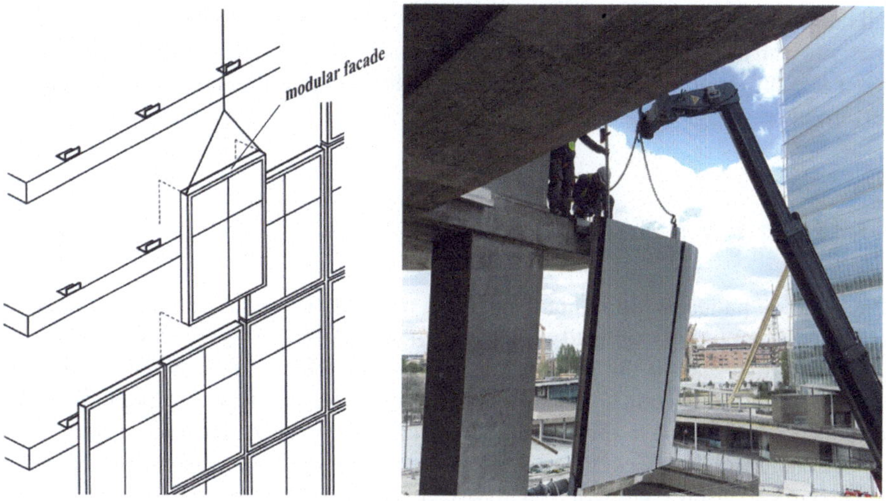

Fig. 5 Installation phases of unitized construction systems. Credits: UNICMI

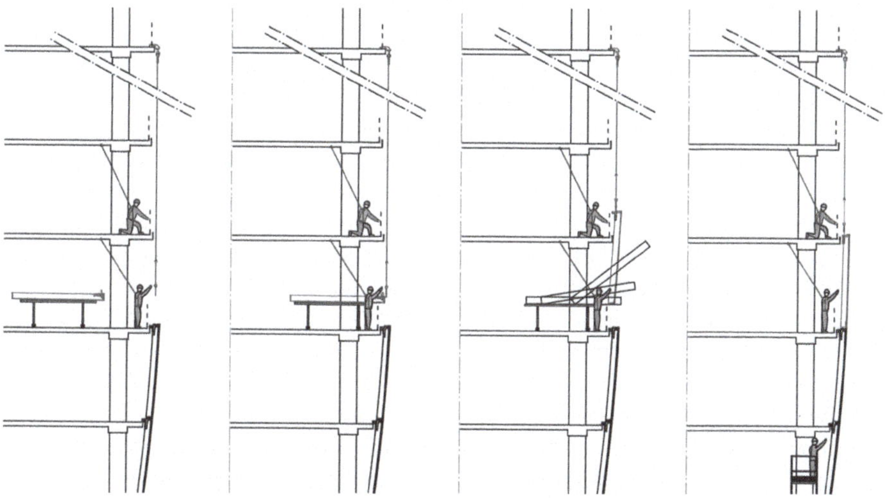

Fig. 6 Installation procedures of unitized construction systems. Credits: UNICMI

rather than on the full variability of the façade. The quality assurance protocol cannot be considered universally standardized, as it must be reviewed and adjusted according to the specific requirements and performance criteria of each project. Increasing architectural complexity, parametric variation, and large-format units may exceed the dimensional assumptions embedded in prior test evidence.

In complex projects, identifying a truly representative module for the entire façade system may be difficult, making the extension of laboratory results to the full building envelope more challenging. Moreover, the standard verification process does not fully capture installation-induced variability, transportation stresses, or in-service differential building movements, which may influence long-term façade performance.

3.3 Point-Fixed Façades System

The cladding panels (monolithic, laminated, or insulating glass units) are supported locally by point fixings, typically consisting of articulated fittings (rotules) and spider connectors that anchor the glass to the supporting substructure. When the dimensions of the glass panels so require, additional intermediate fixing points may be introduced, in addition to those usually located at the corners, in order to limit glass deflections. In these cases, the mechanical behavior shall be assessed by means of dedicated structural calculation analyses [12]. As no continuous framing system is present, joint sealing is subsequently carried out to ensure façade continuity and weather tightness. The joint dimensions shall be calculated considering, for example, cladding movements, thermal expansion, and other relevant effects. Point-fixed glazing systems often rely on project-specific mechanical verification, particularly when glass panels contribute to global stiffness through fins or, for example, rod-supported systems. While EN 13,830 addresses curtain wall performance at system level, it does not explicitly regulate structural glass as a primary load-bearing material. As a result, designers must rely on additional standards, such as EN 16,612 for glass strength assessment, and project-specific mechanical calculations [13]. This fragmented regulatory scenario may lead to uncertainties regarding the effect of post-breakage behavior and/or failure mechanisms (progressive). The installed glazing may be restrained either by fittings that clamp the glass at the corners or by systems that require partial or full drilling of the glass panels (see Fig. 7). These fixing devices are connected to the building's load-bearing structure through vertical supporting members (mullions), which may be manufactured from various materials, including glass fins and may be arranged in trussed or cable-stayed substructures. The mullions are connected to the primary structure by means of dedicated fixing brackets, in a manner like the one adopted for the previously described façade typologies.

The main advantages of point-fixed glass façades is the high architectural quality resulting from maximum transparency due to the absence of conventional framing members; flexibility in the selection of both primary and secondary supporting substructures; applicability in roof and horizontal configurations; satisfactory thermal performance when insulating glass units are used, with a reduction of thermal bridges; good acoustic insulation when combined as well with insulating glazing (the possibility of integration into double-skin façade systems, thereby achieving increased thermal and acoustic performance). The main disadvantages are the need for extreme care during handling and installation of glass panels, owing to the technical refinement of the system; reliance on sealant performance for watertightness (and in joints design), and increased mechanical demands on the glass.

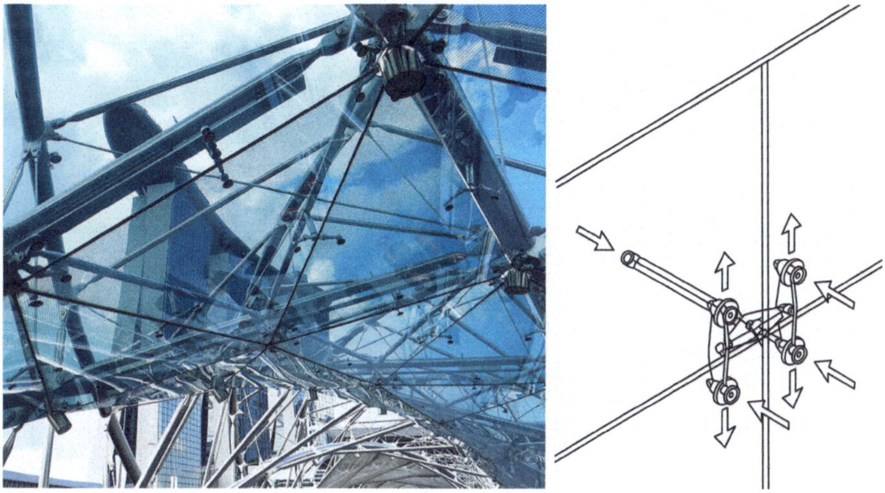

Fig. 7 Complex-shaped glass bridge with a point-fixed glazing system, Helix Bridge, Singapore. Credits: Studio di Ingegneria Rigone

3.4 Structural Sealant Glazing Systems (SSGS)

The American term structural sealant glazing defines a specific façade typology composed of glazed infill panels bonded to the supporting frame using structural silicones, which transfer the loads acting on the façade to the metal structure. In France, the designation vitrage extérieur collé has been preferred in order to avoid confusion with suspended glazing systems which, although sometimes referred to as "structural", do not exploit the structural function of silicone sealants. It is important to clarify from the outset that curtain wall systems employing structural sealant glazing (SSGS) do not have a load-bearing function and do not contribute to the primary structural system of the building [14]. The term structural refers exclusively to the method by which the glass is connected to the subframe (e.g. metal based material) by means of structural sealants applied along the perimeter of the glazing (see Fig. 8).

In the United States, the developed construction typologies are mainly based on two different systems. The first, referred to as strip window or two-side supported systems, employs a structural sealant applied only along two opposite sides (typically the vertical edges), while the remaining sides are mechanically retained. The second type, known as total wall or four-side supported systems, consists of glazing bonded along all four sides to the supporting frame through the use of structural sealant. Structural Sealant Glazing Systems (SSGS) are typically assessed under the European Technical Assessments based on ETAG 002 and EN 13,022 [15, 16].

However, the long-term mechanical behavior of structural silicones, particularly under cyclic thermal loading and environmental ageing, remains critical especially in

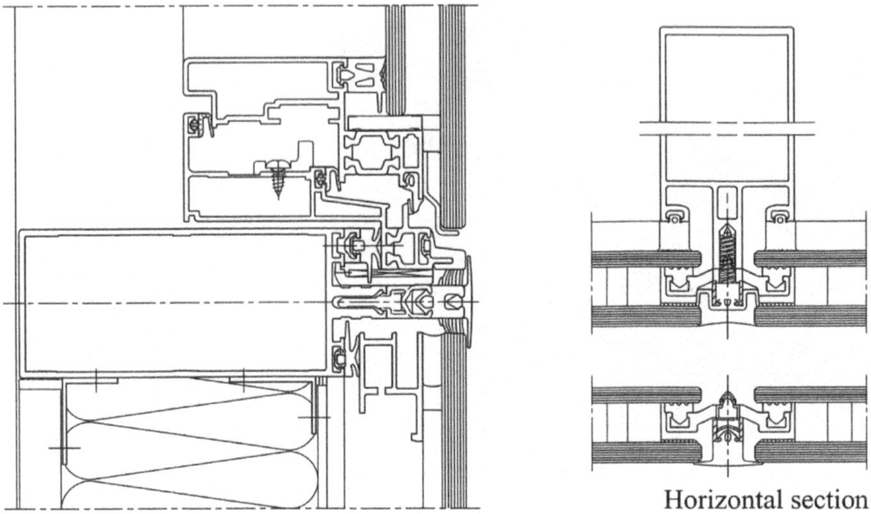

Fig. 8 Stick construction system with structural sealant glazing systems (SSGS). Credits: UNICMI

aggressive exposure conditions such as marine environments or atmospheres characterized by high levels of pollution. Certification procedures verify initial performance under standardized laboratory conditions, but particular conditions, such long-term durability and residual capacity after glass fracture require careful project-specific evaluation. SSGS technology may be adopted both for mullion and transom façades and for unitized or framed façade systems [17]. Owing to the absence of externally exposed metal elements, these façades generally offer favourable thermal and acoustic performance, together with simplified external cleaning operations, including automated maintenance systems. Few practical considerations: when structural silicone is applied to clear transparent glass, it remains visible as a perimeter frame a few millimetres wide. To reduce this effect, ceramic fritting can be applied to the glass edges over a width sufficient to cover the sealant bead. The fritting process consists of applying enamel through a fine-mesh screen onto the glass surface according to the desired pattern; subsequent kiln firing, typically associated with the tempering process, vitrifies the enamel layer.

Alternative SSGS façade solutions adopt a so-called stepped configuration, achieved by offsetting the outer and inner panes of the insulating glass unit. This arrangement provides perimeter protection to the inner pane, while only the outer pane is structurally bonded to the supporting frame. When using SSGS technology, particular attention must be paid to laminated glass units, manufactured by bonding two or more glass panes with an interlayer of polymeric material such as advanced, high-stiffness ionomer interlayers, or PVB (polyvinyl butyral) which, if exposed to prolonged weathering, may develop undesirable yellowing along the glass edges. The silicone sealants employed in SSGS façades vary depending on their function, namely structural bonding, secondary sealing, or weatherproofing.

3.5 Double-Skin Façades System

Double-skin façades (see an example in Fig. 9) consist of an external glazed skin, a ventilated cavity, shading devices, horizontal compartmentation elements, and an internal skin. High energy-efficient building envelopes have a primary role in reaching a high sustainability and improving the indoor comfort [18]. Their operating principle is dynamic insulation through natural or mechanical ventilation within the cavity. In detail, this envelope typology is composed according to the following configuration (from the exterior towards the interior):

- An external screen which, depending on the configuration and the ventilation strategy adopted to meet performance requirements, may be, more or less, permeable to air. This outer skin is generally realized using transparent enclosure systems, allowing the transmission of solar radiation required to activate the intended ventilation or heating mechanisms;
- A ventilated cavity of variable height and depth, depending on building orientation, geometry, exposure, and specific functional requirements, such as internal maintenance and cleaning operations;
- A system for regulating daylight penetration into the internal spaces, typically located within the cavity and designed to be movable and/or adjustable in order to modulate its operation in relation to solar radiation intensity and internal and external environmental conditions;
- A horizontal partition, which may serve multiple functions, including fire compartmentation, vertical subdivision of the cavity, or as a walkway to facilitate maintenance operations;
- An internal skin, more or less transparent, either operable and/or fixed, air-permeable, and constructed using the façade technologies described previously.

The operating principle of double-skin façades is based on dynamic insulation provided by natural or mechanically driven ventilation within the air cavity between the two layers. During the summer season, the primary goal is to reduce indoor discomfort caused by solar heat gains (see Fig. 10).

CFD models are specially designed to evaluate convective processes in buildings by solving the Navier—Stokes equations, since ventilation flows are complex [19]. In the winter season, simulation studies have shown that double-skin façades can reduce heat transmission losses by approximately 20–30% compared to conventional curtain wall systems. Thermal comfort in winter conditions is further enhanced by the higher surface temperatures of the internal façade layers. CFD has been validated as a numerical simulation tool by several authors for natural and mechanical ventilation scenarios [20].

The type of ventilation refers to the driving forces at the origin of the ventilation of the cavity located between the two glazed layers [22]. The need to control airflow rates while ensuring adequate acoustic, thermal, and fire performance has led to the development of different technological solutions (see Fig. 11), which can be

Fig. 9 Double-skin façades system, Hadid Tower (Generali) in Milan during the installation of the outer skin. Credits: Studio di Ingegneria Rigone

classified as follows: Single vertical channel façade: the cavity is continuous both vertically and horizontally.

The main advantages of this solution are its constructional flexibility and minimal compositional constraints on the external skin, allowing maximum envelope transparency. However, it presents disadvantages related to the difficulty of controlling airflow rates due to pressure variability within the cavity, as well as challenges in ventilating internal spaces, since exhaust air from lower floors may enter upper levels. From an acoustic standpoint, vertical and horizontal continuity may create acoustic bridges between adjacent floors and rooms. In the event of fire, the cavity may

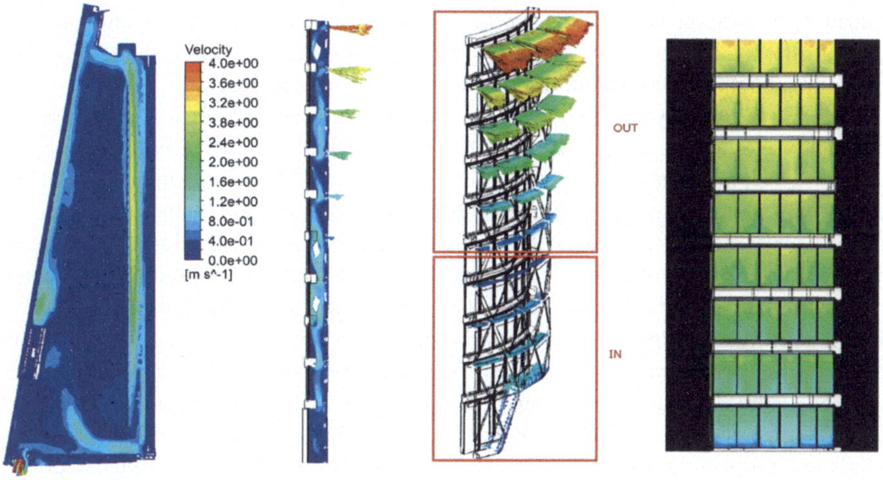

Fig. 10 Double-skin façade system, example of CFD based analysis. Credits: Rigone et al. [21]

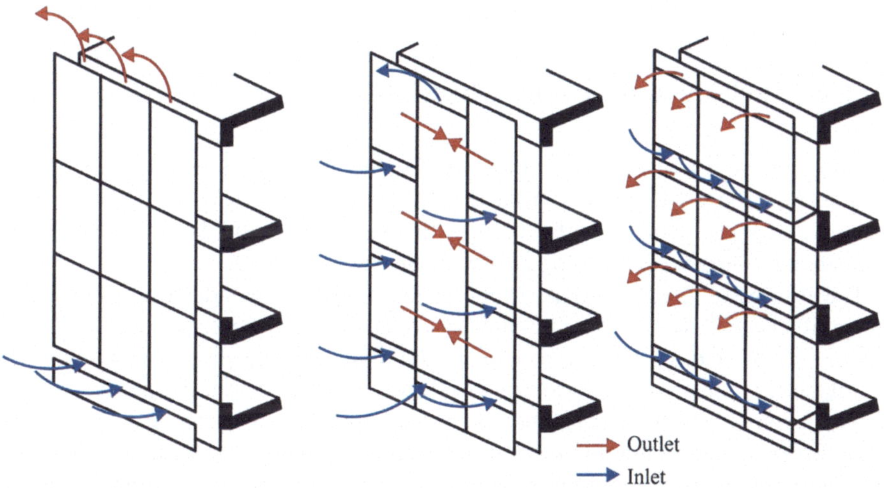

Fig. 11 Double-skin façades, from left to right: vertical channel, dual channels and elements system. Credits: UNICMI

promote the spread of flames and smoke. Dual-channel façade: this system is characterized by the alternation of vertically compartmentalized elements (also referred to as cells), subdivided horizontally at each floor level and equipped with devices for external air inlet, and full-height vertical shafts provided with air inlets at the base and exhaust openings at the top. These shafts collect exhaust air from adjacent cells through lateral openings, with the stack effect inducing air extraction from the individual cells. Compared to the previous solution, this configuration allows improved control of both acoustic and thermal performance; however, lateral openings must be carefully sized to prevent pressure reversal phenomena. In the event of fire, the vertical channels may facilitate the spread of flames and smoke.

Elements façade system: characterized by horizontal and vertical partitions at each floor level and at each mullion, defining individual enclosed cavities equipped with dedicated air inlet and outlet openings. This technology, well suited to unitized curtain wall systems, further reduces the limitations observed in the previous configurations, particularly with respect to acoustic insulation and airflow management. Improved energy performance (see Table 1) is achieved at the expense of greater compositional rigidity of the façade and a partial reduction in overall transparency. Double-skin façades introduce dynamic thermal and airflow behavior that extends beyond static performance verification. Although each skin may be assessed under EN 13,830, the combined environmental performance depends on ventilation strategies, shading control, and maintenance regimes. The exposure conditions of the façade itself represent a crucial aspect to be assessed during the design stage. Current harmonized standards do not fully address the dynamic and operational aspects of such systems, particularly regarding long-term performance under changing climatic conditions.

4 CE Marking and Product Regulation

CE marking is a mandatory certification for the circulation of products within the entire European Economic Area (EEA). It establishes a common regulatory framework by providing manufacturers with shared test methods, procedures, and assessment criteria recognized by all European Member States.

The affixing of the CE mark presupposes the fitness for intended use of construction products, namely all products manufactured to be permanently incorporated into building works. CE marking is mandatory and represents the system with which all manufacturers must comply in order to place their products on the market within the European Union. It confirms that the finished product has declared performance levels for regulated requirements in relation to its intended use. Responsibility for affixing the CE mark lies with the manufacturer, or with its authorized representative established within the EEA. The CE marking may be applied directly to the product, to a label attached to it, to the packaging, or to the accompanying commercial documentation. For curtain wall systems, the relevant harmonized product standard is UNI EN 13,830:2003, which covers stick construction systems; unitized systems and double-skin façades (understood as façades composed of individual double-skin units).

For curtain wall systems employing structural bonding of glazing (SSGS—Structural Sealant Glazing Systems), it is currently possible to obtain a European Technical Assessment (ETA) based on the ETAG 002 guidelines developed by EOTA (European Organization for Technical Assessment). The ETA applies to SSGS curtain wall systems, however, individual curtain wall assemblies incorporating structural

Table 1 Advantages and disadvantages of different double-skin façade typologies

Façade typology	Aesthetic/ compositional constraint	Integrability in existing buildings	Thermal insulation	Natural ventilation	Cross ventilation	Acoustic insulation	Fire performance
Single vertical channel	*	***	***	*	**/***	*	*
Dual-channel	**/***	*	***	*/**	*	**	**
Elements	***	*	***	***	*	***	***

glazing may also be CE marked in accordance with EN 13,830. Although curtain wall systems are defined under EN 13,830 as non-load-bearing envelope elements, contemporary architectural practice increasingly employs glass as a structural material in the form of fins, beams, and columns. In such applications, glass contributes directly to global or local stability, introducing challenges that extend beyond conventional façade product certification. Structural glass design within the European framework currently relies on a fragmented set of standards. In the previous chapter it was mentioned the standard EN 16,612 provides guidance for determining the load resistance of glass panes, while EN 13,022 addresses structural sealant glazing assemblies. However, a fully harmonized Eurocode for structural glass has only recently been under development, and its implementation remains incomplete. The future Eurocode 10 for glass includes in its scope structural glazed elements having a class of consequence higher than CC0. Non-structural elements, typically associated with consequence class CC0, are instead covered by the existing European standard UNI EN 16,612 [23]. The document defines the applicable limit states for glazed elements, including the serviceability limit state (SLS) and the ultimate limit state (ULS), and introduces two additional limit states specific to glass: the fracture limit state (FLS) and the post-fracture limit state (PFLS) [24]. Attention must be paid to post-breakage behavior. Unlike ductile materials, glass exhibits brittle fracture, and its residual load-bearing capacity in laminated configurations depends on the interlayer properties and the specific stratigraphy of the panes, including any thermal treatments. While laboratory tests verify ultimate load resistance under controlled conditions, they do not always take real-world scenarios involving accidental impact, stress concentrations, or cumulative damage.

These considerations highlight a regulatory boundary: when glass assumes a structural function beyond envelope infill, designers must move from product-level compliance toward mechanical reliability-based design, integrating façade standards with broader structural engineering principles. The current regulatory version of EN 13,830:2003 specifies the following essential performance requirements (see Fig. 12): reaction to fire; fire resistance; fire spread; resistance to wind load; resistance to self-weight (permanent load); impact resistance; resistance to horizontal loads; air permeability; watertightness; acoustic insulation; thermal transmittance; water vapor permeability; resistance to thermal shock; durability [25].

A revised version of the standard, published by CEN in 2022, is currently blocked by the European Commission pending citation in the Official Journal of the European Union (OJEU). This issue does not concern curtain wall standards alone. At present, more than 120 harmonized standards are considered not fully compliant with the provisions of Regulation (EU) No. 305/2011 (CPR), which has recently been revised by Regulation (EU) 2024/3110. For the time being, the two regulations coexist, and all standards related to the relevant product standards currently continue to mainly follow the previous version of the regulation. The new regulation will become fully applicable approximately between 2028 and 2030, giving the sectors time to revise their documentation [26, 27].

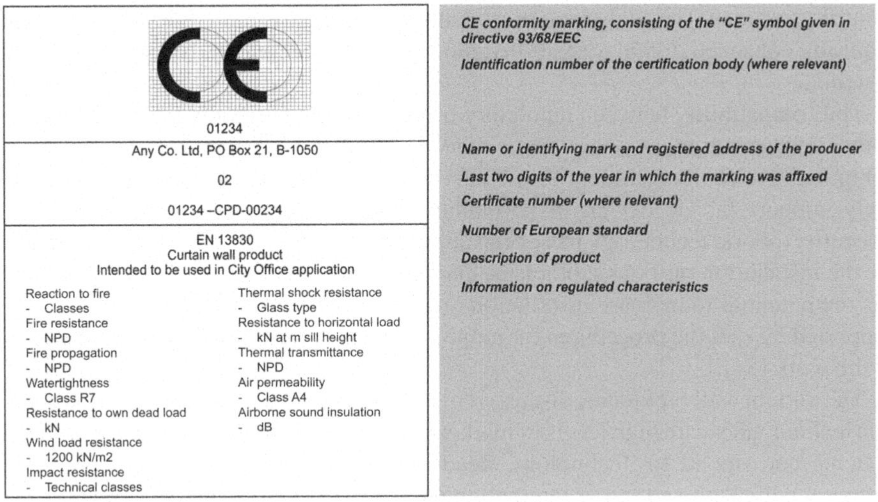

Fig. 12 Example of CE marking information for curtain wall systems to be included in the documentation of the product. Credits: UNICMI

While CE marking establishes harmonized procedures for performance declaration, it is important to recognize the inherent limitations of standardized laboratory testing. Curtain wall systems are typically tested under controlled conditions that simulate wind pressure, air permeability, and watertightness through predefined cycles. However, real environmental exposure may involve combined loading effects, fatigue phenomena, debris impact [28], and increasingly severe climatic events [29, 30]. European wind resistance testing differs significantly from hurricane or debris-impact protocols adopted in other international regulatory contexts. As climate patterns evolve [31], questions arise regarding the representativeness of current load assumptions. Furthermore, certification tests are performed on specific specimens and configurations; extrapolation to complex geometries or customized façade systems must therefore be undertaken with caution.

The growing adoption of composite materials (e.g., Glass Reinforced Concrete, or and Glass Fibre Reinforced Polymer for profiles or main components of the façade architecture), hybrid timber–glass–aluminum systems, and recycled materials introduces variability in mechanical and fire performance properties. Current harmonized standards are primarily tuned for conventional aluminum and more in general metal-based systems. Sustainable use of resources and environmentally friendly products is rapidly becoming increasingly important, supported by more stringent green regulations. Novel materials, such as bio-based polyamides for thermally broken insulted profiles [32], often require European Technical Assessments (ETA), which provide case-by-case approval rather than full harmonization. Fire classification under EN 13,501, long-term durability testing, and ageing behavior represent key areas where sustainability-driven innovation may outpace regulatory adaptation. However, CE marking procedures traditionally assume serial production and standardized quality control processes. Unique geometries challenge the conformity assessment model, requiring enhanced digital traceability, monitoring, and integrated quality assurance systems.

The compatibility between regulatory frameworks and digitally fabricated façade components remains an open question requiring further development [33]. Advanced computational tools, including AI-assisted optimization and digital twins, increasingly support façade design in particular in the scenario of complex building. Recently, robotic technology has been supporting installation activities, particularly for the installation of fixings, brackets, and connectors in curtain wall systems. This has the potential to increase installation quality; nevertheless, at present, it must be supported by specific procedures for monitoring and controlling the different phases of the work [34].

The curtain wall typologies discussed in this paper represent established construction technologies with high levels of market maturity. Conventional stick and unitized systems correspond to Technology Readiness Level (TRL) 9, as they are fully commercially deployed across international markets [35]. Structural Sealant Glazing Systems (SSGS) similarly achieve TRL 8–9, depending on configuration and certification pathway. Structural glass systems incorporating load-bearing fins or beams typically approximately range between TRL 7 and 9, depending on project-specific implementation and regulatory approval. Smart and adaptive façade technologies, including electrochromic glazing and integrated sensor systems, generally range between TRL 6 and 9, reflecting varying degrees of commercial deployment [36]. This classification is provided solely to clarify technological maturity and does not imply endorsement of specific proprietary systems.

5 Installation

Regardless of the system typology, correct installation represents the most critical phase of the process. The design of the façade and its anchorage systems shall take into account the structural tolerances of the building. Tolerances related to straightness and squareness of the façade grid shall always remain within the prescribed limits, ensuring effective fixing and sealing of the infill panels as well as the correct installation and operation of openings, in particular doors. The presence of the specified clearances between coupled frame components (for example, mullion-to-mullion joints in stick systems) shall always be verified, in order to accommodate thermal movements and ensure proper operation of opening elements.

The façade shall be installed in strict compliance with the boundary conditions (hinges, sliders, pads, etc.) defined during detailed design and mechanical analysis (calculation report). The Eurocode 9 (design of aluminum structures), in its Part 1–1 "General rules", identifies in Clause 9 the serviceability limit states [37], with direct reference to the EN 1990 "basis of structural and geotechnical design" identifies in clause 5.4 [38] the serviceability limit states for buildings and, in particular, the clause 1.8 of annex A "application for buildings". All joints (e.g. expanding, movement, and coupling joints) shall be executed in accordance with the dimensions and locations specified in the façade construction drawings.

When dealing with the stack joint gap ($\delta v,tot$) of a unitized system, for example, it is necessary to consider the algebraic sum of the main tolerances, including the maximum deflection of the slab edge ($\delta 1$), thermal expansion ($\delta 2$), manufacturing tolerances ($\delta 3$), and installation tolerances ($\delta 4$). In addition, the joint must be capable of accommodating movements induced by seismic actions, depending on the restraint conditions of the façade module.

The structural tolerances referred to above are normally defined with respect to three reference axes:

X-axis: horizontal, in the plane of the façade, generally positive to the right;

Y-axis: vertical, in the plane of the façade, generally positive upwards;

Z-axis: horizontal, perpendicular to the façade plane, generally positive towards the interior of the building.

Tolerances generally apply to the significant points of façade to structure interaction, namely those used for anchorage fixings, those that may cause physical interference, those that limit installation clearances, or those that could prevent proper operation of openings (see Fig. 13). At present, no national or European technical standards define construction tolerances of building structures specifically suitable for the correct installation of curtain wall systems. Reference is therefore commonly made to voluntary best-practice guidelines, such as the technical document UX 60, guide to the Installation of curtain wall systems, published by UNICMI (Italian Association of Metal Construction, Envelope and Fenestration Industries). According to UX 60, any significant point of the building structure interacting with the façade shall not deviate from its theoretical position by more than $\pm$ 25 mm along each of the three axes defined above [39]. These values can, of course, vary from project to project, particularly when considering cantilevered structures, where specific tolerances can be defined in the project specifications. In addition, the following more restrictive local tolerances apply: center-to-center distance between two successive columns or load-bearing walls (*X* and *Z* axes): $\pm$ 20 mm; height between two successive floor slabs (*Y*-axis): $\pm$ 20 mm; vertical deviation between two successive floor slabs (*Z*-axis): $\pm$ 20 mm; level difference between corresponding points on the same floor slab (*Y*-axis): $\pm$ 10 mm; vertical deviation between points along the same column (*Z*-axis): $\pm$ 10 mm, etc.

The first group of three tolerances applies not only to axes and levels of successive structural elements, but also to distances between any two significant points belonging to them.

5.1 Tolerances in Curtain Wall Installation

It is the responsibility of the façade supplier to verify, prior to installation, that the structural tolerances do not exceed the adjustment capacity of the anchorage systems, that cast-in elements are correctly positioned (see Fig. 14), and that no part

Fig. 13 Typical connection of a curtain wall bracket installed on metal and concrete supports; the bracket design allows position adjustment and tolerance correction through a slotted hole. Credits: Studio di Ingegneria Rigone

of the structure interferes with proper façade installation, with particular regard to the clearance between structural edges and the façade (e.g. using clash detection tools in BIM).

Other important aspects are related to the geometry and/or challenging shape of the façade and the building itself, which may represent current limits for curtain

Fig. 14 Cast-in channels: examples of remedial works following wrong (**a**, **b**) specified installation tolerances (**c**, **d**). Credits: Studio di Ingegneria Rigone

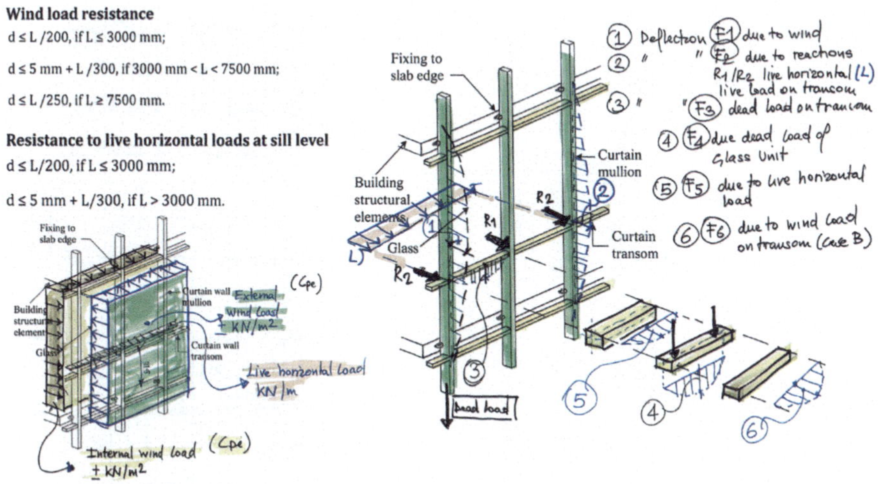

Fig. 15 Displacement of curtain walling framing according to the product standard norm. Credits: Paolo Rigone

wall systems as defined by standards (e.g., cantilevered structures with post-tension systems or façades inclined inward/outward up to 15°). Under these conditions, it is normal to expect "large" vertical displacements from the structure, particularly due to shrinkage in concrete elements. This is a typical issue that arises when aiming for high transparency with large glass units or when using very lightweight structural and substructural systems.

Figure 15 illustrates a typical approach to addressing this problem. Standards define expected displacements for the façade frame under wind loads (calculated taking into account the internal and external pressure coefficients for each zone [40]) and/or horizontal live loads and set displacement limits to ensure that the curtain wall maintains its performance, including air permeability, watertightness, and overall serviceability.

However, complex projects often require moving beyond standard commercial systems profiles toward customized solutions. Challenging building shapes, such as differing floor-to-floor heights or cantilevered floors next to standard floors, can increase differential displacements beyond typical expectations. This issue is further emphasized when stack joints of the façade system are located at sill level rather than floor level, making them more visible and less easily mitigated with flashing or other finishing solutions. In such cases, the system must also be tested in the laboratory to verify that it meets the performance requirements specified for the bespoke project, considering both environmental conditions and anticipated building movements.

Careful attention must also be paid to the interface and displacement accommodation between the structure and the curtain wall. Joint dimensions presented in early design drawings may be undersized relative to actual structural displacements and construction tolerances encountered during detailed design (see Fig. 16). In these

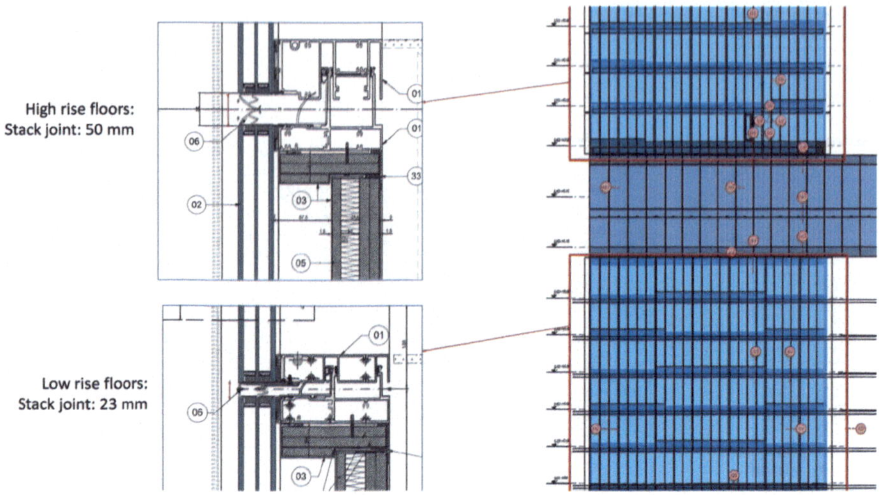

Fig. 16 Different stack joints to accommodate very wide displacements. Credits: Studio di Ingegneria Rigone

situations, the framing must be modified to provide a custom solution fully compatible with the structural movements. the performance of a façade is not only related to a good design, but also to the quality of construction and installation. In order to evaluate a façade real performance, especially in case of complex ones, laboratory tests and on-site tests on full-scale mock-ups are today essential [41]. The standard EN 13,051 defines the method to be used to check for any points of water leakage that may occur in curtain walling when installed on a building [42]. Consequently, commercially available framing profiles may be insufficient, requiring the production of custom made geometries to accommodate the tolerances associated with the façade system joints.

The façade shall be installed in accordance with its theoretical position as defined in the design drawings and with respect to the reference lines set out at each floor level. Cast-in elements for façades anchoring may be positioned within a tolerance of $\pm$ 10 mm parallel to the reference line and $\pm$ 5 mm perpendicular to the reference line.

The Building Envelope Contractor shall survey the primary structure for line, level and position of fixing points prior to erection of the envelope and give notice to the main contractor if it is not possible to erect the envelope as planned [43].

Any potential non-conformity shall be promptly reported to the client (often represented by the general contractor, as façade works are typically subcontracted), and no modifications shall be carried out without written approval. Installation shall be suspended in the affected areas until such non-conformities have been resolved. With the introduction of the harmonized standard EN 1090-1, with specific reference to the execution of steel structural components (and therefore in relation to the technical standard EN 1090-2), has effectively provided a tool that highlights the need to "coordinate" requirements and tolerances that were already relevant to the activity of the structural designer and, consequently, to the manufacturer [44]. The evaluation criteria for the various requirements related to the manufacturing process of steel structural components are contained in the technical standard EN 1090-2, including essential tolerances, manufacturing tolerances, and functional tolerances [45].

In addition, the façade shall meet aesthetic requirements of verticality and alignment of continuous lines, both vertically and horizontally (see Fig. 17). A line is considered vertical if its points do not deviate by more than ± 5 mm from the theoretical vertical established by a plumb line; this value is reduced to ± 2 mm between points within the same story up to a height of 3.50 m, and to ± 3 mm for stories exceeding 3.50 m. Also, the ± 5 mm tolerance shall be doubled for buildings exceeding ten stories in height; if façade installation begins before completion of the entire structural frame, the maximum allowable vertical deviation shall be increased to ± 10 mm, and to ± 20 mm for buildings over ten stories. Again, a line is considered horizontal if its points do not deviate from the theoretical horizontal, established by a level, by more than ± 5 mm, with a stricter limit of ± 2 mm between points belonging to the same façade module. Finally, where the façade is intersected by structural expansion joints, each façade portion shall be considered structurally independent from the others.

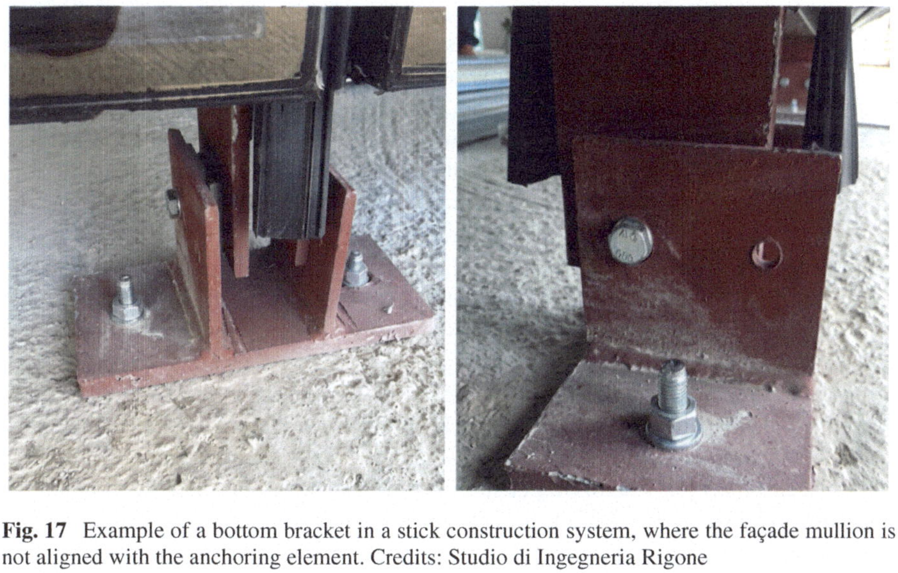

Fig. 17 Example of a bottom bracket in a stick construction system, where the façade mullion is not aligned with the anchoring element. Credits: Studio di Ingegneria Rigone

6 Conclusion

As discussed in this manuscript, a thorough understanding of the classification of curtain wall systems, including stick systems, unitized façades, and structural sealant glazing, is essential for anticipating installation challenges and selecting the appropriate assembly strategy. Each system type presents unique tolerances, movement allowances, and fixing requirements that directly influence the achievable performance.

However, contemporary façade engineering extends beyond conventional typologies. the increasing use of structural glass, hybrid materials, digital fabrication methods, and adaptive envelope technologies requires a critical reassessment of how current European product standards address innovation and long-term performance. It was noted the importance of CE marking and regulatory compliance, which define the essential performance criteria for curtain walls, such as mechanical resistance, watertightness, air permeability, thermal and acoustic performance. While the harmonized framework provided by the norm EN 13,830 and related standards ensures a common basis for performance declaration, certification remains primarily based on standardized laboratory testing under controlled conditions. as building geometries become more complex and climatic events more severe, questions arise regarding the extent to which current test protocols fully represent in-service conditions and in particular related to the extreme-event scenarios. Finally, the installation phase of curtain wall systems represents a critical step in ensuring that the designed performance of the building envelope is fully achieved. As emphasized throughout this chapter, strict adherence to structural tolerances, alignment, and joint specifications is essential. Detailed design generally relies on standard façade systems. However, the growing complexity of projects and the challenges related to the construction phase increasingly require the development of custom solutions. Custom façade systems must be tested both in the laboratory and on-site, as it is difficult or, in many cases, impossible, to extend the application and validity of existing system test certificates. The joints of a custom façade system must be designed in accordance with the expected structural displacements and construction tolerances. Even minor deviations in the positioning of anchoring elements, frame profiles, or panels can compromise key performance criteria, insulation, mechanical behavior, and overall durability behave of the building envelope. In addition, the interface between mechanical design, product certification protocols, and on-site execution becomes increasingly critical in projects involving customized or non-series-produced components, where the extension of existing test evidence is challenging. Future regulatory revision will need to address these dimensions in order to align certification procedures with the quick transformation of façade technology and their components.

References

1. EN 13830 (2020) Curtain walling, product standard. European Committee for Standardization
2. UNI 11173 Windows, Doors and curtain walling, selection criteria based on air permeability, watertightness, resistance to wind. Italian National Standards Body
3. EN 12154 Curtain walling, Watertightness, performance requirements and classification. European Committee for Standardization
4. EN 12152 Curtain walling, Air permeability, performance requirements and classification
5. EN 12179 Curtain walling, Resistance to wind load, test method. European Committee for Standardization
6. EN 13116 Curtain walling, Resistance to wind load, performance requirements. European Committee for Standardization
7. EN 12631 Thermal performance of curtain walling, calculation of thermal transmittance
8. EN 14019 Curtain Walling, Impact resistance, performance requirements. European Committee for Standardization
9. EN 13119 Curtain walling, Terminology. European committee for standardization
10. Morris F (2013) Definition and types of curtain walls. Curtain wall systems: a primer, pp 3–7
11. UNICMI (2018) UX60 guide to the installation of curtain walls
12. UNICMI (2013) UX70 mechanical design of curtain walls
13. EN 16612 Glass in building, determination of the lateral load resistance of glass panes by calculation
14. CEN/CENELEC (1994) Mandate m/108, curtain walling, 04/33 external walls (including cladding), internal walls and partitions
15. EN 13022-1 Glass in building, structural sealant glazing, Part 1: glass products for structural sealant glazing systems for supported and unsupported monolithic and multiple glazing
16. EN 13022-2 Glass in building, structural sealant glazing, Part 2: Assembly rules
17. ETAG 002 (2012) Guideline for European technical approval for structural sealant glazing kits (SSGK) EOTA
18. Baldinelli G (2008) Double skin façades for warm climate regions: analysis of a solution with an integrated movable shading system. Build Environ 44:1107–1118
19. Manz H, Schaelin A, Simmler H (2004) Airflow patterns and thermal behavior of mechanically ventilated glass double façades. Build Environ 39:1023–1033
20. Ramponi R, Blocken B (2012) CFD simulation of cross-ventilation flow for different isolated building configurations: validation with wind tunnel measurements and analysis of physical and numerical diffusion effects. J Wind Eng Ind Aerodyn 104:408–418
21. Rigone P, Mazzucchelli ES, Giussani P, Ancillotti A (2021) Double skin façades: Externally and naturally ventilated small and big cavities, a case study in Milan. In: 16th Conference on advanced building skins. Advanced Building Skins GmbH, pp 44–53
22. Loncour X, Deneyer A, Blasco M, Flamant G, Wouters P (2005) Contributed Report. Ventilated Double Skins Façades, Classification & illustration of façade concepts. Belgian Building Research Institute
23. EN 19100-1 (2024) Eurocode 10, Design of glass structures Part 1: General rules. European Committee for Standardization (CEN), Brussels
24. Guerra V, Giussani P, Rigone P (2025) The new Eurocode 10 for the design of glass structures: the prEN 19100 series and the application to glass elements loaded orthogonally to their plane. Costruzioni Metalliche: 01–14
25. EN 13830 (2003) Curtain walling, Product standard
26. European Commission (2011) Regulation (EU) No 305/2011 of the European Parliament and of the Council
27. European Commission (2014) Regulation (EU) 2024/3110 of the European Parliament and of the Council
28. Technical Note N.4 Simulated windborne debris impact testing of building envelope components

29. Limonta S, Giussani P, Rigone P (2022) Resilient facades behavior and performance of dry exterior wall systems in case of extreme windstorms. Costruzioni Metalliche, pp78–90
30. Mejorin A (2018) Exceptional atmospheric events resilience of the curtain wall. Techne, pp 323–330
31. Powar O, Jayachandran A (2021) Learning from façade failures over urban landscape: aftermath of cyclone Vardah. Nat Hazards Rev 22(1):04020049
32. Technoform Bautec GmbH (2018) Bio-based polyamides for insulating profiles. https://www.technoform.com/it. Last Accessed on Feb 15 2026
33. Azcárate-Aguerre JF, Klein T, Konstantinou T, Veerman M (2022) Facades-as-a-Service: The role of technology in the circular servitisation of the building envelope. Appl Sci 12(3):1267
34. Iturralde K, Feucht M, Illner D, Hu R, Linner T, Bock T, Eskudero I, Rodriguez M, Gorrotxategi J, Izard J-B, Astudillo J, Santos J, Gouttefarde M, Fabritius M, Martin C, Henninge T, Nornes S, Jacobsen Y, Elia L (2022) Cable-driven parallel robot for curtain wall module installation. Autom Constr 138:104235
35. Mankins JC (1995) Technology readiness levels: a white paper. Advanced concepts office, office of space access and technology, National Aeronautics and Space Administration (NASA), Washington, DC, USA
36. Yfanti S, Sakkas N (2024) Technology readiness levels (TRLs) in the era of co-creation. Appl Syst Innov 7:32
37. EN 1999-1-1 Eurocode 9, design of aluminum structures, part 1-1: General rules
38. EN (1990) Eurocode, basis of structural and geotechnical design
39. UNICMI (2005) UX42 Guide to the installation of windows and doors
40. CNR-DT 207/2008 (2010) Guide for the assessment of wind actions and effects on structures, CNR, Advisory Committee on Technical Recommendations for Construction, Rome
41. Giussani P, Rigone P (2022) The performance characteristics of curtain walls: from on-site checks to final acceptance tests. Costruzioni Metalliche, pp 44–60
42. EN 13051 Curtain Walling, watertightness site test
43. CWCT (2005) Part 7, Robustness, durability, tolerances and workmanship, standard for systemized building envelopes
44. EN 1090-1 Execution of steel structures and aluminium structures, part 1, requirements for conformity assessment of structural components
45. EN 1090-2 Execution of steel structures and aluminium structures, part 2, technical requirements for steel structures

Next-Generation Timber Façades: Performance-Based Design and Smart Technology Integration

Domas Valiukas, Ricardo Seijas, and Laura Craft

Abstract Sustainability imperatives and technological innovation are converging to reshape building façade design. Timber has become central to a new generation of low-carbon envelope systems that combine biobased materials with digital intelligence. This chapter presents the development of Staticus' new Timber-Hybrid Unitized Façade (HUF). It examines key implementation challenges, namely structural validation, hygrothermal behavior, and fire performance, and highlights the enabling role of testing for new material implementation. It also considers how architectural aesthetics and sustainability regulations have driven the development of the Timber-HUF system. The subjects of performance testing, embodied carbon, and digital monitoring are explored, including details of the Textiltorget project in Stockholm, the first implementation of Staticus' Timber-HUF. Key results achieved on the Textiltorget project include an up to 48% reduction in embodied carbon compared to a standard aluminum unitized façade system, and better thermal performance, with a U-value of 0.46 W/m^{2}K achieved. Equilibrium moisture content was maintained within safe thresholds, typically below 18%, while the treated timber used achieved a combustibility Class B rating. Additionally, this chapter examines Staticus' willow-hybrid prototype development, which demonstrates the feasibility of regenerative materials in façade applications. A number of other innovation milestones were achieved during the research into the Timber-HUF system. A Predictive Performance Tool that utilizes sensors to record metrics such as temperature and humidity has been developed. The data gathered will allow analysis by machine learning algorithms to forecast performance and support predictive maintenance.

Keywords Façades · Timber-hybrid unitized façades · Biobased materials · Regenerative materials · Predictive performance · Circularity · Lifecycle management · Digital twins

---

D. Valiukas · R. Seijas (✉) · L. Craft
Staticus, Vilnius, Lithuania
e-mail: Ricardo.Seijas@staticus.com

A. Mejorin et al. (eds.), *Performance-Based Façade Design*, Springer Tracts in Civil Engineering, https://doi.org/10.1007/978-3-032-28835-6_10

Nomenclature

Timber-HUF System	Timber-hybrid unitized façade system
Glulam	Glue laminated timber
CLT	Cross laminated timber
DfMA	Design for Manufacturing and Assembly
Biophilic Design	Designed aimed at increasing connection to nature and nature-like design in order to improve health and wellbeing
EPBD	Energy performance of buildings directive
WWR	Window wall ration
FSA	Façade surface area
LCA	Life cycle assessment

1 Introduction

Façades serve as the boundary between architecture and environment, mediating the exchange of light, heat, sound, and moisture. The façade industry is undergoing transformation driven by carbon neutrality commitments and digitalization of building performance, namely a shift from design intent to verified performance data.

Aluminum curtain walls have dominated commercial buildings over the last century, offering lightweight strength, rapid assembly, and manufacturing precision. However, aluminum production is among the most carbon-intensive industrial processes, with global emissions averaging 15 kgCO_2e/kg aluminum, varying between 6 and 18 kgCO_2e/kg [1].

While Europe achieves lower emissions through renewable electricity and higher recycled content levels [2], constraints limit the large-scale uptake of low-carbon aluminum.

Timber offers a compelling alternative: engineered timber products require minimal production energy, store biogenic carbon, and source from sustainably managed forests. Glulam and CLT have transformed structural construction, and extending their use to façades more broadly presents new sustainability opportunities. However, timber's natural variability, hygroscopic nature, and combustibility present implementation challenges.

Staticus, a Northern European unitized curtain wall façade contractor, has approached this challenge systematically through research, prototyping, and collaboration. Its incorporation of timber has evolved from large format prefabricated timber façades to advanced timber hybrid unitized curtain wall systems. This progression extends toward adaptive building systems that coordinate façade intelligence with mechanical systems, circular economy frameworks enabled by digital material passports, and regenerative material ecosystems where construction participates in ecosystem restoration.

This chapter examines: the evolution from functional prefabrication to sustainable engineering, including material optimization and smart technology integration (Sect. 2); Staticus' approach and methodology for the Timber-HUF's development (Sect. 3); the results and outcomes of its research, including a case study of the Textiltorget project (Sect. 4); a discussion of lessons learned and system refinements (Sect. 5); the technological readiness level and market status of the Timber-HUF (Sect. 6); conclusions (Sect. 7); and future directions toward adaptive systems, circular economy, and performance-based regulation (Sect. 8).

2 Theoretical Framework: The Evolution of Timber Façades, Material Optimization, and Smart Technology Integration

2.1 From Functional Prefabrication to Sustainable Engineering

Over the past 15 years, Staticus has delivered timber façades across the Nordic and Baltic regions. Figs. 1 and 2 show example projects where Staticus delivered timber façade systems, though notably none show visible timber elements, as these early systems were designed with the timber façade fully concealed.

In the early 2010s, the company developed its first modular timber façade units for markets where thermal performance and cost efficiency were the primary drivers. These early systems featured timber-framed panels with aluminum adapters for connection points and glazing interfaces, typically constructed from spruce, larch, or pine glulam sections. The modular units were installed on projects with a higher ratio of opaque to glazing area, where the architecture prioritized performance over high transparency or frame slenderness, and where there were no requirements for visible timber, enabling a more cost-effective solution.

Fig. 1 Early examples of timber façade systems developed by Staticus

Fig. 2 Another example of a modular timber façade system developed by Staticus

The systems comprised timber framing, aluminum adapters, mineral wool insulation, drywall, and vapor barriers positioned to the rear of the timber structure. Elements were manufactured under controlled factory conditions and following Design for Manufacturing and Assembly (DfMA) principles. These principles improved production efficiency and minimized material waste. The large-format timber units were fully assembled at Staticus' facility before delivery, which reduced exposure to harsh Nordic weather conditions and minimized on-site labor requirements.

While this early generation of modular timber façade units was successfully implemented across numerous projects in Scandinavia—including Solseilet and UNN Pasienthotell, both in Tromsø, Norway—several design and performance considerations limited large-scale adoption, including:

- **Wide visual frame joints**: The location of the external aluminum adapters, attached to the outer side of the long side of the glulam section, combined with aluminum window sub-framing on the inner side, created wide joints between modules. Frame and joint widths reached 250–300 mm for fixed glazing and up to 450 mm for operable elements. While these dimensions could be concealed in designs with large opaque areas, they made the units unsuitable for projects where transparency and frame slenderness were architectural priorities, thus essentially limiting adoption to buildings with lower window-to-wall ratios (WWR).
- **Concealed timber**: Since the primary driver in these early systems was thermal performance and there were no aesthetic drivers for timber visibility, the glulam framing was concealed within the façade system. This enabled more cost-effective timber material to be selected.

The development of these early systems laid the foundation for the next generation of timber façade systems. They demonstrated that it was possible to meet stringent performance targets for airtightness and thermal transmittance using large-scale, prefabricated timber façade modules, thus proving the technical viability of this approach.

2.2 A Shifting Design Landscape

In the late-2010s, architectural design intent and policy frameworks converged to redefine façade design priorities. Two parallel market drivers emerged that would redefine the demand for timber façade systems: evolving architectural aesthetics and increasingly stringent sustainability regulations.

Evolving architectural aesthetics. Contemporary architectural practice has driven demand for a next generation of timber façades characterized by two key aesthetic shifts:

- **Increased glazing and narrow frames**: A return to curtain wall design principles with slender framing profiles and higher window-to-wall ratios (WWR), prioritizing transparency and visual lightness over the opaque-dominant compositions that suited first-generation systems.
- **Exposed timber and biophilic design**: A renewed interest in biobased materials and natural aesthetics, with architects specifying visible, high-quality timber to leverage its tactile warmth and biophilic design. This shift positioned timber not as a hidden structural material, but as a primary architectural finish (see Fig. 3) that contributes to occupant wellbeing and environmental connection.

Evolving sustainability regulations. The EU Green Deal, adopted in 2019, established a comprehensive framework for achieving climate neutrality by 2050. Buildings, which are responsible for approximately 40% of EU energy consumption and 36% of greenhouse gas emissions [3], with a further 5–12% attributed to the embodied energy of materials used for building construction [4], were identified as a critical industry contributor. Two key regulatory instruments have emerged to drive decarbonization in the building and construction sector: the EU Taxonomy for Sustainable Activities and the revised Energy Performance of Buildings Directive (EPBD).

The EU Taxonomy establishes technical screening criteria that define when economic activities, including construction, can be considered environmentally sustainable. Under these criteria, new buildings must provide lifecycle assessments, including a Global Warming Potential (GWP) disclosure. The revised EPBD, which entered into force in 2024, mandates that all new buildings must meet carbon limits, aligned to zero-emission pathways from 2030 onwards, with public buildings required to meet this standard from 2028. Critically, the directive requires member states to establish maximum thresholds for the GWP of new buildings across their full lifecycle, from raw material extraction through end-of-life.

Fig. 3 The new architectural trend of using timber as a primary architectural finish that enhances occupant wellbeing. Photos ©Staticus, ©Andrius Gudelis. Reprinted with permission

In response to these EU-level mandates, member states have progressively implemented national frameworks for embodied carbon accounting and limits. The Netherlands was the first to act, introducing mandatory environmental performance limits (MPG) from January 2018, with binding caps tightened further in 2021 [5]. France launched its RE2020 regulation in January 2022, establishing maximum emissions thresholds that progressively tighten from 640 kg CO_2e/m^2 for single-family homes in 2022 to 415 kg CO_2e/m^2 by 2031 [6]. Denmark implemented whole-life carbon limits for large buildings under BR18, with significantly stricter requirements (7.1 kg CO_2e/m^2/year) entering force from July 2025 for nearly all new builds above 150 m^2 [7].

Figure 4 illustrates the timeline of climate declaration and limit value integration across the Nordic countries against the EPBD as documented by Nordic Sustainable Construction (as of September 2025). It shows the progressive roadmaps Nordic countries have established for implementing limit targets. The trajectory is clear: binding carbon intensity limits, which effectively set maximum allowable emissions per square meter that buildings cannot exceed, are rapidly becoming the regulatory norm across Europe, with other member states expected to follow suit ahead of the EU's 2030 deadline.

Additionally, some jurisdictions have introduced minimum requirements for biobased materials as a complementary strategy for reducing embodied carbon. Since November 2024, all public buildings in Lithuania, where Staticus is headquartered, must comprise a minimum of 50% timber or other organic construction materials [8]. This regulatory requirement has triggered a marked increase in both research into and market demand for biobased construction systems, creating immediate commercial drivers for high-performance timber façade solutions.

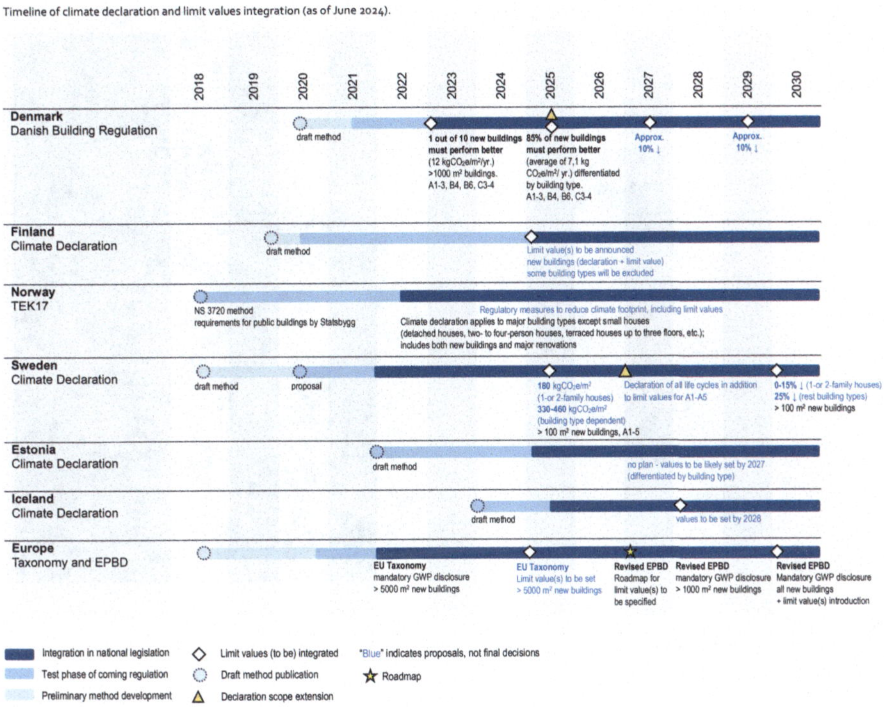

Fig. 4 A graphic timeline showing climate declaration and limit value integration in different Nordic countries since 2016. Nordic Sustainable Construction. © (2024) *Nordic Sustainable Construction: Overview of Nordic climate declarations*. Available at https://www.nordicsustainableconstruction.com/knowledge/2024/ Reprinted with permission

This concurrent shift in both architectural and regulatory priorities has reframed the role timber plays in façades. It is no longer merely a structural alternative or a material hidden behind cladding, but a material of strategic regulatory and commercial importance. It contributes to compliance with carbon limits while simultaneously realizing biophilic design principles and contemporary aesthetic expectations for transparency and visual refinement.

2.3 *Material Optimization and Regenerative Alternatives*

Resource constraint. Timber provides substantial carbon benefits. However, it remains a limited biogenic resource, and global demand for timber continues to grow, creating increasing pressure on forest ecosystems and agricultural land (as illustrated in Fig. 5).

If the global construction industry were to transition entirely to timber using conventional long-rotation species, such as spruce, larch or pine with growth cycles

Fig. 5 A birds-eye view of two adjacent land plots, one heavily forested and the other covered in dense vegetation. © Shutterstock. Available at Shutterstock.com. Image ID: 1274894119. Licensed under Shutterstock Standard License

of 20–30 years, without parallel advances in forestry management or regenerative alternatives, the consequences would be severe. A 2023 World Resources Institute analysis projects that providing 50% of new construction with wood between 2010 and 2050 would require annual industrial wood harvests to more than double and to average 40% higher than baseline [9]. Such expansion would create acute competition for agricultural land, significant strain on ecological systems, and potential biodiversity loss.

Evidently, such expansion is neither ecologically nor economically viable. This realization led Staticus to conclude that timber construction and traditional forestry alone cannot solve the construction industry's sustainability challenge. The company therefore explored fast-growing regenerative species that have short growth cycles and can sequester carbon at higher rates than conventional softwoods.

2.4 *Exploring Regenerative Biomaterials: Willow as a Case Study*

To address land use constraints and the limitations of long-rotation timber crops, Staticus investigated the feasibility of fast-growing, regenerative alternatives. Various short-rotation coppice species were evaluated, with willow (Salix spp.) emerging as a leading alternative for replacing conventional softwoods in façade applications.

Willow, sometimes referred to as "the bamboo of the North," is native to Europe and thrives in temperate and sub-arctic climates of the Northern Hemisphere, making it well suited to the Nordic and Baltic regions (see Fig. 6). It offers a range of advantages, including:

- **Rapid growth and short rotation cycles**: Willow is typically grown on 2–5-year rotations, with guidelines reporting growth up to 4 m in the first year after cutback and 6–8 m by the third year [10]. Shorter rotation cycles produce stems of 20 cm diameter with straight lengths of 30–60 cm, which may be appropriate for specific façade components or laminated products [11]. Longer rotation willow crops (up to 8 years) can achieve stem diameters exceeding 40 cm with straight lengths up to 2 m, making them suitable for structural applications.
- **Accelerated carbon sequestration**: Willow plantations generally achieve faster biomass accumulation and higher carbon uptake rates than conventional spruce or pine plantations on comparable sites. This stems from willow's rapid growth in short-rotation coppice systems which allows multiple harvests and continuous sequestration.
- **Regenerative coppicing**: Following harvest, willow regenerates naturally by sprouting new shoots directly from the remaining stump. This eliminates the need for replanting, reduces soil disturbance, and preserves topsoil structure. The regenerative growth cycle fosters sub-surface microbial biodiversity and supports long-term soil carbon storage.
- **Biodiversity benefits**: Willow plantations support diverse ecosystems, providing habitat for pollinating insects, birds, and other wildlife while maintaining agricultural productivity on marginal land. Additionally, willow possesses phytoremediation properties, making it capable of removing heavy metals from contaminated soils and contributing to land restoration.

Engineering challenges and material characterization. Despite its environmental and agricultural advantages, willow has not yet been widely implemented as a construction material. Its primary applications have historically been limited to external fencing, noise barriers, landscape screening, or incineration for energy recovery. Key implementation challenges requiring further investigation include:

- **Moisture content management**: Willow exhibits a moisture content of 50–55% at harvest and after drying processes (e.g. kiln) the moisture content is reduced to 18–25% [12]. This is high compared to the 12–15% moisture content of traditional glulam inputs [13]. This elevated moisture content necessitates advanced drying protocols and controlled conditioning to achieve the stability required for façade construction (see Fig. 7). Further research is needed to assess optimal drying methods and their implications for large-scale implementation.

Fig. 6 A plantation of growing willow trees, together with a graphic illustrating different stages of the tree's growth cycle. © Pilebyg, © SHL. Reprinted with permission

Fig. 7 Contains two photographs of willow at different stages of processing. © MOOW, © Staticus. Reprinted with permission

- **Dimensional limitations**: Natural stem geometry imposes constraints on available dimensions. While selective breeding programs have improved stem straightness and reduced kinks and knots, these dimensional limitations influence how willow can be processed and integrated into composite structural systems.

2.5 *Smart Technology Integration*

The drive for digital integration. While material innovation addresses embodied carbon, digitalization targets carbon across the entire building lifecycle. As buildings

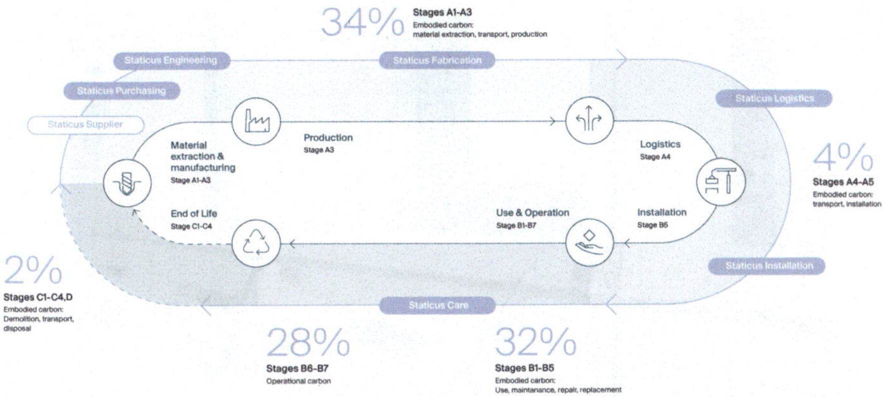

Fig. 8 A graphic illustrating the carbon generated at different stages of a 2030 office building's carbon lifecycle based on the LETI Embodied Carbon Primer, 2020 [15] combined with parts of Staticus' value chain and service offering. Illustration created by the authors

become more energy efficient and electricity generation decarbonizes, the proportion of whole-life carbon attributed to operational energy is declining significantly. Today, that ratio has shifted from operational carbon accounting for two-thirds of emissions in 2020 to a 50:50 split between operational and embodied carbon [15]. By 2030, operational carbon is projected to account for only one-third of total emissions [15], with upfront embodied carbon and embodied carbon during use stages, including maintenance, repair, and replacement, each representing approximately one-third of whole-life carbon.

This shift means that embodied carbon during the use stages, including maintenance and replacement, are now equally important as upfront embodied carbon (see Fig. 8). Façade systems, which are subjected to continuous environmental exposure, represent a critical opportunity for intervention. Predictive intelligence offers a pathway to extend service life, reduce maintenance requirements, and improve resource efficiency across the operational lifespan of façades.

Staticus recognized that the same precision modelling techniques it was using for structural validation and hygrothermal analysis could also be deployed for real-time monitoring of façade performance. Building on work done during a Norway Grants-funded research project, the company has developed a range of smart-technology initiatives (including the system of sensors illustrated in Fig. 9) with the aim of transforming façades from passive enclosures into active data ecosystems capable of informing both design optimization and building operations.

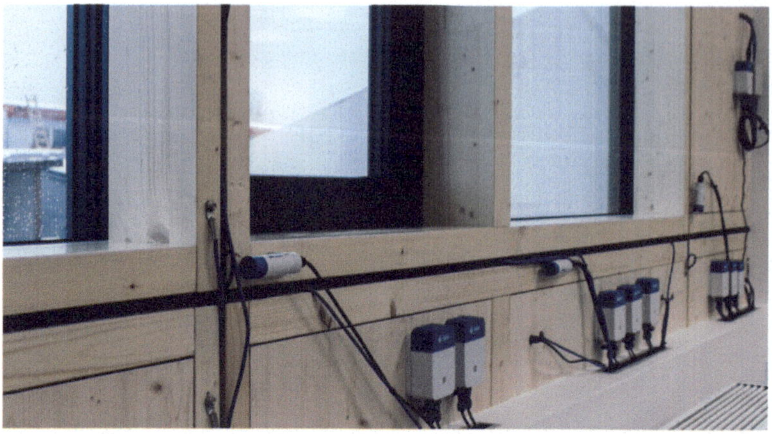

Fig. 9 A photograph of the system of Internet of Things (IoT) sensors Staticus is developing to monitor façade conditions and then relay data to digital simulation models

3 Approach and Methodology

3.1 A Multi-faceted Methodological Approach

Staticus employed a multi-faceted approach to the development of its Timber-HUF system, and adjacent research and development activities. This ensured the end outcomes both contribute to the façade industry's overall development and address specific market needs.

One key element of Staticus' methodological approach to this task is a commitment to modeling. In the early design stages, engineers deployed numerical modeling and empirical validation to test specific hypotheses. Structural models developed in the laboratory allow for calibration of finite element simulations. This step is crucial in the establishment of a virtual twin in a finite element environment, which will be the basis for future developments of the unitized facade elements.

For hygrothermal performance modeling, transient-state analysis was employed, rather than steady-state analysis. Transient-state analysis acknowledges that temperature peaks can differ in duration, and this temporal variation impacts interior surface temperatures, which directly influence condensation risk. Furthermore, transient state analysis is less conservative regarding the engineering judgment and provides a more accurate prediction of actual internal surface temperatures, leading to more optimized solutions.

This focus on modeling was supported by an investment into digital tools to gather critical data on the Timber-HUF's performance. A digital workflow was created which linked façade geometry, material quantities, and carbon calculations within a single model, enabling real-time optimization of environmental performance. In terms of calculations for embodied carbon, the Centre for Windows and Cladding Technology's (CWCT) methodology was used.

Intensive testing of components and mock-ups was also critical in both the initial development of prototypes, and in the deployment of the Timber-HUF system in real world projects. Connection testing by Kaunas University of Technology (KTU) and Staticus assessed the shear, bending, and tensile strength at key joints during the initial development of the Timber-HUF system. Stiftelsen for Industriell og Teknisk Forskning (SINTEF) established the main principles for the digital twin models, taking physical testing data into the finite element environment. Accelerated weather chamber testing was used to validate the initial modelling on hygrothermal performance.

For the system's first deployment in a real project, a large-scale performance mock-up went through full EN-standard testing, covering air permeability, water-tightness, wind load resistance, and impact resistance. Extreme weather testing was also used to validate the transient hygrothermal modeling approach and confirm design assumptions. Once production was underway, a comprehensive Quality Control and Quality Assurance (QC/QA) program specific to the Timber-HUF system was implemented, including systematic measurement of timber moisture content at multiple stages.

Finally, a critical component of Staticus' approach was to collaborate with multiple industry experts. The Timber-HUF system was developed in partnership with Oslo Metropolitan University (OsloMet), SINTEF, and KTU. Further material research into the potential use of willow was conducted together with Schmidt Hammer Lassen Architects, Pilebyg (a Northern Danish willow grower and producer), and the design company Made out of Willow (MOOW).

The following section details the approach and process adopted during the design of the Timber-HUF system. It covers the initial design, research into its connections and hygrothermal performance, material optimization while maintaining equilibrium moisture content within safe thresholds, fire performance, and embodied carbon and life cycle impact. A case study of the Textiltorget project, the first real world use of the Timber-HUF system, provides examples of the methodologies used in practice. Finally, a brief summary of the ongoing research into the use of willow and digital tools developed together with the Timber-HUF is provided.

3.2 Staticus' Timber Hybrid Unitized Façade (HUF) System

Design concept. To respond to evolving market drivers for increasing transparency, natural aesthetics, and future carbon regulations, Staticus initiated a multi-year R&D project under the Norway Grants scheme. In 2021, the company partnered with OsloMet, SINTEF, and KTU. The main objectives of this project were to bring a timber-hybrid unitized façade (Timber-HUF) to market and develop a more environmentally friendly automated façade system that is integrated into a building's control systems.

The Timber-HUF system is a fully unitized system constructed from a composite timber-aluminum frame comprising:

- **Internal timber frame**: Timber sections, including pine, larch, and spruce, replace traditional curtain wall aluminum sections on the inside of the weather line, serving as the primary structural load path in non-load bearing structures.
- **External aluminum adaptor**: A thermally broken aluminum adaptor creates a pressure equalized chamber with water and air seal performance line. The adaptor is sourced from low-carbon, high recycled content aluminum (> 75% post-consumer recycled content). By placing the aluminum adaptors in the exterior zone, higher water and airtightness performance is ensured, along with improved service life, ease of replacement and durability.
- **Narrow frame design**: The location of the external aluminum adaptors, attached to the short outer side of the glulam section, enables a slim frame and joint design of only 90–100 mm, comparable to conventional aluminum curtain walls, making the system suitable for projects where transparency and architectural slenderness are priorities.
- **Unitized curtain wall**: The system's modular units are fully unitized, with glazing, insulation, vapor barriers, and mechanical fixings all prefabricated under high-quality factory conditions. In terms of installation, the Timber-HUF system follows the same principles as a traditional aluminum unitized curtain wall, minimizing site time, and increased safety conditions during construction.

Structural engineering and connection testing. Developing this Timber-HUF system presented a range of engineering challenges. Timber and aluminum have differences in their modulus of elasticity (~ 10 GPa for timber vs. 70 GPa for aluminum), coefficient of thermal expansion (5×10^{-6} 1/K for timber along the grain vs. 24×10^{-6} 1/K for aluminum, cross section tolerances, sensitivity to moisture content, and non-uniform strength distribution along longitudinal axis. Because the Timber-HUF system is a complex composite system, it was critical to guarantee adequate load transfer through all material connections and joints while accounting for these differences.

Both numerical modelling and empirical validation were needed to ensure a consistent load transfer across these materials.

Staticus conducted extensive connection testing in collaboration with KTU and OsloMet, which assessed shear, bending, and tensile strength at key joints. Finite element simulations allowed the structural model developed in the laboratory to be calibrated. All connections of the Timber-HUF system were tested to calibrate and verify the numerical modelling of L-connections and T-connections.

Full load-displacement profiles were plotted to investigate the behavior of connections under progressive loading. The experimental elastic limits were determined from these graphs, enhancing understanding of bearing capacity in the fully assembled frame. Finite element simulations were then calibrated using this empirical data, allowing the structural model developed in the laboratory to accurately predict real-world behavior.

In addition to structural testing of joineries and load transferring connections, full performance mock-up testing was carried out in accordance with European standards for curtain wall performance (EN 13830), meeting compliance for air permeability, watertightness, and wind and impact resistance (see Table 1).

Hygrothermal performance modelling. Hygrothermal performance, the movement of heat and moisture through buildings, is critical for high energy efficiency, occupant comfort, and long-term durability. The Timber-HUF system generally demonstrates improved thermal performance due to timber's much lower thermal conductivity compared to aluminum (0.13 W/mK for softwood timber vs. 160 W/mK for aluminum) [16].

However, the key challenge to overcome was assessing timber's hygrothermal behavior, given its natural tendency to absorb and retain moisture over the service life of the façade.

Staticus' engineering team recognized that steady-state models, which assume constant boundary conditions, could not capture the real behavior of a timber-hybrid façade, especially in regions where humidity levels and temperatures fluctuate significantly, such as the Nordics. While numerical simulations can be used to study hygrothermal behavior, they often yield conservative results because of uncertainties in the boundary conditions and material properties.

Steady-state versus transient-state analysis. Steady-state analysis is highly sensitive to temperature gradients and identifies the location of the dew point within the assembly at a given moment. However, it cannot account for the thermal mass of materials or the temporal dynamics of temperature and moisture cycling.

Transient-state analysis, by contrast, recognizes that temperature peaks may differ in duration but not necessarily in magnitude, and that this temporal variation has a significant impact on interior surface temperatures, which directly influence condensation risk. Because exterior temperatures fluctuate continuously, transient analysis provides a more accurate prediction of actual surface temperatures, aligning more closely with experimental test results.

Staticus therefore employed transient-state analysis to simulate coupled heat and moisture transfer through the Timber-HUF components over time. A parametric study

Table 1 Results of laboratory testing in accordance with EN 13830

Test	Standard	Classification/declared value
Air permeability	EN 12152	AE/900 pa
Water tightness	EN 12154	RE/1200 pa
Wind resistance	EN 12179	± 2000 pa (serviceability) ± 3000 pa (safety)
Impact resistance	EN 14019	Class I4/E4
Acoustic performance	EN ISO 10140 Part 2 EN ISO 10848 Part 2	Rw (C; Ctr) = 40 (− 2; − 7) dB Dn, f, w (C;Ctr) = 56 (− 1; − 4) dB

based on different variables was conducted across dynamic boundary conditions reflecting both interior climate control and external weather exposure.

Design impact and material optimization. Based on the results of these transient hygrothermal analyses, engineers determined the strategic placement of vapor barriers, breathable membranes, and drainage cavities. This eliminated the need for additional thermal isolation layers and overcame the problem of conservative overdesign that had hampered previous timber façade applications due to lack of reliable performance data.

The system was optimized to maintain equilibrium moisture content within safe thresholds (typically below 18% to prevent biological degradation). This allowed the amount of vapor resistant materials to be reduced by 10–15% without compromising durability or performance. Critically, the analysis demonstrated that properly designed ventilation cavities and drainage pathways could manage moisture without requiring impermeable vapor barriers that might trap moisture within the assembly.

Accelerated weather chamber testing, which reproduces annual cycles of condensation and drying in controlled conditions, was used to validate this analytical approach. No critical moisture accumulation or biological degradation was observed even after exposure equivalent to several years of simulated cycles, confirming that the design achieves long-term resilience.

Fire performance and safety design. One of the largest barriers to the broader use of timber in façades remains the ongoing scrutiny over material combustibility. Untreated glulam is classified as Class D-s2, d0 [14] under European standard EN 13501-1, meaning it is combustible but with limited contribution to fire spread due to minimal falling burning materials.

In order to meet the stringent fire safety requirements of projects, a series of fire protection interventions were investigated as part of the Timber-HUF system development, including:

- **Waterborne intumescent coatings**: Application of coating systems combined with intumescent lacquer that expands under heat, forming an insulating char barrier. This waterborne intumescent treatment improves the fire classification of the timber to Class B-s1, d0, significantly reducing its combustibility while maintaining the material's natural appearance.
- **Non-combustible external aluminum adaptors**: The external aluminum components shield the timber frame from direct exposure to fire, providing an additional layer of passive fire protection.
- **Fire-stop materials at joints**: Integration of fire-stop materials at horizontal and vertical joints prevents the spreading of fire and smoke between façade modules and floors.

Full-scale tests of the Timber-HUF system under EN 1364-4 configuration type 5 are required to confirm its fire performance in relation to fire spread between floors. The treated timber, with combustibility Class B rating, positions the Timber-HUF system suitability for medium-rise commercial buildings under most European jurisdictions.

Embodied carbon and lifecycle impact. In response to the evolving sustainability regulations and increasing requirements for whole lifecycle carbon assessments with binding carbon thresholds, the façade industry has adopted standardized methodologies to quantify the embodied carbon of façades across the building lifecycle. In addition to European Standards EN 15804 and EN 15978, the facade industry uses the Centre for Windows and Cladding Technology (CWCT) embodied carbon methodology [17] because it translates these generic standards into façade-specific guidance. The CWCT methodology provides consistent system boundaries, default assumptions, and data requirements tailored to the technical complexity of curtain wall systems, enabling comparable project-specific results.

The CWCT methodology was used for calculating the upfront embodied carbon emissions of Staticus' Timber-HUF system and compared against Staticus' range of low-carbon façade systems. Assessments were calculated for façade modules measuring 3.7 m × 1.4 m with 35% window-to-wall ratio (WWR), double glazing (or low carbon glazing triple glazed unit), and an aluminum sheet spandrel. Upfront embodied carbon calculated from material extraction and manufacturing through to site installation (LCA stages A1-A5) and results were recorded as a carbon intensity per façade surface area (FSA):

- Standard aluminum unitized facade system: ~ 250 $kgCO_2e/m^2$
- Premium Timber-HUF system: ~ 115 $kgCO_2e/m^2$
- Premium + Timber-HUF system: ~ 105 $kgCO_2e/m^2$

3.3 *Case Study: Textiltorget, Stockholm*

Project overview. The Timber-HUF system was implemented on its first project, the Textiltorget office and retail redevelopment in Stockholm, Sweden (see Fig. 10). Completed in 2023, this project was designed by LINK Arkitektur for client Fabege AB, representing a significant milestone in demonstrating the commercial viability of timber-hybrid unitized façade systems.

Fig. 10 Visuals of the completed Textiltorget project featuring Staticus' Timber-HUF façade system. Photos © Andrius Gudelis. Reprinted with permission

Project summary:

- **Category**: Office and retail redevelopment
- **Façade area**: 1400 m^2
- **System**: Timber-HUF (Timber-Hybrid Unitized Façade)
- **Completion year**: 2023
- **Upfront embodied carbon**: 156 $kgCO_2e/m^2$

Staticus' Timber-HUF system was installed on upper floors as part of the building renovation. Due to the location of the Timber-HUF system on levels 4–6, the inclusion of combustible materials in façades was permissible under Swedish building regulations [18]. One corner of the building was retrofitted with an HUF system commencing on the first floor. Taking into consideration the sprinkler system installed within the premises, the overall design was deemed acceptable by the fire engineering assessment.

The upfront embodied carbon intensity calculated for the project is 156 $kgCO_2e/m^2$, which represents an increase from the baseline presented in Sect. 3.2 (115 $kgCO_2e/m^2$). This is due to project-specific factors including a different window-to-wall ratio and insulated glass unit build-ups. The upfront embodied carbon intensity was calculated for stages A1-A5 in compliance with the CWCT methodology [17].

Design, production, and installation. For the design stage of the Textiltorget project, Staticus' engineers applied a digital workflow that linked façade geometry, material quantities, and carbon calculations within a single model. This approach enabled real-time optimization of environmental performance alongside structural and thermal requirements. Each façade unit was designed as a repeatable but configurable component, optimized for efficient transportation and streamlined installation sequencing (completed Timber-HUF units are shown in Fig. 11).

Manufacturing was carried out at Staticus' production facility in Vilnius, Lithuania, under controlled environmental conditions. Precision CNC machining ensured tight tolerances for the critical hybrid timber-aluminum connections, maintaining the dimensional accuracy essential for unitized curtain wall performance.

Fig. 11 Timber-HUF elements used for the Textiltorget project at Staticus' production facility

A comprehensive Quality Control and Quality Assurance (QC/QA) program specific to the Timber-HUF system was implemented throughout production. This program included systematic measurement of timber moisture content at multiple stages: prior to assembly, after completion of factory assembly, following shipping, and after installation on site. Special precautions were taken to mitigate excessive water entering the unitized façade elements. Each installed element was covered with a water resisting membrane to secure the elements and to drain the rainwater outward. This rigorous monitoring ensured that moisture levels remained within specified limits (typically 12–15%) throughout the supply chain, preventing dimensional instability or moisture-related degradation.

All façade components, including glazing units, thermal breaks, vapor barriers, gaskets, and mechanical fixings, were pre-assembled in controlled factory conditions. These controlled conditions ensured consistent quality control, eliminated weather exposure during assembly, and simplified on-site installation.

Once transported to the site in Stockholm, the unitized approach enabled rapid installation. The system achieved installation rates of up to 70 m^2 per day. This accelerated installation schedule ensured project deadlines were met, minimized disruption to existing building tenants during the renovation, and reduced on-site construction emissions by limiting the duration of crane operation and site activity.

Curtain wall testing. Prior to installation, a large-scale performance mock-up was constructed (see Fig. 12) and subjected to full EN-standard testing to assess compliance with curtain wall performance requirements. The testing program included:

- Air permeability (EN 12152): Testing under positive and negative pressure differentials to verify air-tightness performance,
- Watertightness (EN 12154): Both static and dynamic water testing to simulate wind-driven rain and verify water penetration resistance,
- Wind load resistance (EN 13116): Structural testing under design wind pressures to confirm load-bearing capacity and deflection limits,
- Impact resistance (EN 14019): Hard body and soft body impact testing to assess resistance to accidental damage.

Extreme weather testing. In parallel with performance mock-up testing, extreme weather testing was conducted using a climatic double chamber (see Fig. 13), where both interior and exterior environmental conditions were hydrothermally controlled to simulate extreme environmental conditions.

This validation of the transient hygrothermal modelling approach confirmed design assumptions: no critical condensation accumulation or biological degradation (rot or mold) was observed, even under simulated climate extremes. The results verified that the vapor barrier detailing, drainage cavity design, and material selection successfully manage moisture mitigation without compromising long-term durability.

Fig. 12 Elements of the Timber-HUF used for the Textiltorget project mounted on Staticus' internal testing rig for performance testing

3.4 A Willow-Hybrid Prototype: Development and Methodology

To investigate the practical application of willow as an alternative to softwood timbers in the Timber-HUF system, Staticus partnered with Schmidt Hammer Lassen Architects, Pilebyg (a Northern Danish willow grower and producer), and the design company Made out of Willow (MOOW). The consortium received an Implement Materials grant from the Danish Association of Architects, Arkitektforeningen, to develop and test a willow-hybrid façade prototype.

The partners represent the full value chain from growers through manufacturers, fabricators, and architects. Over the past 30 years, Pilebyg has selectively bred

Fig. 13 The Timber-HUF used for the Textiltorget project undergoing extreme weather testing

specific willow varieties to minimize kinks and knots while maximizing stem straightness, thereby addressing a key challenge that has historically prevented willow's use as a construction material.

A system mock-up was successfully produced (see Fig. 14), demonstrating the feasibility of manufacturing willow-based façade components using existing production infrastructure with appropriate process modifications.

Fig. 14 A willow-hybrid façade prototype developed by the consortium

3.5 The Predictive Performance Tool: Development and Methodology

As part of the Norway Grants project and the Staticus Care strategic initiative, a Predictive Performance Tool was developed. It is designed to forecast façade behavior and pre-empt performance degradation. It achieves this by creating machine learning models focused on the detection of performance patterns. The system operates as follows:

- Embedded sensors (see Fig. 15) record temperature, humidity, differential pressure, and timber moisture content at key interface zones.
- Signals from these sensors are logged via Internet of Things (IoT) gateways and then transmitted to a central analytics server.
- Machine-learning algorithms can then identify anomalies such as localized moisture spikes or increased condensation risk.

For more than two years, the prototype façade at Staticus' Vilnius facility (see Fig. 15) has been collecting data continuously. Analysis revealed patterns of diurnal moisture fluctuation correlated with external humidity and solar radiation. These discoveries will lead to engineering and material refinements of Timber-HUF elements.

When integrated into project workflows, the predictive performance tool provides multiple benefits:

Fig. 15 Two photographs of a prototype of Staticus' Timber-HUF with a network of sensors integrated. One photograph shows the prototype's exterior while the second shows the sensor system from the interior

- Predictive maintenance allows building managers to detect anomalies earlier and proactively act, preventing costly interventions later on.
- Live data can be used to validate analytical models and support the ongoing development of lighter, more efficient designs.

Datasets from multiple projects can be fed into a growing digital knowledge base for future designs. As part of the Norway Grants program, Staticus developed a Predictive Performance Tool designed to forecast façade behavior and pre-empt performance degradation. The project was driven by three core research questions:

- Can a building's energy efficiency be improved by understanding façade performance under varying indoor and outdoor conditions?
- Can predictive and preventive maintenance strategies be developed based on monitoring and forecasting façade performance?
- Can the material requirements for a façade be reduced based on knowledge of its expected performance over its operational lifetime?

The tool was specifically developed to address condensation and mold prevention in the Timber-HUF system, with performance calibrated against live data from sensors installed on a physical performance mock-up. The development methodology proceeded in three stages:

Stage 1: Risk spot analysis. A detailed hygrothermal risk assessment was conducted to identify locations within the timber façade assembly with elevated humidity or moisture risk. This analysis informed the strategic placement of sensors at critical interface zones, including timber-to-aluminum junctions, ventilation cavity boundaries, and areas adjacent to thermal bridges.

Stage 2: Performance mock-up construction and sensor installation. The Timber-HUF system performance mock-up was constructed and installed adjacent to a conditioned space at Staticus' Vilnius facility. A network of embedded sensors was integrated into the assembly to record:

- Internal and external air temperature,
- Differential pressure across the façade assembly,
- Relative humidity at key interface zones,
- Surface moisture levels,
- Timber moisture content within the structural framing.

Signals from these sensors are logged via Internet of Things (IoT) gateways and transmitted to a data warehouse for continuous monitoring and analysis.

Stage 3: Data collection and pattern recognition. For more than two years, the prototype façade has been collecting data continuously under real-world environmental conditions. The next phase of the project will use machine-learning algorithms to detect patterns, enabling the system to identify anomalies such as localized moisture spikes, increased condensation risk, or deviations from expected thermal performance (the data inputs for this process are shown in Fig. 16).

When integrated into project workflows, the predictive performance tool provides multiple benefits:

- **Predictive maintenance**: Building managers can detect anomalies earlier and take proactive action, preventing costly reactive interventions and extending component service life.

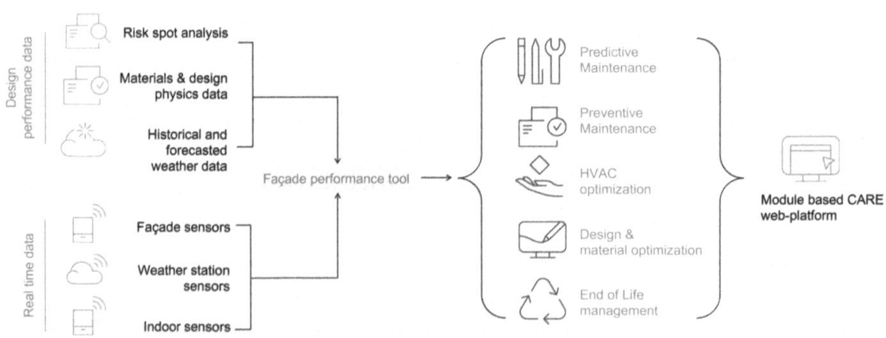

Fig. 16 A flow diagram showing the data inputs the façade performance tool receives and where its outputs can be applied. Figure created by the authors

- **Design validation**: Live data validates analytical models and supports the ongoing development of lighter, more material-efficient façade designs with confidence in long-term performance.
- **Knowledge base development**: Performance datasets from multiple projects can be aggregated into a growing digital knowledge base, establishing empirical benchmarks for future timber façade designs.

Integration with Building Management Systems (BMS). The next step in this smart technology evolution is to couple the predictive tool with Building Management Systems (BMS). Through open communication protocols such as BACnet and MQTT, façade performance data can be integrated directly into building-wide control strategies, informing HVAC operation and indoor climate control in real time.

For example, when surface temperatures or humidity levels at critical façade interfaces approach condensation thresholds, the BMS can automatically adjust ventilation rates or modify internal temperature set-points to maintain safe conditions. This coordinated approach maintains occupant comfort without resorting to energy-intensive dehumidification strategies.

By integrating predictive façade performance models with BMS platforms, it becomes possible to optimize the coordination between the building envelope and mechanical systems, thereby improving energy performance, enhancing operational efficiency, and supporting more accurate end-of-life condition assessments. The system can track critical metrics such as the number of times design service loads are exceeded and cumulative exposure to elevated moisture levels, enabling data-driven decisions about component replacement and lifecycle management.

In this way, façades are transformed from static barriers into responsive environmental interfaces that dynamically coordinate with mechanical systems to optimize energy use and occupant comfort.

Digital surrogate models. Once sufficient empirical data has been collected, machine-learning algorithms can generate a digital surrogate model. This surrogate model will predict façade behavior under varied future conditions without continuous sensor feedback, thus dramatically reducing monitoring costs and complexity while eliminating the need for physical sensors in live project installations.

Furthermore, digital surrogates can be applied during design stages to simulate future climate scenarios, enabling designers to future-proof buildings for climate adaptation and resilience. Currently, façade systems are designed based on the probability of exceeding limit states using historical climate data. However, as temperatures rise, the number of extreme heat days increases, and overall humidity levels change, designing in order to anticipate future weather conditions becomes essential.

Following conventional design processes, increased limit states typically equate to greater material mass. Digital surrogate models offer an alternative approach: rather than designing static façades with oversized material specifications, designers can take an adaptive and integrated approach. This involves utilizing the adaptability of HVAC systems to adjust internal conditions in response to external stresses, guided by predictive façade performance models. For instance, surrogate models could predict moisture dynamics under projected 2050 humidity and temperature profiles, facilitating the design of façades optimized for future climate conditions rather than historical norms.

This approach enables material reduction while maintaining or improving performance, directly addressing the research question of whether material requirements can be reduced based on knowledge of expected performance over a façade's operational lifetime.

Staticus' XD Façade Management Platform: closing the digital loop. Staticus has developed the XD Façade Management Platform, a digital platform built on a Building Information Modelling (BIM) backbone with intuitive dashboards. The tool was originally developed to track project delivery data and resources across the entire construction process, including technical and workshop drawings, production rates, logistics processes, installation progress, quality assurance protocols, and payment plan management.

The XD platform now also serves as the operational interface for Staticus' internal data (see Fig. 17). Recent upgrades have added a carbon tracking functionality, and the next generation will enable live tracking. The embodied carbon of every component is recorded at the production stage and updated during transport, with data automatically aggregated into a project-wide carbon ledger.

Staticus recently completed its first XD mock-up for carbon tracking and is currently developing as-built documentation and material passport integration. These will enable full traceability of material composition, origin, and environmental impact across the façade lifecycle, supporting circular economy principles and facilitating future material recovery and reuse.

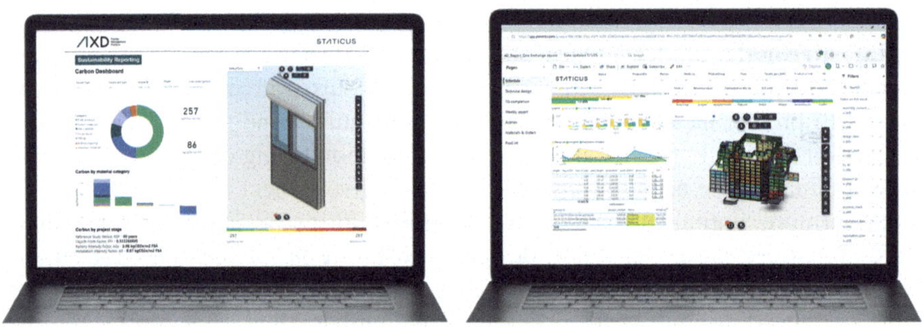

Fig. 17 Two screenshots that illustrate the functionality of Staticus' XD Façade management platform

4 Results and Outcomes

The outcomes from of the research into and development of the Timber-HUF, together with the complementary exploration of willow as a façade material and digital solutions for performance and maintenance optimization, potentially impact façade design, production, installation, and maintenance in multiple ways.

This section details the key results achieved across this research, including successful management of moisture content, substantial reduction of embodied carbon when compared to standard aluminum systems, and the development of a willow-based prototype.

4.1 Hygrothermal Performance

Optimization of the Timber-HUF during the development phase enabled the maintenance of equilibrium moisture content at typically 18% or lower, thus keeping the hygrothermal performance within safe thresholds and preventing biological degradation.

During the Textiltorget project, careful and thorough monitoring throughout the value chain ensured moisture levels within specified limits, which were typically 12–15%. This ensured there was no dimensional instability or moisture-related degradation.

During testing in an accelerated weather chamber, no critical moisture accumulation or biological degradation was observed. This included after exposure equivalent to several years of simulated cycles.

4.2 Material Optimization

As a result of this strong hygrothermal performance, it was possible to optimize the amount of vapor resistant materials needed. A 10–15% reduction in such material was achieved without any comprising of durability or performance.

4.3 Fire Safety Performance

The application of coating systems combined with intumescent lacquer elevated the classification of the timber from Class D-s2, d0 for untreated glulam, under European standard EN 13501-1, to Class B-s1. This approach reduced combustibility without impacting the material's natural aesthetic.

An additional layer of passive fire protection was achieved via external aluminium components, which shield the timber frame from direct exposure to fire, while the integration of fire-stop materials at horizontal and vertical joints prevents first and smoke spreading between façade modules and floors.

4.4 Embodied Carbon and Lifecycle Impact—Results

As illustrated in Fig. 18, Staticus' Timber-HUF systems represent a 54 and 58% reduction, respectively, when compared to a Standard aluminum unitized façade system.

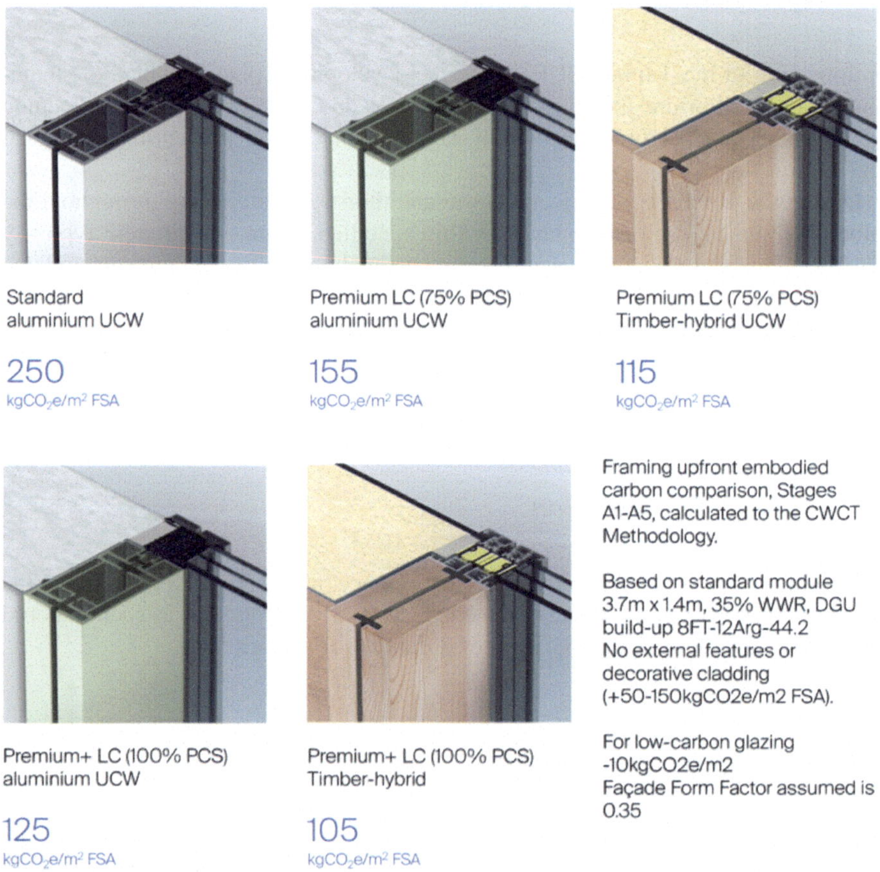

Fig. 18 A graphic representation of carbon intensity per façade surface area for different façade systems, including two models of the Timber-HUF system

When comparing Staticus' Premium and Premium + low carbon aluminum unitized façade system to their equivalent Premium and Premium + Timber-HUF systems, ~ 26 and 16% improvements are observed, respectively. Aligned with the CWCT methodology, these results exclude the biogenic carbon storage of timber. When this is accounted for, the effective carbon footprint becomes even lower.

In the case of Textiltorget, the upfront embodied carbon intensity calculated for the project is 156 $kgCO_2e/m^2$, which represents a slight increase from the baseline presented in Fig. 18 (115 $kgCO_2e/m^2$). This is due to project-specific factors including a different window-to-wall ratio and insulated glass unit build-ups. The upfront embodied carbon intensity was calculated for stages A1–A5 in compliance with the CWCT methodology [17].

In addition, the Timber-HUF system's modularity supports disassembly and reuse, a key principle of circularity. Because components can be refurbished and redeployed, service life can be extended, and impacts at end-of-life stages (C1–C4) are reduced.

It is noted that the façade module test case assessed represents a lean façade design. For designs with external features and additional cladding design assumptions (including stone, terracotta, GRC, and brick slip), the carbon intensity of unitized façade systems will increase. Depending on specific design requirements, an additional 50–200 $kgCO_2e/m^2$ FSA can be anticipated.

4.5 Textiltorget Project Outcomes

The deployment of Staticus' Timber-HUF system in the Textiltorget project confirmed the engineered performance values developed during the research and development phase, demonstrating that the system delivers measurable benefits across environmental, thermal, acoustic, and structural performance metrics. These benefits include areas such as:

- **Carbon reduction**: The system achieved a reduction of over 40% in embodied carbon compared to an equivalent aluminum curtain wall baseline, confirming the carbon calculations developed during the design phase and validating the environmental value proposition of timber-hybrid façade systems.
- **Thermal performance**: Enhanced thermal insulation was achieved, with the installed system delivering a U-value of 0.46 W/m^2K. This represents an improvement over conventional aluminum curtain walls (typically 0.70 W/m^2K), reducing operational heating energy and contributing to improved energy efficiency certification for the renovated building.
- **Architectural drivers**: The Timber-HUF system's minimal connection joint width of 90–100 mm enabled visual parity with conventional aluminum façades, meeting the architectural intent for a high transparency, contemporary appearance while incorporating the natural aesthetic of timber.

- **Acoustic performance**: Improved acoustic isolation, as compared to aluminum systems, was achieved due to timber's higher sound dampening properties, which correlate with the material's mass and viscoelastic characteristics.

4.6 A Willow-Hybrid Prototype: Outcomes

The consortium constructed a scaled façade unit that combines willow framing with the Timber-HUF aluminum adapter system. This prototype integrates the structural, thermal, and moisture management strategies developed for conventional timber while adapting connection details and fabrication processes to accommodate willow's specific material characteristics. A system mock-up was successfully produced, demonstrating the feasibility of manufacturing willow-based façade components using existing production infrastructure with appropriate process modifications.

The next phase will undertake extensive product and system testing to evaluate long-term durability, the adhesion and performance of protective coatings, hygrothermal behavior under cyclic environmental exposure, and biological resistance. If these durability tests prove successful, willow could establish a new closed-loop material stream for façade systems. The material renews quickly, operates as carbon-negative over its lifecycle when accounting for both sequestration and embodied emissions, and can be cultivated close to manufacturing hubs, reducing transportation impacts.

Implications for sustainable construction. The willow research program represents a fundamental shift in how the construction industry might approach material sourcing. Demonstrating the industrial application of fast-growing regenerative crops can advance agroforestry integration and reposition façade manufacturing as a participant in ecosystem restoration rather than resource extraction. If willow-based systems prove commercially viable, they could be cultivated on marginal agricultural land, flood-prone areas, or previously contaminated sites, generating construction materials while simultaneously providing carbon sequestration, soil remediation, and biodiversity habitat.

Regenerative alternatives to timber offer a pathway toward construction systems that operate within ecological boundaries while meeting global building demand. Rather than simply substituting one material for another, it suggests a systemic transformation in how buildings source, utilize, and ultimately return materials to biological cycles.

5 Discussion

5.1 Lessons Learned and System Refinements

The Textiltorget project generated valuable practical insights that informed ongoing refinement of the Timber-HUF system. Key learnings and optimizations included:

- **Vapor barrier detailing**: Fine-tuning of vapor barrier laps, penetrations, and termination details based on installation feedback, improving ease of assembly and reducing potential points of failure.
- **Connection optimization**: Further reduction in the use of aluminum connectors through improved interlocking timber joinery designs, which would reduce material costs and embodied carbon while maintaining structural performance.
- **Assembly protocols**: Streamlining factory assembly sequences based on production experience, improving manufacturing efficiency, and reducing labor hours per module.
- **Logistics coordination**: Refinement of packaging, transportation, and on-site handling procedures to minimize risk of damage and optimize installation sequencing.

These refinements are being incorporated into subsequent design iterations and will inform future implementations of the Timber-HUF system. The successful completion of Textiltorget established proof of concept for timber-hybrid unitized façades in the Nordic market, demonstrating that these systems can meet the rigorous performance, aesthetic, and logistical requirements of commercial construction while delivering significant environmental benefits. The project serves as a reference case for architects, developers, and regulatory authorities evaluating the viability of biobased façade systems for urban building applications.

6 Technology Readiness Level and Market Status

The Timber-HUF system has been implemented in a real-life project in Stockholm. The building is currently fully operational, which supports its classification at Technology Readiness Level (TRL) 9, indicating that the system has been successfully proven in an operational environment.

The smart monitoring component associated with the Timber-HUF system is currently assessed at Technology Readiness Level (TRL) 6, indicating a system prototype that has been demonstrated in a relevant environment. At this stage, a fully functional prototype has been developed and is being tested in a simulated operational setting through Staticus' IoT mock-up installed at its facilities. This mock-up continuously collects performance data from a Timber-HUF assembly, providing a practical basis for evaluating façade behavior under relevant environmental conditions and supporting ongoing hygrothermal performance analysis.

The current level of maturity demonstrates that the concept has progressed beyond laboratory-only investigation and into prototype validation in a relevant environment.

7 Conclusions

Staticus' trajectory, from early prefabricated timber modules through the Timber-HUF system to regenerative materials research and intelligent systems integration, illustrates construction's transformation: buildings as active participants in ecological cycles, dynamically responsive through coordinated systems, and transparent in resource flows and performance.

7.1 Converging Imperatives: Performance, Carbon, and Circularity

Regulatory developments (Sect. 2), such as the EU Green Deal through national embodied carbon limits, establish binding constraints making conventional aluminum-intensive façades untenable. The Timber-HUF system's 54–58% embodied carbon reduction (Sect. 4) provides a commercially viable response without compromising performance.

Rigorous validation through structural testing, hygrothermal modelling, fire design, and full-scale mock-ups (Sect. 3) has established performance parity. Textiltorget (Sect. 3) confirmed results in practice. The unitized configuration and disassembly protocols position the system within circular economy frameworks, with modularity enabling material reuse and closed-loop resource management (Sect. 8).

7.2 Digital Intelligence as Performance Enabler and Systems Integrator

Predictive tools and sensor networks (Sect. 3) fundamentally altered the design equation. Empirical performance data calibrating models against real-world sensors reduced design uncertainty and eliminated unnecessary material. Transient hygrothermal analysis (Sect. 3) enabled 10–15% reductions in vapor-resistant materials without compromising durability.

Sensor infrastructure enables the coordination of sophisticated façade-mechanical systems (Sect. 3). This represents a paradigm shift from independent envelopes and services to coupled systems coordinating dynamically. Rather than the façade physically adapting, building systems responding to façade data create adaptive behavior. Digital surrogate models (Sect. 3) will enable designing for projected

2050 climate conditions through adaptive systems coordination rather than oversized specifications.

7.3 Regenerative Materials: Beyond Carbon Reduction

Research into willow (Sect. 3) exemplifies the shift from minimizing carbon to contributing to ecosystem restoration. Willow's rapid growth (2–5 years vs. 20–30 years), regenerative coppicing, and phytoremediation address resource constraints identified in World Resources Institute projections [9]. The successful willow-hybrid mock-up (Sect. 4) demonstrates feasibility. If durability testing confirms viability, regionally sourced regenerative materials could establish closed-loop bio-industrial systems, thus generating materials, restoring ecosystems, and creating rural opportunities simultaneously.

7.4 Performance-Based Regulation

Regulatory frameworks (Sects. 2 and 8) reveal an industry in transition with binding carbon limits becoming the norm. Façade contractors must be prepared for performance-based regulation, where buildings demonstrate compliance through measured outcomes. Exposed timber framing reflects biophilic design principles while adaptive systems coordination enables dynamic environmental quality management, demonstrating that sustainability and human wellbeing are complementary.

7.5 Toward Regenerative, Intelligent, Human-Centric Architecture

Construction stands at an inflection point. The pathway forward integrates biobased materials reducing embodied carbon, digital intelligence enabling material optimization and systems coordination, and circular design extending service life. This is evolving façades from passive barriers into active environmental interfaces.

The challenge is scaling innovations across industry, developing regenerative supply chains, establishing data standards enabling digital twins at portfolio scale, and restructuring business models that align incentives with circularity. The integration of rigorous analysis, validation testing, digital monitoring, and systems integration has created a blueprint for any building system confronting carbon reduction, circularity, and performance optimization imperatives.

As embodied carbon increasingly dominates lifecycle impacts, façade design can become a strategic intervention with decades-long consequences. This is the foundation that has been established through Timber-HUF system development, regenerative materials research, and digital intelligence integration.

8 Future Directions: Toward Adaptive and Regenerative Façades

Building on the Timber-HUF system and regenerative materials research, this section examines emerging directions that extend Staticus' current work, combining material innovation, digital intelligence, and systems thinking to address carbon reduction and circular resource management.

8.1 Adaptive Performance Envelopes

Predictive performance tools and BMS integration (Sect. 3) enable façade intelligence to actively coordinate with mechanical systems, optimizing energy performance, indoor environmental quality, and occupant comfort in response to dynamic conditions.

Current building design treats façades and mechanical systems separately: façades meet static specifications while HVAC systems handle worst-case loads with limited integration of in-use façade behavior. This can result in oversized mechanical systems and missed coordination opportunities.

The Timber-HUF system's monitoring infrastructure, which measures temperature, humidity, differential pressure, and thermal conditions at critical interfaces, generates data streams that, when integrated with BMS platforms, enable coordinated control strategies neither system could achieve independently.

Predictive HVAC coordination. Coupling façade sensor and weather data with occupancy schedules enables proactive environmental control. When sensors detect condensation conditions, the BMS incrementally adjusts temperature setpoints or ventilation rates to maintain safe dewpoint margins without energy-intensive dehumidification. Solar radiation sensors data combined with HVAC usage enable the BMS to predict peak cooling loads and pre-cool thermal mass during off-peak hours, shifting energy consumption and reducing costs. Differential pressure sensors inform intelligent ventilation control, modulating fresh air intake based on actual infiltration rather than design assumptions.

Dynamic response and climate adaptation. Façade intelligence enables nuanced indoor environmental control through dynamic setpoint strategies informed by real-time façade behavior. During mild conditions, the system widens acceptable temperature bands; during extreme weather, it tightens control to protect occupant comfort and façade durability. This is all done without excessive energy consumption.

As extreme weather events increase across Northern Europe, this integrated approach represents architectural resilience: material-efficient façades aided with intelligent building services that adapt to actual conditions rather than oversized material specifications. The timber remains unchanged; it is the system that responds.

Energy performance optimization. Façade-HVAC integration delivers measurable benefits including reduced heating and cooling energy consumption through optimized setpoint management, minimized peak electrical demand through load-shifting, extended mechanical equipment service life through reduced runtime, and improved indoor environmental quality without energy penalties. This addresses operational carbon, which accounts for approximately one-third of whole-life emissions through 2030, while timber-for-aluminum substitution addresses embodied carbon, creating comprehensive lifecycle impact reduction.

8.2 *Circular and Traceable Manufacturing*

The Timber-HUF system's modularity and Staticus' XD Façade Management Platform position timber-hybrid façades as exemplar circular construction. Realizing circularity requires comprehensive material traceability, standardized disassembly protocols, and economic models incentivizing reuse over disposal.

Digital material passports. Building on XD platform carbon tracking (Sect. 3.5), Staticus is developing comprehensive digital material passports for each Timber-HUF system documenting:

- material composition and origin (timber species, grade, moisture content, forest certification; aluminum specifications and recycled content);
- manufacturing data (production batches, quality control results, applied coatings);
- installation and operational history (as-built location, sensor performance data, inspection records);
- performance validation (cumulative thermal performance, moisture exposure cycles, design loads exceeded);
- and end-of-life recovery instructions (disassembly sequences, fastener specifications, material separation guidance).

This transforms façade modules into "banks" of materials with known provenance, documented performance history, and verified condition, which thus enables informed reuse, refurbishment, or recovery decisions.

Design for disassembly and reuse. The unitized system configuration supports disassembly, but circularity in practice requires development to prioritize material recovery and longevity. The main structural connections utilize mechanical fasteners, enabling non-destructive disassembly. Timber and aluminum components are designed for clean separation, with sealants and gaskets specified for removability. Connection geometries are standardized across Staticus' production, enabling modules to be reconfigured for different building contexts, while material components can be replaced without affecting adjacent units, enabling selective refurbishment.

Emerging business models. Staticus is exploring façade-as-a-service models where owners lease rather than purchase systems, with Staticus retaining ownership and responsibility for maintenance, monitoring, refurbishment, and recovery. This aligns economic incentives with circularity: manufacturers benefit from durable design and material recovery while owners gain predictable lifecycle costs and guaranteed end-of-life take-back. The XD platform and digital passports provide essential tracking infrastructure, demonstrating how digital tools enable fundamental restructuring of business relationships and resource flows from linear "take-make-dispose" economics to circular systems where materials retain value across multiple use cycles.

8.3 Regenerative Material Ecosystems

The willow research (Sects. 3 and 4) demonstrates construction's potential as an active participant in ecosystem restoration rather than merely reducing harm.

Expanding available regenerative materials. While willow is Staticus' primary focus, other species warrant investigation. Cold-hardy bamboo offers exceptional strength-to-weight ratios with 3–5 year growth cycles. Hemp fiber composites from 4-month rotations produce insulation while improving soil structure. Mycelium-based materials grown on agricultural waste produce near-zero carbon insulation, though durability challenges remain. Methodologies developed for willow, namely drying protocols, protective coatings, and hybrid connections, provide transferable frameworks for other regenerative species.

Agro-industrial integration. The willow consortium (Pilebyg, MOOW, Schmidt Hammer Lassen, Staticus) demonstrates vertically integrated value chains enabling regional bio-industrial clusters. Cultivation on marginal land, such as flood-prone areas, contaminated sites, and abandoned parcels, generates materials without displacing food production. Biomass follows cascading use: primary stems become structural framing; secondary growth becomes particle board; and residues are composted or used for biogas. Regional cultivation (within 200–500 km) minimizes transport emissions while creating rural economic opportunities and achieving carbon-negative supply chains where biogenic sequestration exceeds processing emissions.

Biodiversity and ecosystem services. Willow plantations provide ecosystem benefits aligning with EU Biodiversity Strategy by supporting pollinators, providing wildlife habitat, filtering agricultural runoff, stabilizing stream banks, and removing heavy metals through phytoremediation. Integrating cultivation into façade supply chains reframes manufacturing as ecological practice generating carbon sequestration, water filtration, and biodiversity habitat as co-products.

Scalability. Willow's 2–5 year rotations enable 4–10 harvests within one spruce rotation timeframe, producing comparable biomass from less land. Cultivation on marginal land avoids competition with food production. If durability testing confirms viability and cultivation scales industrially, regenerative materials could supply substantial façade requirements without the ecological pressures of conventional forestry expansion.

8.4 Data-Driven Sustainability Policy

Current regulations rely on prescriptive limits (maximum embodied carbon intensities, minimum biobased content). Staticus' digital infrastructure (machine learning models and predictive tools) enables performance-based approaches based on measured outcomes rather than design assumptions.

From prescriptive to performance-based compliance. Traditional regulations specify materials and assemblies. In contrast, performance-based codes define desired outcomes and accept any valid demonstration. Timber-HUF sensor networks generate empirical performance data verifying compliance: measured thermal performance, verified airtightness through continuous monitoring, durability validation through moisture sensors, and occupant comfort verification through surface temperature monitoring. This addresses the "performance gap" between predicted and actual building performance.

Regulatory innovation and policy feedback loops. As digital twins proliferate, regulators gain visibility into actual building performance at portfolio scale. Aggregated data informs evidence-based policy, identifying which strategies deliver outcomes and which fall short. Performance data establishes empirical benchmarks demonstrating carbon reductions and durability. Sensor data that reveals unexpected issues triggers design refinements and updated guidance. Buildings with continuous monitoring qualify for streamlined approvals, as empirical data provides stronger compliance evidence than pre-construction calculations.

Enabling innovation. Performance-based frameworks enable novel solutions that prescriptive codes might prohibit. The willow-hybrid façade lacks established standards. Under prescriptive regulations, approval could require years of standardization. Under performance-based codes, the consortium demonstrates compliance through laboratory testing, full-scale mock-ups, instrumented pilots, and validated digital twin simulations, thus accelerating market adoption that is critical to meeting 2030 and 2050 climate targets.

8.5 Human-Centric Design

Technical achievements must ultimately serve human occupants. Natural materials, responsive systems, and evidence-based design enhance wellbeing, productivity, and connection to nature.

Material experience and sensory engagement. Exposed timber (Sect. 2) aligns with biophilic design principles emphasizing natural materials in occupied spaces. The Timber-HUF system's exposed framing provides visual warmth and tactile qualities associated with natural materials without compromising technical performance, thus demonstrating that sustainability and human-centric design are mutually reinforcing.

Dynamic environmental quality. Integrated façade-HVAC coordination (Sect. 3.5) enables control of indoor environmental parameters, including daylight optimization while preventing glare, thermal stability, air quality management based on actual occupancy, and humidity control maintaining comfortable ranges.

Transparency and connection to exterior. Higher window-to-wall ratios and slender profiles address needs for exterior connection and daylight access. The Timber-HUF system's 90–100 mm joint widths achieve visual parity with aluminum curtain walls while incorporating visible timber elements.

Occupant agency. Digital monitoring enables transparency through real-time environmental data. Local environmental control and performance feedback engage occupants as active participants in building operation, creating participatory relationships rather than passive conditioning.

Acknowledgements
Authors:

Domas Valiukas, System Design Lead, Staticus.
Ricardo Seijas, IT Project Manager, Staticus.
Laura Craft, Head of Sustainability, Staticus.

Additional Contributors

Aulikki Sonntag, Chief Business Development Officer, Staticus.
Ieva Dauderytė, Head of Marketing, Staticus.
Simona Varkojūtė, Marketing and Communication Project Manager, Staticus.
Gediminas Putnikas, Head of Concept Design, Staticus.
Eimantas Tinginys, Façade System Designer, Staticus.
Edgaras Pangonis, Head of Technical Design, Staticus.

Research Partners

OsloMet, Norway.
SINTEF, Norway.

Kaunas University of Technology (KTU), Lithuania.
Schmidt Hammer Lassen Architects (SHL Architects), Denmark.
Pilebyg, Denmark.
Made out of Willow (MOOW), Denmark.

References

1. Jones C et al. (2025) Circular ecology, inventory of carbon & energy (ICE), version 4.1. Available at: https://circularecology.com/embodied-carbon-footprint-database.html. Last accessed on 18 Dec 2025
2. European Aluminum (n.d.) A low carbon footprint. [online] European Aluminum. Available at: https://european-aluminum.eu/projets/a-low-carbon-footprint/. Last accessed on 18 Dec 2025
3. European Commission (2020) In focus: energy efficiency in buildings. [online] Available at: https://commission.europa.eu/news-and-media/news/focus-energy-efficiency-buildings-2020-02-17_en. Last accessed on 18 Dec 2025
4. European Commission (2022) Buildings and construction. [online] single-market-economy.ec.europa.eu. Available at: https://single-market-economy.ec.europa.eu/industry/sustainability/buildings-and-construction_en. Last accessed on 18 Dec 2025
5. Climate Laws (2018) The Environment Buildings Decree of the Netherlands (Besluit bouwwerken leefomgeving, Bbl) - Climate Change Laws of the World. [online] Available at: https://climate-laws.org/document/the-environment-buildings-decree-of-the-netherlands-besluit-bouwwerken-leefomgeving-bbl_1057. Last accessed on 18 Dec 2025
6. Ryan W (2025) RE2020 updates 2025: what changes now. [online] EPD Guide. Available at: https://epd.guide/environmental-regulations-and-laws/re2020-updates-2025-what-changes-now. Last accessed on 18 Dec 2025
7. Nordic Sustainable Construction (2024) Danish political agreement tightens the limit values for new buildings and extends the impact. [online] Available at: https://www.nordicsustainableconstruction.com/news/2024/june/tillaegsaftale-paa-engelsk. Last accessed 18 Dec 2025
8. Klimato Kaita (2022) Net-zero government roadmap. [online] Available at: https://klimatokaita.lt/climate-change-mitigation/net-zero-government-roadmap/. Last accessed on 18 Dec 2025
9. Searchinger T, Peng L, Waite R, Zionts J (2023) Wood is not the climate-friendly building material some claim it to be. [online] Available at: https://www.wri.org/insights/mass-timber-wood-construction-climate-change. Last accessed on 18 Dec 2025
10. Envirocrops (2024) Short rotation coppice willow best practice guidelines. [online] Available at: https://envirocrops.com/resource/short-rotation-coppice-willow-best-practice-guidelines. Last accessed on 18 Dec 2025
11. Pilebyg and MOOW (2025) Findings from as yet unpublished pilot trials in 2025
12. Kofman PD (2012) Storage of short rotation coppice willow fuel. Coford Connects, Harvesting/transport no. 30. [online] Available at: https://www.coford.ie/media/coford/content/publications/projectreports/cofordconnects/HAR30_LR.PDF. Last accessed on 18 Dec 2025
13. European Committee for Standardization (2003) EN 1991-1-5 English version Eurocode 1: actions on structures - Part 1–5: general actions - thermal actions. [online] Available at: https://www.phd.eng.br/wp-content/uploads/2015/12/en.1991.1.5.2003.pdf. Last accessed on 18 Dec 2025
14. European Committee for Standardization (2013) EN 14080 English version Timber structures—Glued laminated timber and glued solid timber - requirements. [online] Available at: https://eximcorp.co.in/pdf/en-14080-2013-timber-structures-glued-laminated-timber.pdf. Accessed on 18 Dec 2025

15. London Energy Transformation Initiative, Various Contributors (2024) LETI embodied carbon primer, supplementary guidance to the climate emergency design guide. [online] Available at: https://www.leti.uk/_files/ugd/252d09_8ceffcbcafdb43cf8a19ab9af5073b92.pdf. Last accessed on 18 Dec 2025
16. ISO (2017) ISO 10077-2 3rd edition, thermal performance of windows, doors and shutters—calculation of thermal transmittance—part 2: numerical method for frames. [online] Available at: https://www.iso.org/standard/64995.html. Last accessed on 18 Dec 2025
17. CWCT (2022) How to calculate the embodied carbon of facades: a methodology. [online]. Available at: https://www.cwct.co.uk/pages/embodied-carbon-methodology-for-facades. Last accessed on 18 Dec 2025
18. Swedish National Board of Housing, Building and Planning—Boverket (2022) Application of the European construction standards, EKS 12. [online] Available at: https://www.boverket.se/en/start/publications/2023/application-of-the-european-construction-standards-eks-12/. Last accessed on 18 Dec 2025

Coat-ing

Pauline Bellec and Robert-Jan van Santen

Abstract External thermal insulation is facing a challenge in Europe, how to conceal low cost to low carbon. Existing solutions, tries to tackle each time one of those issues, but none have been able to include both. Staying in the norms and Eurocode makes it difficult to change the traditional industrialized way of insulating. This chapter presents a lightweight solution based on only three materials, combining a loose insulation with a prestress perforated membrane-façade to a framework, integrating the windows as well as the structure, named coat-ing. However no existing European standards or technical reference cover the case of loose insulation behind a fabric-membrane. Therefore, dedicated research and tests are being conducted to understand how this performing façade will react to existing wall. Research was carried around the granular pressure, membrane deformation, wind-induced pressure, acoustic performance, fire resistance and hygrothermal performance. Using a loose granulated insulation, shows how the pressure exerted against the existing wall and the membrane is governed by the Janssen effect, with roughness of the surface the main variable. A loose insulation with a perforated membrane will act as a sound absorbing system for the outside and perform great attenuation at the same time. The total system also attenuates about 78% of exterior wind pressure. Hygrothermal performance of the chosen insulation enable to avoid the use of unnecessary elements such as rainscreen. However, the fire tests are being the most challenge one, as direct contact between the insulation and the fabric accelerate the façade degradation. A broader conclusion from this chapter, is expressing how innovative building solution must define their own performance criteria as almost no standards cover them.

Keywords ETI · Perforated membrane-façade · Expanded granulated cork · Insulation · Low-carbon

---

P. Bellec (✉)
Coat-ing, Champs-sur-Marne, France
e-mail: p.bellec@coat-ing.eu

R.-J. van Santen
VS-A.EU Group, Hong-Kong, China
e-mail: rjvs@vs-a.group

A. Mejorin et al. (eds.), *Performance-Based Façade Design*, Springer Tracts in Civil Engineering, https://doi.org/10.1007/978-3-032-28835-6_11

1 Introduction

Buildings are the largest energy consumers in Europe, and approximately 75% [1] of the existing European building stock is considered poorly insulated and is now a major priority to ensure both energy savings and indoor comfort. With the introduction of the Energy Performance of Buildings Directive (EPBD), the European Union is setting targets for improving the energy efficiency.

Energy losses in buildings arise from multiple factors. Studies estimate that 30–60% of energy losses occur through the façades (e.g., Fig. 1). Depending on building compactness, scale, wind exposure … these losses are most of the time associated with poor performance of doors and windows, and defects leading to undesired air leakage. Additional sources of energy loss include natural or mechanical ventilation, the roof and the area in touch with the ground.

Climate conditions also play an important role in how those challenge manifest across Europe. The Köppen climate classification (e.g., Fig. 2) highlighted geographical regions where these challenges are preeminent. In regions where heating system compensated for poorly insulated envelopes, thermal insulation was often neglected. However, following the oil crisis in 1973, and the awareness of greenhouse gas emission (GHG), improving thermal insulation of buildings has been identified as a must.

Additionally, obsolete buildings have been preferably demolished, and salvaged materials were reused in new construction. At the same time, the widespread use of non-reusable material, combined with products designed with planned obsolescence, supported an industry generating substantial GHG. This approach is no longer viable and enhancing existing buildings is now a priority.

Three main façade renovation strategies can be identified: internal insulation (ITI), external insulation (ETI) and off-site insulation panels. ITI are often cost-effective,

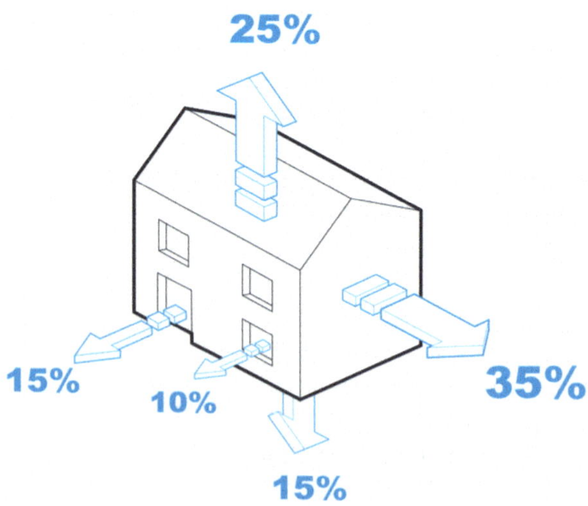

Fig. 1 House heat loss diagram (© coat-ing)

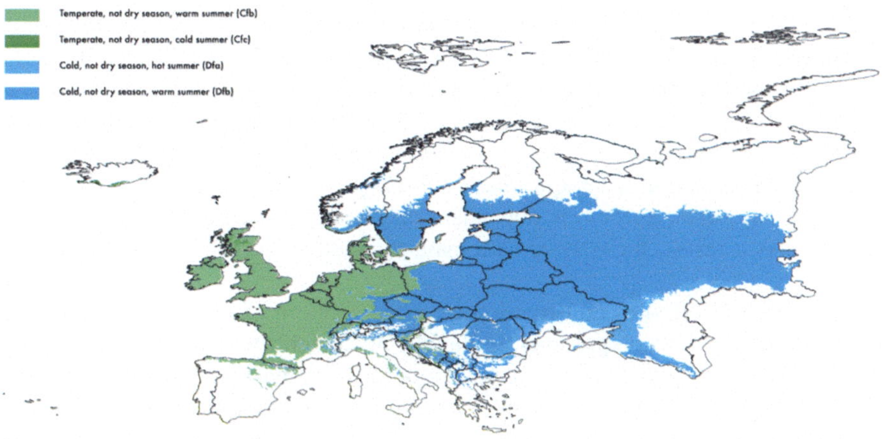

Fig. 2 Köppen map (© coat-ing)

but leading to a reduction of usable floor area, with the risk of introducing condensation risks and thermal bridge. ETI requires a complex on-site installation and remain sensitive to execution defects, especially when the facades have discontinuities (junctions and openings). Off-site façade, mainly introduce for the new construction, can be complex for renovation due to connections difficulties, higher weight and cost.

This chapter will look at an innovative façade renovation system based on a prestressed perforated membrane-façade system membrane and a loose insulation material. The system aims to combine the benefits of the on-site and off-site as well as façade cladding and window.

The performance of this system will be analysed through a succession of experimental investigations. The use of a membrane-façade and a loose insulation, needed to be tested with a variety of unconventional experimentation.

2 Background and Theoretical Framework

Applying internal thermal insulation (ITI) was characteristic of early renovation practice. The limitations are numerous: a reduction in the usable floor surface, the need to move the radiators (typically positioned beneath windows), and the lack of thermal bridging treatment at critical points, such as slab edges. Occupants often felt a summer discomfort, and numerous defects have been reported due to the occurrence of condensation. Nevertheless, this approach has gradually evolved, sometimes incorporating high-tech and efficient vacuum insulation panels. It continues to be widely adopted due to its cost-effectiveness and is often considered the only viable option when the architectural character of the façade must be preserved.

External Thermal Insulation (ETI) can be implemented using either rainscreen cladding systems or composite systems (ETICS), in which a finishing layer is applied directly over the thermal insulation that is mechanically fixed to the masonry. This method requires the installation of external scaffolding, and it is significantly more complex as most of the components are directly exposed to weather conditions.

All façade discontinuities, such as junctions, corners and openings, create a risk of performance discontinuity, potential weak points. Improvements of the available systems and installation techniques have made it possible to address many of these challenges. However, success is dependent on the experience and qualification of installers. As a result, these systems remain vulnerable to defects and executions failures.

The rise of off-site construction, carried out in factories inspired the EnergiesSprong [2] ("energy leap" in Dutch) movement, which originated in the Netherlands. This approach involves assembling large parts of façade in factory, using standardized technical systems, enabling consistent quality, high production rates with the ambition of reducing the cost. However, off-site insulation construction is not as efficient as off-site construction for new buildings.

These limitations enhance the need of thinking of an additional system, simplifying constructability, sustainability and adaptability to existing building. The performance of façades systems depends on many physical mechanisms, including heat transfer, air leakage, moisture transport, mechanical loading …

Moisture/condensation/vapor diffusion

It is commonly said that moisture or water vapor migrates from the inside to the outside. This varies depending on climatic conditions and the type of occupancy, this transfer, which obeys the laws of physics, is not a problem. However, it is becoming one when thermal insulation is vulnerable to water, and when, at certain times the water vapor content becomes saturated (HR = 100%), is condensing and introducing water into the insulation itself. Adding a supplementary component, a vapor barrier, can solve this problem, but special care must be taken to ensure that there are no discontinuities that could be fatal [3].

Mechanic loading/wind pressure

This mainly concerns self-weight, seismic resistance, and wind resistance.

Any material applied to a facade generates an additional vertical load called deadload that must be supported by the existing structure [4]. This load can be applied to a surface (bonded products) of an existing wall. The added structure may also be a self-supporting construction, with the weight transferred downwards to be transferred to a cantilevered frame or a specific beam, as is common with perforated membrane-façade system components.

Seismic resistance is becoming an increasingly frequent requirement: the fixings of the added assembly must then withstand horizontal loads while remaining functional when the substrate begins to move [5].

Wind is particularly intense, mainly in suction, in the immediate vicinity of corners, parapets, and protruded elements. Impacts caused by wind-borne debris or human activity near the facade must also be considered [6].

Acoustics
Acoustic performance is measured in three different ways.

It is common to speak of sound attenuation to characterize the reduction of noise from a wall located between a noisy exterior space and the interior. This characteristic is applicable in both directions.

The term "sound absorption" is less frequently used when discussing facades, as they are highly reverberant (between 97 and 98% of sound is reflected by a concrete wall or a glass panel). This is why cities are traditionally noisy places.

Even less often is the noise generated by the facade itself: for example, by differential movements between parts that move or expand in opposite directions, or by wind that causes certain components to vibrate, such as perforated sheet metal or tubular profiles with open ends.

Fire performances
Fire resistance is a characteristic largely defined by regulations, both in terms of requirements and compliance methods. This includes:

- The fire reaction of materials, classified from non-combustible to increasingly flammable according to the Euroclass system (EN 13501-1:2018 [7]).
- The fires resistance of structural elements (e.g., a wall), acting as effective fire barriers for a given time (EN 13501-2:2016 [8] and EN 1363-1:2020 [9]).
- Project related geometries preventing the spread of fire from one part to another, from one floor to another, or from one property to its neighbour (EN ISO 13785-1:2002 [10] and EN ISO 13785-2:2002 [11]).

3 System Description

Coat-ing is a lightweight, low carbon façade renovation system based on three components:

- A unique extruded aluminium profile, acting as both a structural subframe and integrated window frame.
- A pre-stressed, micro-perforated façade membrane-façade system, acting as a lost formwork, ensuring durability and breathability.
- A loose-fill insulation made of expanded granulated cork, offering thermal and acoustic performance.

With the use of a perforated membrane-façade system and expanded granulated cork, comes a challenge of experiencing the façade in a new way. The installation system and its behaviour will be explained.

Step 1: Survey and preparation

The existing building can be 3D-scanned and generate its own accurate execution model with all required information to set the details and manufacture all the components (Fig. 3).

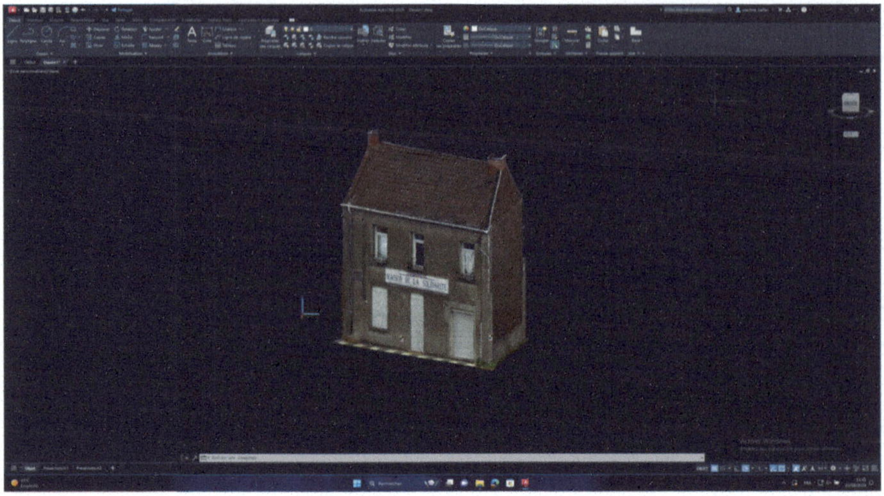

Fig. 3 3D scan example of a house for the second prototype, 2024 (©coat-ing)

Step 2: Mounting the aluminium framework

Extruded aluminium profiles are installed along the perimeter of the façades and around windows and door openings, defining the new envelope structure. As no connection between windows other than the perforated membrane-façade system, the pre-assembled frames can be installed without any other constraints than matching the edge of the opening in the wall, necessary for the air-tight sealing (Fig. 4).

Step 3: Installing the perforated membrane-façade system

The microperforated perforated membrane-façade system is fixed and flattened onto the aluminium framework, forming the external visible façade (Fig. 5).

Step 4: Insufflation of cork granules

Expanded granulated cork is blown in the cavity from the top, at multiple points. The cavity is filled uniformly, and gentle vibration is applied to improve distribution, especially at the underside of windows. Additionally, the cork can be easily removed, when necessary.

Step 5: Finishing touches

Once the insulation is completed, the excess of perforated membrane-façade system is trimmed, while some interior finishing details are applied (Figs. 6 and 7).

Expanded granulated cork

Expanded granulated cork is a transformed product of cork. Cork is harvested every six to eight years from the bark of cork oak trees. A less premium cork is harvested form the upper layer of the tree, it is less uniform and contains more imperfections. This portion is primarily used for expanded cork. The cork is ground into granules,

Fig. 4 Installation of the aluminium profile on the facade and around the window on the first coat-ing demonstrator, 2026 (©coat-ing)

used for lower quality cork stopper, insulation and decorative applications. All the scrap from the industry is ground and process through different treatments.

To produce expanded cork, raw granules are poured in an autoclave and heated to approximately 300 °C for 10–15 min, depending on the humidity. During this process, the granules release water and natural resin which act as binder and form block. The blocks are then cut into panels of various thickness, while offcuts are turned into loose granules (Fig. 8).

Aluminium a strategic choice for coat-ing

Aluminium's lightweight nature and extrudability showcase it as a common material for façade detailing. With the coat-ing system aluminium is used for a double task: insulating the wall and integrating the framework of the window at the same time (Fig. 9).

The extrusion process generates almost no material loss.

Untreated aluminium

Within an approach of environmental sobriety, untreated aluminium profiles, i.e. without any protective surface treatment, can be used. Alloys 6060 T6 and 6082 T6 naturally develop a protective alumina layer (Al_2O_3), ensuring excellent corrosion resistance, even in outdoors conditions.

Fig. 5 Installation of the façade perforated membrane-façade system on the first demonstrator, 2026 (©coat-ing)

Studies on the ageing of aluminium [12] exposed between 13 and 16 years in various environments (urban, marine, industrial) show a corrosion depth of less than 0.1 mm, with a very stable behaviour over time (Fig. 10).

By maintaining the aluminium untreated, the surface ages, and evolves naturally over time, losing its shiny appearance in favour of a uniform grey patina. This transformation does not alter its mechanical properties or longevity and reduces unwanted solar reflection.

The choice of 6060 T6 for standard extruded profiles, and 6082 T6 for mechanically stressed parts, allow the framework to combine:

- lightness, resistance, durability,
- Simplicity of manufacture (extrusion, machining)
- environmental coherence (recyclability, sobriety)

Although aluminium is used, the selected alloys try to meet both the structural and environmental requirements, supporting a durable, lightweight and responsible approach to building renovation.

Prestressed perforated membrane-façade system

The perforated membrane-façade system used by coat-ing had to meet several technical requirements:

Fig. 6 Photo of the first demonstrator, 2026 (© coat-ing)

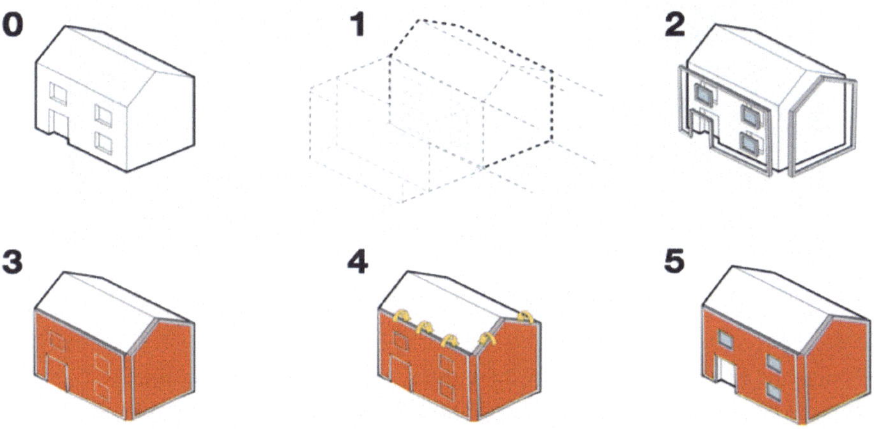

Fig. 7 Coat-ing step by step (© coat-ing)

Fig. 8 Expansion process of cork, raw granules are poured in the autoclave (©coat-ing)

- It must be fire-resistant, to comply with façade fire safety requirements.
- It must be perforated to prevent condensation within the insulation layer.
- It must have long-term mechanical stability. To prevent deformation that could lead to uneven cork distribution, leading to thermal bridging. And to avoid aesthetic degradation, caused by a loose membrane-façade. Observations have shown numerous cases in which façade membranes lost their tension over time (Fig. 11).

Fig. 9 Library of dies at Lucé by Hydro (© coat-ing)

Fig. 10 Example of raw aluminium on the Grand Palais in Lille by OMA, inaugurate in 1994, the aluminium still shows great aesthetical quality despite being untreated (© coat-ing)

Following extensive research and review of available solutions, the Frontside 381 print, a pre-stressed and micro-perforated perforated membrane-façade system developed by the French company Serge Ferrari, was selected. It holds internationally recognised certifications demonstrating compliance with applicable codes and mandatory requirements, including a Euroclass B-s2,d0 fire rating obtained through standalone component testing. As discussed in Sect. 4, the behaviour of the membrane under fire conditions within the assembled system differs from its standalone classification, a distinction that has direct implications for system-level fire design.

Key characteristics of this membrane include:

- A bi-axial tensioned structure providing a very high dimensional stability,
- Guaranteed colourfastness over 10 years, thanks to a proprietary printing process by *Printable,* based in the Netherlands.
- *Euroclass B-s2, d0* fire rating, ensuring low flammability and smoke development, one of the highest ratings for composite membranes used in architecture.

Window renovation

Renovation of window opening involves four keys' aspects that impact interior daylight comfort:

- *Replacement of the window frame*: a high-performing frame has usually bulkier profiles, which reduce the clear glazed area.
- *Replacement of the glazing*: the transition from single to high-performance insulating glazing improves the thermal performance but results again in reduced light transmittance.

Fig. 11 Renovation of EDF facility, such loose tension is not noticeable on the coat-ing system. Lyon, France by Jean-Paul Viguier et associés 2009–2011 (© coat-ing)

- *Addition of ETI*: the depth of the window reveals increases, creating a '*tunnel effect*', reducing both solar exposure and natural light as well as the outward views.
- *Addition of a roller shutter:* reduces the height of the window opening (about 20 cm), even though daylight has better penetration into the room through the upper part.

There is a broad consensus that the visual and daylight comfort of renovated dwellings has significantly diminished. While this is clearly not the wish of architects, it is a consequence of technical solutions applied independently to address each issue in isolation. An alternative approach was therefore required.

Integrating the structural framework to the window framework provides improvement across each of the four-point described above:

- *Replacement of the window frame*; window frames are installed on the exterior face of the wall. No frame elements remain visible from the interior, allowing the entire opening to function as a glazed area. This approach enables an increase of 25–30% in the glazed area.
- *Replacement of the glazing*; although as discussed earlier, insulating glazing is less transparent, performance can be slightly improved by using extra-clear glass for the inner pane (a gain of 1–2%). This configuration, frequently implemented by VS-A.EU, also improves the *Sgs coefficient* (solar protection performance of the combined glazing and interior shading system
- *Addition of ETI*; the 'tunnel effect' is eliminated by a chamfered design of the window, replacing sharp angles with bevelled surfaces. This allows a maximum penetration of natural light. In this configuration, ETI no longer obstructs both daylight and outward view.

  Additon of a roller shutter; when requested by occupants, the roller shutter is positioned in front of the ETI rather than within the window reveal. This eliminates the thermal bridge associated with shutter boxes, while requiring the design of an integrated coherent enclosure.

4 Experimental Methodology

Given the absence of existing analytical models applicable for textile-membrane façade systems filled with granular insulation, a series of dedicated experimental protocols were developed to characterize the mechanical, acoustic and fire behavior of the system. They were performed by certified laboratories when applicable, while others were carried out by VS-A.HK in Honk-Kong and by coat-ing itself with the help of Chalmers University in Sweden.

- Mechanical behavior of granulated expanded cork
- Membrane façade deformation
- Wind-induced pressure transmission
- Acoustic performance for sound absorption and transmission

- Fire behavior
- Hydrothermal performance

4.1 Mechanical Behaviour of Cork Granules

One of biggest challenge with the system, is to understand the lateral pressure exerted by the granulated expanded cork on the façade membrane. Indeed, unlike liquids, granular materials do not transmit pressure hydrostatically. The friction between grains against confining walls causes pressure to saturate with depth rather than increasing linearly. Therefore coat-ing looked at the approach known as the *Janssen effect*, for the application of grain silos.

The Janssen model describes vertical pressure in a granular column as:

$$p_z(z) = \lambda \rho g \left(1 - \frac{\exp(H - z)}{\lambda}\right) \quad (1)$$

$$\lambda = L_x L_y / \left(2\mu \left(K_y L_x + K_x L_y\right)\right. \quad (2)$$

For a circular section of radius R, ρ is the grain density, K is the Janssen coefficient, and μ is the grain-wall friction coefficient.

The apparent mass measured at the base of a silo follows:

$$m_{app} = m_c \left(1 - \exp\left(-\frac{m_{th}}{m_c}\right)\right. \quad (3)$$

where m_c is the saturation mass:

$$m_c = \lambda \rho g \quad (4)$$

The internal friction angle, φ, of expanded granulated cork was measured by the natural angle method: a free pile of cork granules was formed and the natural slope angle measured, giving φ = 35.09° (e.g., Fig. 12). From this, the at-rest Janssen coefficient was estimated as:

$K_0 = 1 - \sin\varphi = 0.38$ with active and passive bounds $K_a = 4.2$ and $K_b = 0.24$.

To verify the *Janssen effect*, a first experiment had to be done using a circular tube (e.g., Fig. 13), acting like a silo. Enabling to confirm the general validity of the Janssen effect. The tube was of two diameters (36 and 60 mm) and of different materials (cardboard vs. perforated membrane-façade system). The apparent mass curves showed a fast initial rise followed by a strong deceleration, like the Janssen shape. The friction coefficient bounded at $< 0.3 < \mu < 0.56$ (Rankine theory). And normalized results were found for the 36 and 60 mm confirming the models.

Fig. 12 Natural angle method measurement for cork (© VS-A.EU)

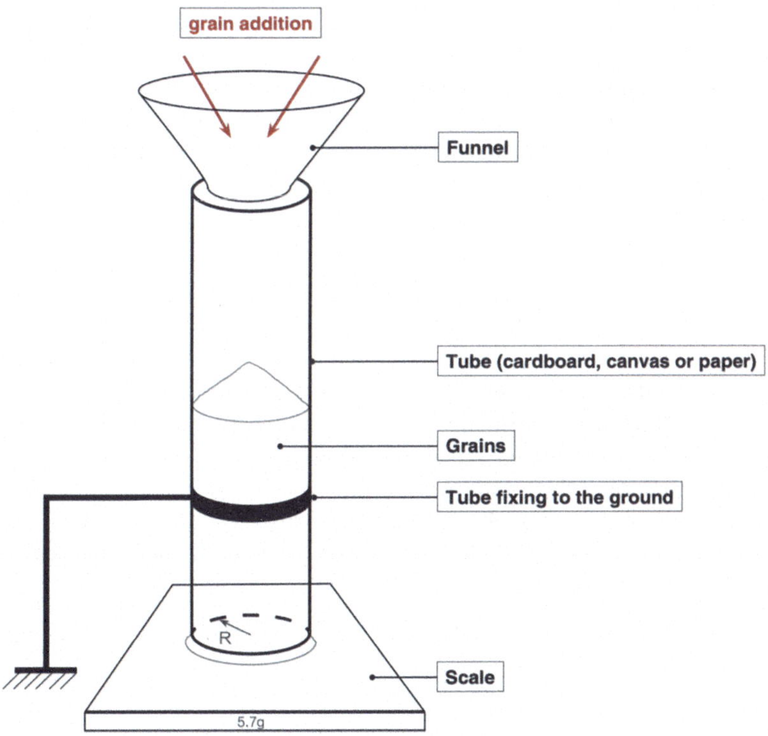

Fig. 13 Experimental process on a circular tube (© VS-A.EU)

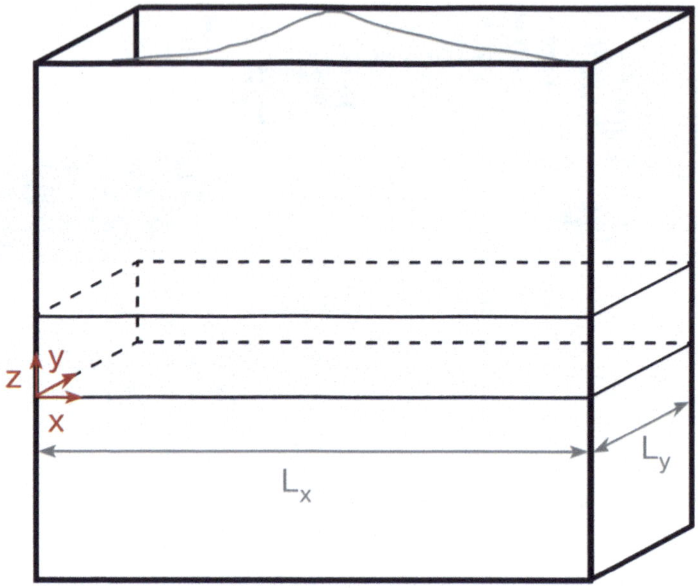

Fig. 14 Experimental process on a rectangular section (© VS-A.EU)

Following those results, a second series of tests were conducted, using rectangular sections (e.g., Fig. 14) with again two materials (glass and perforated membrane-façade system), to understand the roughness of the wall and the apparent lateral pressure on the wall. To do so, assumptions had to be done, the same expression were used except for λ:

$$\lambda = \frac{L_y}{2\mu K} \ when \ L_x \gg L_y \tag{5}$$

The results showed the Janssen effect was still working but showed a difference with the material used for the experiment. The change in wall material and therefore in the grain-wall friction coefficient has a considerable effect on the apparent mass and therefore the pressure of the grain on the wall.

Controlled vibrations were also added in a second test after filling, simulating the in-situ vibration used during the insufflation of the cork during the installation process.

4.2 Membrane façade Deformation Load under Varying Uniform Load

This experience meant to study the tensile properties of a perforated membrane-façade system sample, from the Frontside 381 of Serge Ferrari, by measuring its deformation under different loads (e.g., Fig. 15). The determination of the deflection of the perforated membrane-façade system was tested under incremental loads, made from rice grains uniformly spread other the surface, ranging from 0 to 13 kg, with a 2.5 kg increase after each test (e.g., Table 1). A sample of 0.88 m^2 was build and position horizontally. The study and data collected provides insights into how the perforated membrane-façade system stretches under tension, allowing the calculation of tensile stress and strain.

The deflection formula used for the test was this followed one:

$$Deflection = D1 - average\ D2 \quad (6)$$

Each measurement was repeated three times for accuracy.

Fig. 15 Experimental bench of the membrane façade under varying uniform load (© VS-A.EU)

Table 1 Table of measurement under varying uniform load on 4 different samples

Weight (kg)	0	2.5	5	7.5	10	13
Total pressure (DaN/m^2)	0.61	3.40	6.19	8.97	11.76	15.11
Deflection measurement 1.1 (mm)	10.35	18.41	20.59	26.04	29.46	31.45
Deflection measurement 1.2 (mm)	16.96	23.56	26.41	29.02	30.59	33.71
Deflection measurement 2.1 (mm)	2.44	10.77	15.26	18.32	20.95	25.02
Deflection measurement 2.2 (mm)	4.98	13.38	16.99	19.66	23.37	26.90

4.3 Wind-Induced Pressure Transmission

For this second experimentation, the aims were to investigates the influence of expanded granulated cork on wind induced pressure within a membrane-façade system. The study is based on a comparative experimental approach, combining a reference configuration with several test configuration to assess the impact of cavity depth and insulation material on pressure transmission.

A control configuration is defined as an air-filled cavity without perforated membrane-façade system, which serves as a baseline for characterizing the exterior wind pressure behaviour in the absence of a textile interface. Building upon this reference, six experimental configurations are investigated. These configurations consist of a perforated membrane-façade system facade positioned at distances of 100, 200, and 400 mm from the exterior wall, with the cavity either filled with air or with cork insulation. This approach makes it possible to isolate the respective effects of cavity depth and insulation material under identical geometric conditions (Fig. 16).

Pressure measurements are performed using differential and exterior pressure sensors to capture the pressure dynamics acting on the system under controlled wind conditions. The experimental protocol allows the identification of both overpressure (push) and depression (pull) phases generated during the tests. This will enable to obtain the differential pressure applied to the perforated membrane-façade system for different cavity depths (10, 20, and 40 cm). The cavity will be either air-filled or cork-filled, and the ratio between the differential pressure acting on the perforated membrane-façade system and the corresponding exterior pressure during both pull (depression) and push (overpressure) phases.

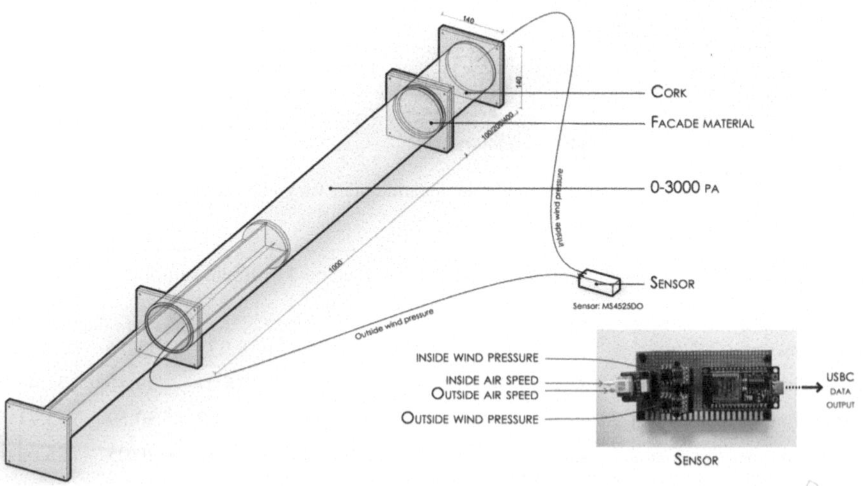

Fig. 16 Experimental bench system built for the wind induced pressure transmission (© VS-A.EU)

Table 2 Measurement for the pull phase or depression

Configuration	ΔP min (Pa)	Outside pressure (Pa)	Transmission pull (%)
10 cm—Air	−537	+10,612	5.0
10 cm—Cork	−624	+7442	8.4
20 cm—Air	−773	−3600	21.5
20 cm—Cork	−1744	+7455	15.8
40 cm—Air	−551	+6163	8.9
40 cm—Cork	−1342	+7312	18.4

Table 3 Measurement for the push phase or overpressure

Configuration	ΔP min (Pa)	Outside pressure (Pa)	Transmission pull (%)
10 cm—Air	+750	+10,612	7.0
10 cm—Cork	+347	+4888	7.1
20 cm—Air	+426	+11,254	3.8
20 cm—Cork	+473	+7455	6.3
40 cm—Air	+179	+6163	2.9
40 cm—Cork	+2178	+7312	29.8

Those result, will enable to have a better understanding on how wind influence the perforated façade (Tables 2 and 3).

For each configuration, the pressure transmission ratio is calculated by identifying the extreme value of differential pressure (maximum for the push phase, minimum for the pull phase) and dividing its absolute value by the exterior pressure measured at the same instant (same data line). The transmission ratio is expressed as a percentage according to:

$$Transmission\ (\%) = \frac{Differential\ Pressure^{*}}{Outside\ Pressure}\ 100 \qquad (7)$$

4.4 Acoustic Performance for Sound Absorption ISO 354

The experimentation was carried by Matelys Research Lab. The investigation aimed to determine the acoustic absorption of the solution, intrinsic acoustic parameters of a granulated insulation with a perforated membrane-façade system Frontside 381 by Serge Ferrari with a 28% opening, under diffuse field conditions.

The acoustic absorption characterization was carried on a study of three insulation materials:

- Expanded granulated cork non-compacted (e.g., Table 4)

Table 4 Acoustic parameters for non-compacted granulated cork (79.2 ± 1.9 kg m^{-3})

Static airflow resistivity (σ)	3600 N s m^{-4}
Open porosity (φ)	0.66
High-frequency tortuosity limit (α∞)	1.83
Thermal characteristic length (Λ′)	387 μm

Table 5 Acoustic parameters for compacted granulated cork (92.8 ± 1.1 kg m^{-3})

Static airflow resistivity (σ)	10,600 N s m^{-4}
Open porosity (φ)	0.62
High-frequency tortuosity limit (α∞)	2.14
Thermal characteristic length (Λ′)	236 μm

Table 6 Acoustic parameters for polystyrene (11.5 ± 0.1 kg m^{-3})

Static airflow resistivity (σ)	3100 N s m^{-4}
Open porosity (φ)	0.49
High-frequency tortuosity limit (α∞)	1.44
Thermal characteristic length (Λ′)	518 μm

- Expanded granulated cork compacted (e.g., Table 5)
- Expanded polystyrene (e.g., Table 6)

The choice of these three materials was decided to build a comparative analysis with another mainstream existing thermal insulation material, polystyrene.

The characterization was performed using the standing wave impedance tube (Ø 44.45 mm, frequency range 250–4400 Hz), following the methodology described by the ISO 10534-2 [13].

Following the acoustic absorption (α) formula:

$$\alpha = E_a/E_i \tag{8}$$

Because all three materials consist of loose granules, conventional three-microphone measurements were not applicable, again. Instead, measurements using upstream microphones only were performed to estimate the dynamic density and bulk modulus of each material.

From the measurements, several parameters were extracted: static airflow resistivity (σ), open porosity (φ), high-frequency tortuosity limit (α∞), viscous characteristic length (Λ), and thermal characteristic length (Λ′). These parameters are needed for the Johnson-Champoux-Allard dissipation model [14].

In addition, the perforated membrane-façade system was also studied with a 2 × 3 mm^2 perforation with a 28% opening, was characterized as an acoustic facing layer, using the surface impedance method described by Jaouen and Bécot [15].

The diffuse field absorption simulations were performed using the Transfer Matrix Method with the *Alpha Cell v16–2024 software*, for a material thickness of 25, 50, 100 and 200 mm with and without the perforated membrane-façade system.

4.5 Acoustic Performance for Sound Insulation

This study aimed to understand the acoustic attenuation of the system on the indoor comfort. It was applied to three different types of wall configuration according to the existing European stocks: a 200 mm hollow brick wall, a 140 mm concrete wall and 140 mm concrete wall with an interior lining made from 45 mm of glass wool and 12.5 mm of plasterboard. Simulations were carried once again using the Transfer Matrix Method and the *Alpha Cell software* developed by Matelys. The acoustic parameters were the same one as the one used in *3.4 Acoustic performance for sound absorption*. The three insulation materials tested were, once again, non-compact expanded granulated cork, compact granulated cork and polystyrene.

4.6 Fire Performance Test—SBI Test EN 13823

The reaction to fire behaviour of coat-ing was assessed through a Single Burning Item (SBI) test conducted by RISE Research Institutes of Sweden AB, an ISO/IEC 17025 accredited laboratory. The test was performed for informatorily purposes according to EN 13823:2020 + A1:2022.

The specimen consisted of the full coat-ing system: the Frontside 381 façade membrane (0.97 mm thick, 550 g/m^2), 140 mm of expanded granulated cork (density 73 kg/m^3 as measured), and a 2 mm aluminium casing. The specimen was mounted directly onto plasterboard complying with EN 13238 (e.g., Fig. 17). The cork granulate was conditioned at 23 ± 2 °C and 50 ± 5% relative humidity until constant mass was achieved.

The SBI method exposes the specimen to a flame source in a corner configuration. The key parameters recorded include the Fire Growth Rate index (FIGRA), Total Heat Release over 600 s (THR_{600s}), Smoke Growth Rate index (SMOGRA), Total Smoke Production over 600 s (TSP_{600s}), and the presence of flaming droplets or particles.

It should be noted that this test was carried as a first try. Indeed, it had to be aborted before completion pursuant to EN 13823 clause 8.5 (e.g., Fig. 18). As explicitly stated by RISE, the results cannot be used for product classification or regulatory approval. They are reported here as preliminary informatorily data only.

Fig. 17 Experimental bench at RISE for SBI before the test (© coat-ing)

4.7 *Moisture and Hygrothermal Behaviour*

Moisture behaviour was investigated using the Saturated Salt Solution Method under a controlled temperature (20 °C) and relative humidity conditions. The samples were sealed in an environment with 93% RH with KNO_3 salt, and they were weighted at different time (e.g., Fig. 19). The samples were initially dried and weighed at time

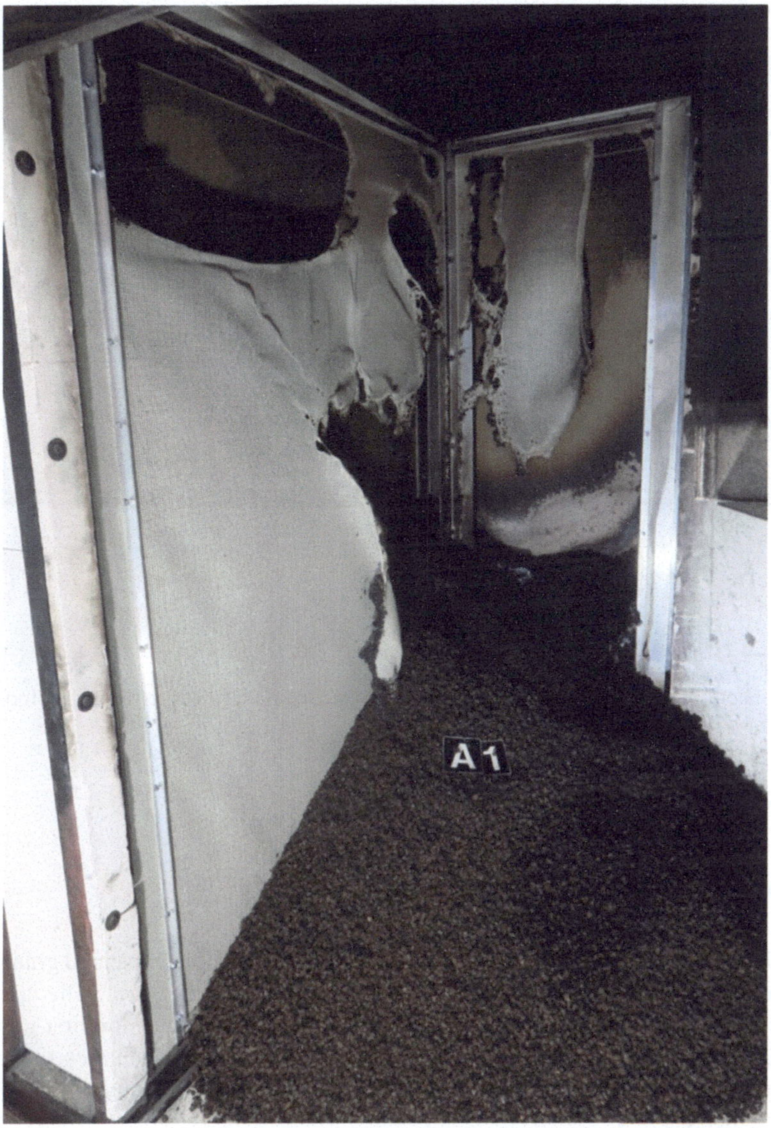

Fig. 18 Experimental bench at RISE for SBI after the test (© coat-ing)

zero. The absorption was tracked by weighing the samples at regular intervals (from 30 minutes up to 336 h). No forced air circulation was applied.

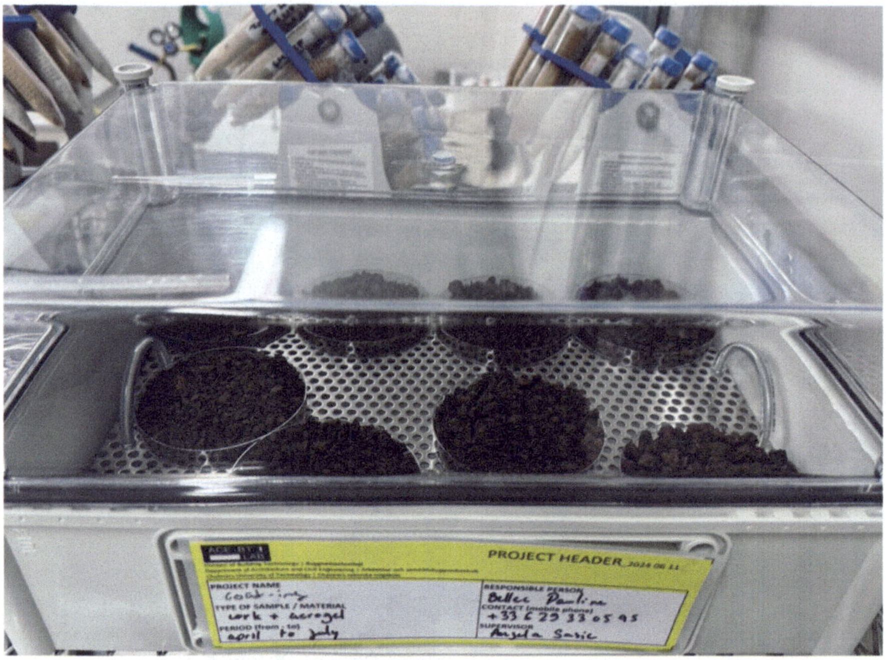

Fig. 19 Photo of the experimental bench for the Saturated Salt Solution Method performed at Chalmers University (© coat-ing)

5 Experimental Results and Analysis

5.1 *Mechanical Behaviour of Cork Granules*

Circular tube experiment

The silo filling tests approach with circular tubes confirmed that expanded granulated cork behaves as a granular material governed by the *Janssen effect*. The apparent mass curves showed a two-phase response: a rapid initial rise as the first granules are added, followed by a strong deceleration when the mass saturates toward an asymptotic limit. A contrast if compared to the behaviour of liquid with a linear pressure increase.

Regression of the experimental data using an $A(1 - \exp(-x/A))$ fit gives good agreement across all tube materials tested. Normalizing the results by tube radius showed a close agreement between the 36 and 60 mm tubes, further validating the model. From these fits, the product $K\mu \approx 0.134$ was derived for paper-walled tubes, and Rankine theory bounded the friction coefficient between cork and paper at $0.03 < \mu < 0.56$.

However, no massive difference was found between the cardboard, paper and canvas, the results were too close to each other.

Table 7 Normalized apparent mass M_c for polycarbonate and perforated membrane

Material	M_c (g)
Polycarbonate	15.68
Perforated membrane-façade system	5.56

Rectangular section experiments

The Janssen effect is confirmed for rectangular geometry. The apparent mass curves follow the same saturating exponential shape as in the circular case, and normalizing results by section dimensions shows close agreement between the 84 × 32 mm and 105 × 62 mm glass-walled sections with a relative difference of 12.1% at saturation. This confirms that the Janssen formula extended to rectangular sections is consistent and reliable across different cavity widths.

The critical finding emerges when comparing the same rectangular section (84 × 32 mm) with two different wall materials: smooth glass and the Frontside 381 perforated membrane-façade system. The relative difference in normalized apparent mass reaches 64.5%. Wall material is therefore the dominant variable, not cavity dimensions (e.g., Table 7).

This result can be explained by the micro-perforations of the perforated membrane-façade system: the holes increase the surface roughness and therefore the grain-wall friction coefficient significantly compared to smooth glass.

The perforated membrane-façade system friction coefficient is 2.82 times higher than that of the smooth wall.

These results suggest that the microperforated perforated membrane-façade system transfers a larger proportion of the cork's weight to the perimeter aluminium profiles and the backing wall, reducing lateral pressure on the membrane, and avoiding massive deformation of the perforated membrane-façade system as shown on chap. *5.2 Membrane façade deformation under varying loads.*

Effect of vibration and additional rate

The vibration tests show that imposed shocks cause an initial spike in apparent mass (e.g., Fig. 20), followed by a relaxation that stabilizes at approximately 1.3–1.5 times the Janssen asymptotic limit. The direction of change, varies between tests, reflecting the sensitivity of the result to the packing state of the grains now of vibration. Similarly, adding cork all at once produces higher apparent mass than incremental addition, because rapid filling creates a more compact packing and therefore a lower K value.

These last two findings highlight an important limitation: the Janssen coefficient K cannot be determined independently from these experiments because it is inseparable from the packing state, which itself depends on installation conditions. An absolute value for the grain-perforated membrane-façade system friction coefficient therefore cannot be extracted from this study alone. A full in-situ characterization is necessary to determine the precise pressure state in coat-ing system under service conditions.

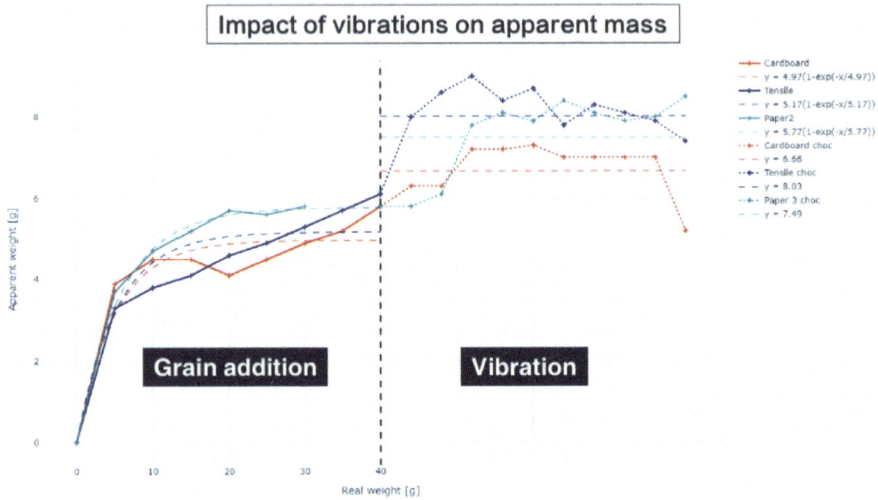

Fig. 20 Impact of vibrations on apparent mass (© VS-A.EU)

Because of inability to proceed further in this direction so far. An alternative approach was proposed measuring the deformation of the membrane-fabric façade under varying uniform loads.

5.2 Membrane façade Deformation Load under Varying Uniform Load

Both samples of the Frontside 381 fabric (1.1, 1.2, 2.1 and 2.2) exhibit a nonlinear stiffening response. The largest deflection increment occurs between 0 and 2.5 kg, with a net deflection of about 8 mm. Beyond this point, each additional 2.5 kg increment produces progressively smaller deformation. This behaviour can be explained using a prestress fabric such as the Frontside 381 fabric. In other words, the more the fabric deflects, the more it resists further deflection, showing a benefit for long-term deformation under repeated critical load, such as wind load.

At maximum test load (13 kg, 15.11 DaN/m^2), net deflection reaches approximately 21 mm and 22 mm for the two samples respectively, with a difference of less than 5%. This consistency validates the reliability of the Frontside 381 membrane as a structurally predictable material for façade applications.

It is important to note that this test was conducted with the fabric positioned horizontally, with the load applied perpendicularly and uniformly over the entire surface by gravity. It is a worst-case scenario, as in reality the system is vertically.

5.3 Wind-Induced Pressure Transmission

The pressure transmission measurements an attenuation of the wind-induced pressure. Across almost all configurations tested, the differential pressure acting on the membrane-fabric represents less than 22 of the exterior wind pressure, meaning 78% of the wind load is being dissipated through the thickness of the system. In other words, the perforation of the fabric as well as the air between granules and the tortuosity of the granules itself (see Sect. 3.4), allowing air to flow through it, equalizing pressure progressively across the system.

Regarding the Eurocode EN1991-1-4, in Sect. 7.2.10 [6], it recommends calculating wind force on the outer layer $cp, net = \frac{2}{3}cpe$ for overpressure and $cp, net = \frac{1}{3}cpe$ for depression. Meaning the outer skin take 67% of the full exterior wind pressure under overpressure and 33% under depression. These values represent conservative design assumptions. With the experimental values, the actual pressure transmitted to the perforated membrane-façade is lower than these values. The value remaining mostly under 22% in both overpressure and depression.

These results confirm that wind pressure is not a critical for coat-ing and prove how a perforated membrane-fabric can help. A dedicated wind tunnel test with instrumented full-scale specimens remains necessary to fully characterize the system's behaviour across a broader range of wind speeds and directions before regulatory certification.

5.4 Acoustic Performance for Sound Absorption ISO 354

The characterization confirms distinct acoustic parameters, comparing the expanded cork to the polystyrene. A noticeable difference was shown with compacted cork, the static airflow is 292% more important than for polystyrene, compaction is critical for the airflow resistivity. This difference as well as the tortuosity have a noticeable impact on the viscous dissipation and lengthen the acoustic path through the material. Therefore, in diffuse condition, a noticeable hierarchy is expressed, expanded cork outperform polystyrene, especially when compacted.

The fabric, however, has very minor influence, with a slight absorption peak on the lower frequencies. These results are good news for coat-ing system as they avoid an important sound reflection, which will compromise the acoustic performance of the solution. In the future any change of façade-membrane will have to be about the same porosity to avoid any changes (Fig. 21).

Looking at how these experiments can be related to real world simulation. Concrete and brick walls are highly reflective surfaces with absorption coefficients close to zero across the audible frequency range. By contrast, coat-ing with cork insulation and membrane-façade achieves absorption coefficients approaching 0.9 at mid-frequencies (e.g., Fig. 22). If set up in urban environments such as streets, schools or public spaces, this shift from near-total reflection to high absorption.

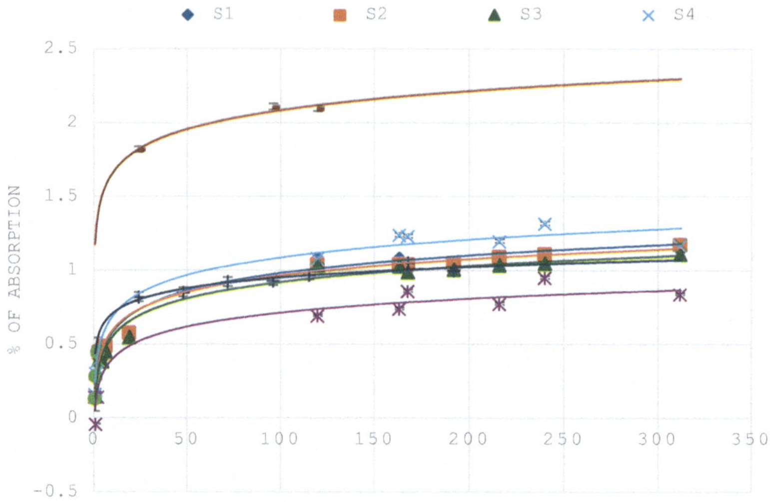

Fig. 21 Percentage of absorption of expanded cork on a 320-hour timeline, on five samples (© coat-ing)

Impacting directly the outdoor comfort for inhabitants. Indeed, road traffic noises, peaks in between 500 and 2000 Hz, precisely the range where cork absorb from 0.7 to 0.9.

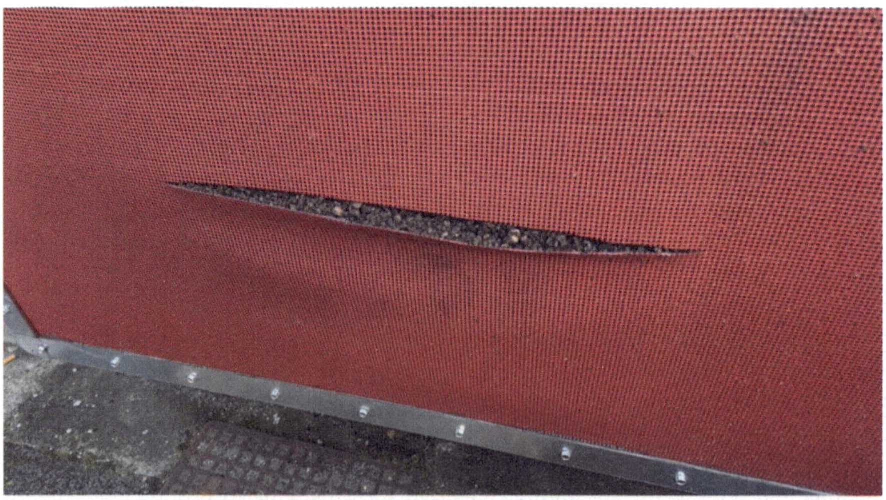

Fig. 22 Photo of a 400 mm cut on the prototype (© coat-ing)

Table 8 Table of the sound transmission for brick wall, bare concrete and concrete with interior lining

Wall materials	R_w (dB)
Brick wall	37
Bare concrete	55
Concrete with interior lining	66

Table 9 Comparative table of the sound transmission R_w [dB] of the different material with polystyrene or cork insulation for 200 mm

	Brick wall	Bare concrete	Concrete with lining
Polystyrene	38.0	55.0	66.0
Non-compacted cork	40.0	58.0	67.0
Compacted cork	44.0	60.0	68.0

5.5 *Acoustic Performance for Sound Insulation*

The three results (e.g., Table 8) confirmed that brick construction is one of the most acoustically vulnerable configurations and therefore the one with the greatest potential for improvement.

Regarding the effect of insulation material, the results were very different regarding the depth, the materials and the compactness.

Polystyrene is negligible insulation on all three wall types, indeed even at 200 mm, the R_w improvement does not exceed 1 dB. Non-compacted cork performs noticeably better, with a gain of +3 dB on the brick base and on bare concrete at 200 mm thickness. Compacted cork, with its significantly higher airflow resistivity, proved to be the best sound insulation: +7 dB on brick and + 5 dB on bare concrete at 200 mm (e.g., Table 9). The ranking compacted cork > non-compacted cork > polystyrene is consistent across all wall configurations and thicknesses.

The effect of thickness is another key finding. Below 200 mm, gains are limited for all three materials regardless of wall type. At 50 and 100 mm, improvements rarely exceed 1–2 dB. Only at 200 mm, a noticeable change appeared, with up to 7 dB attenuation.

5.6 *Fire Performance Test—SBI Test EN 13823*

$$\mathrm{FIGRA}_{0.2}\mathrm{MJ} = 86\ \mathrm{W/s},$$
$$\mathrm{FIGRA}_{0.4}\mathrm{MJ} = 86\ \mathrm{W/s},$$
$$\mathrm{SMOGRA} = 32\ \mathrm{m^2/s^2},$$
$$\mathrm{THR}_{600\mathrm{S}} = 4.1\ \mathrm{MJ},$$
$$\mathrm{TSP}_{600\mathrm{S}} = 170\ \mathrm{m^2}.$$

The test was aborted when the Frontside 381 membrane melted under heat exposure, releasing cork granules onto the sand burner and blocking more than half of it by choking. As stated by RISE, these results cannot be used for product classification or regulatory approval and are reported here as preliminary informatorily data only.

An important distinction must be made when interpreting this result. The Frontside 381 membrane holds a Euroclass B-s2, d0, obtained through testing of the membrane by itself. In the assembled system, however, the membrane is positioned in direct contact with the insulation material. Under heat exposure, the cork may have increased the thermal load on the back face of the membrane, accelerating its degradation.

This highlights a well-documented challenge in façade fire engineering: the fire performance of a composite system cannot be reliably inferred from the sum of its individual component classifications.

Two alternative configurations have been identified for subsequent fire testing. The first substitutes the current facing with a GFM-21 membrane, a glass-fibre composite with 21% open area and a M1 fire classification. The second keeps the Frontside 381 but introduces an air gap and a non-combustible breather membrane (Stamisol A1, Euroclass A1, by Serge Ferrari) between the facing and the cork infill, preventing direct contact at the critical interface identified in the preliminary test. Both configurations are scheduled for spring 2026.

5.7 Moisture and Hygrothermal Behaviour

These results confirm the open porosity and hygroscopicity of loose expanded cork granules. The material can absorb and retain a moderate amount of moisture, without any signs of saturation.

This behavior supports the use of cork granules as an insulation material for coating, while being able to regulate humidity levels and avoid any damages. Additionally, the relatively fast stabilization suggests that cork does not trap moisture excessively and can therefore adapt to fluctuations.

6 Prototypes and Demonstrators

After the validation phase through performance and specific characteristic testing, many demonstrators have been built and are under planning.

In July 2024, a first prototype of 8 m^2 including a window was built in the city of Pont-à-Vendin in North of France.

The frames have been constructed from cut and welded steel sheets. The perforated façade-membrane and cork perfectly reflected the project at this stage. The installation was completed in one day with three people, but the loose cork, with a minimum grain diameter of 3 mm, frequently protruded through the membrane fabric-facade. A minimum grain size of 5 mm for future projects has therefore been preferred. The prototype remained on site for over a year, without any deterioration, changes, or aging to report. Removal took four hours, and the cork is being reused for other prototypes.

Several people expressed concern about the risk of knife attacks. No standard exists for such a test, neither regarding the method nor the expected criteria. Only the EN 3497:2018 + A1:2021 [16] exist, measuring the impact of a stell ball onto the tested specimen. During the dismantling of the prototype, a horizontal incision of 40 cm long. The perforated membrane, despite being slightly taut, does not open, preventing any of the insulating from escaping. When the two sides are pulled apart, a small amount of cork falls out, but stops very quickly; the effect of the relieving arches and the friction with the perforated fabric mean that the weight is transferred to membrane (e.g., Fig. 22).

In January 2026, a training facility has been conjointly open with the contractor Cabre and the company coat-ing, to train teams on the installation and dismantling of the system.

The training centre wall has a more elongated proportion, and the absence of windows means that the span of the perforated membrane-façade system is significantly greater. To keep the deformation within reasonable limits, an intermediate support was needed. To simplify a lengthy process, a steel sinusoidal frame, a rebars for reinforced concrete comprising three longitudinal bars connected by two wires bent into sinusoidal shapes, welded at each intersection, was chosen. A *screw-clamp* is then clipping the perforated membrane to the sinusoidal frame. The *screw-clamp* is the subject of a specific patent.

At the same time multiple methods for installing the perforated membrane-façade system were tested. During that installation, roughly half the weight of the facade was transferred through the perforated membrane-façade system's attachment to the top rail. The number of fasteners was increased consequently.

7 Technology Readiness Level (TRL)

Since the creation of coat-ing in July 2023, the system has been progressing on the technology readiness scale. Many prototypes have been installed and uninstalled for the last couple of years, and a first paid project is undergoing in March 2026. The total installed area of coat-ing so far is around 200 m^2. This stage has allowed each prototype to be validated and modified before the next one, with a clear assessment after each one.

The aluminium profile since end of 2025, are being manufacture in Belgium, confirming the feasability of the system. And fire testing is undergoing in Sweden with the RISE institute. The current installed prototypes and projects are being exposed to real climatic conditions, giving real time feedback on how the different materials react.

Regarding the industrial production, real-building prototypes and demonstrators and a first paid project, the system is reaching a TRL7-8. A few more tests and demonstrators will be needed to fully complete TRL8 in the coming months.

8 Conclusion

This chapter has presented a succession of experimental characterization for an external façade renovation system. Almost no established normative framework, was able to enclose the system. Custom made and dedicated protocols had to be developed, across mechanical, acoustic, hygrothermal, and fire performance domains. The principal findings demonstrate that granular pressure in membrane-enclosed cavities is governed by the *Janssen effect* and dominated by wall roughness rather than cavity depth. That the use of perforated membrane-façade combine with expanded cork enables to have sound absorbing system on the outside and up to − 7 dB sound attenuation in the inside. It also confirmed, how hygrothermal properties will not fragilize the system despite using a perforated membrane directly in contact with the insulation. Moreover, the membrane-cork system attenuates over 78% of exterior wind pressure on the fabric, remaining below the conservative thresholds of EN 1991-1-4.

However, fire performance remains the critical unresolved challenge. The preliminary SBI test identified direct contact between the polymer membrane and the cork infill as the likely cause of accelerated membrane failure, a system-level effect not captured by standalone component certification. Two redesigned configurations have been identified for subsequent testing: a *GFM-21*, by Serge Ferrari, glass-fibre based membrane with M1 fire classification, and a configuration retaining the original facing membrane with a decoupled air gap and a *Stamisol A1*, by Serge Ferrari, Euroclass A1 breather membrane in between the facing and the cork.

Beyond the results, a structural and normative limitation was reached. Existing standards are product-oriented, written to codify known solutions and define minimum acceptable performance for established categories. A prestressed perforated membrane enclosing a granular natural insulation is a fundamentally new system and does not fit any existing category. Therefore, it must define its own performance criteria and demonstrate them through purpose-built testing. This is not exceptional: the absence of harmonized insulation standards for ETICS across Europe illustrates the same gap. Standards describe what already exists. Innovation, by definition, precedes them.

Through iterative prototype deployment, coat-ing is now reaching TRL7-8, with an objective of completing TRL8 by the end of 2026. Completing fire classification, long-term in-situ monitoring, and full-scale wind tunnel testing remain the necessary steps toward regulatory market entry. If the building sector is to accommodate the scale of renovation that European energy policy demands, performance-based regulatory pathways will be essential for systems like this to reach their potential.

References

1. European Parliament and Council of the European Union (2024) Directive EU/2024/1275 on the energy performance of buildings (recast). Official Journal of the European Union L 2024/1275. https://eur-lex.europa.eu/legal-content/EN/TXT/?uri=OJ:L_202401275. Last accessed 10 Apr 2026
2. Energiesprong (2010) Energiesprong Homepage. https://www.energiesprong.org/. Last accessed 10 Apr 2026
3. ISO 13788:2012 Hygrothermal performance of building components and building elements—Internal surface temperature to avoid critical surface humidity and interstitial condensation—Calculation methods. ISO, Geneva
4. EN 1991-1-1:2002 Eurocode 1: Actions on structures—Part 1-1: General actions—Densities, self-weight, imposed loads for buildings. CEN, Brussels
5. EN 1998-1:2004 Eurocode 8: Design of structures for earthquake resistance—Part 1: General rules, seismic actions and rules for buildings. CEN, Brussels
6. EN 1991-1-4:2005 Eurocode 1: Actions on structures—Part 1-4: General actions—Wind actions. CEN, Brussels
7. EN 13501-1:2018 Fire classification of construction products and building elements—Part 1: Classification using data from reaction to fire tests. CEN, Brussels
8. EN 13501-2:2016 Fire classification of construction products and building elements—Part 2: Classification using data from fire resistance tests. CEN, Brussels
9. EN 1363-1:2020 Fire resistance tests—Part 1: General requirements. CEN, Brussels
10. EN ISO 13785-1:2002 Reaction to fire tests for facades—Part 1: Intermediate-scale test. ISO, Geneva
11. EN ISO 13785-2:2002 Reaction to fire tests for facades—Part 2: Large-scale test. ISO, Geneva
12. De La Fuente D, Otero-Huerta E, Morcillo M (2007) Studies of long-term weathering of aluminium in the atmosphere. Corros Sci 49(7):3134–3148. https://doi.org/10.1016/j.corsci.2007.01.006
13. ISO 10534-2:2023 Acoustics—Determination of acoustic properties in impedance tubes—Part 2: Two-microphone technique for normal sound absorption coefficient and normal surface impedance. ISO, Geneva
14. Johnson DL, Koplik J, Dashen R (1987) Theory of dynamic permeability and tortuosity in fluid-saturated porous media. J Fluid Mech 176:379–402. https://doi.org/10.1017/S0022112087000727
15. Jaouen L, Bécot FX (2011) Acoustical characterization of perforated facings. J Acoust Soc Am 129(3):1400–1406. https://doi.org/10.1121/1.3552887
16. EN 13497:2018+A1:2021 Thermal insulation products for building applications—Determination of resistance to impact. CEN, Brussels

Delivery—Built Applications and Case Studies

Stainless Steel Façade Panels in Silicon Valley

Stephen Katz

Abstract Evaluation of façade panel material options requires careful consideration for performance, aesthetics, and site-specific parameters. Stainless steel as a material choice is often not considered due to assumptions regarding initial material and fabrication costs. This is despite the advantages stainless steel can have over more widely used façade panel materials for durability and sustainability aspects. Opportunities and challenges of successfully integrating stainless steel into a project as well as considerations for working with owners, contractors, and consultants are part of the complete picture. This paper examines stainless steel as a primary opaque façade panel for a high-performance office building in Silicon Valley, using the 200 Park project in downtown San Jose, California as a case study. Parametric solar studies drove window-to-wall ratios (WWR) by orientation, and a vertical distribution of embossed stainless steel panels reduced solar heat gain while maintaining daylight and visual clarity. A visual mock-up and glare analysis informed the finish selection to balance diffuse reflectance with brightness and mitigate urban glare. An embodied carbon comparison of three curtain wall variants showed that the stainless-clad system was within ~ 4.2% GWP/m^2 of a baseline version while outperforming aluminum sheet alternatives on the opaque panel components (~10% lower GWP/m^2), supporting stainless steel's durability-driven lifecycle value. The results indicate stainless steel panels can meet facade performance, aesthetic, and resilience goals with manageable carbon tradeoffs when paired with orientation-specific window-to-wall ratios and rigorous finish selection.

Keywords Stainless steel · Life cycle assessment · Details · Performance · Silicon Valley · Solar heat gain control · Parametric modelling · Glare analysis · Embodied carbon · Unitized curtain wall

---

S. Katz (✉)
AIA, LEED AP BD+C, Gensler, Chicago, IL, USA
e-mail: stephen_katz@gensler.com

A. Mejorin et al. (eds.), *Performance-Based Façade Design*, Springer Tracts in Civil Engineering, https://doi.org/10.1007/978-3-032-28835-6_12

1 Introduction

1.1 Theoretical Framework/Background

Project Description

Located in the heart of Silicon Valley, 200 Park Avenue is a 19 story office building of nearly one million square feet. Designed from the inside out, the building form is a flexible container for office uses broken down by carving solar canyons into the mass and introducing daylight deep into the floor plate. Exterior bridges are inserted within the canyons to extend the workplace outdoors and self-shade the building. The warm and inviting building lobby is directly connected to newly redeveloped Park Avenue with a generous bike room and shower facilities as well as additional leasable retail areas, vehicle port- cochere, and other building support infrastructure. In addition to the active ground level, the building contains four levels of subterranean parking and 3 levels of above grade parking (convertible to future office space), a tenant amenity floor consisting of a fitness center and social lounge, rooftop mechanical spaces, and fifteen floors of flexible office space.

As the first new speculatively developed urban office building built in years, the project contributes to the revitalization of downtown San Jose. The project is sited on the redeveloped Park Avenue which connects San Jose's Plaza de Cesar Chavez on the east with the San Jose Performing Arts Center and Guadalupe River Park to the west. The design of the ground plane continues the language of Park Avenue's lush plantings and extends vegetation into building massing notches and even further into the building lobby, taking back the streetscape from vehicles and emphasizing experiences for bikers and pedestrians alike. The project additionally provides a through block pedestrian-oriented Paseo, activated by a commissioned art installation, strengthening connections to the existing urban fabric (Fig. 1).

Fig. 1 Silicon Valley map (OpenStreetMap), San Jose map (Gensler)

Silicon Valley Market Context

Silicon Valley is one of the world's most important hubs for technology, computing and innovation. There are over 2,000 technology companies in this area with a market cap in billions of dollars. The Silicon Valley corridor extends from San Francisco to the north, down the south coast of San Francisco Bay to San Jose. This includes the communities of Redwood City, Menlo Park, Palo Alto, Mountain View, Sunnyvale, Cupertino, and Santa Clara. Instantly recognizable companies such as Apple, Google, Oracle, HP, Adobe, Meta, IBM, Netflix, Cisco, Nvidea, and Intel, either have their headquarters in Silicon Valley or have a major presence and office building footprint. This area is home to several top universities and research institutions including Stanford University, the University of California at Berkeley, Santa Clara University, and NASA's Ames Research Center.

Technology companies in Silicon Valley have been at the forefront of driving changes to office buildings and workplace culture. Open work areas, amenities rich environments, free food, health clubs, ping pong tables, casual attire, and people working outdoors are some of the office building trends which had their start with Silicon Valley technology and start-up companies. As the office building market matured over the years, the emphasis of physical space and outward image building shifted from a somewhat chaotic atmosphere to a highly designed and detailed building with a tightly choreographed environment. The ultimate expression of this could be seen in the Foster + Partners designed Apple Park Ring Building in Cupertino. Technology company offices went from purposely designed disorder to intentionally designed hyper-order with a particular obsession on pure form and detailing (Fig. 2).

The 200 Park project was developed by the Jay Paul Company as a speculative office building starting in late 2018. The Tech sector economy was incredibly strong with Silicon Valley vacancy rates between 5% and 7%. Demand for Class A office space in this area was intense with chronic shortages of top-level buildings. It was in this competitive market that the developer was looking to secure tenants who shared their vision for high quality design and were willing to pay a premium to be in a building which showcased excellence. An amenity rich environment was part of the equation, and 200 Park delivered with major outdoor terraces, a fitness center,

Fig. 2 Facebook offices (Modlar), Apple Park Ring Building (Carles Rabada)

a well-appointed bike room, and indoor parking. 200 Park's finely crafted stainless steel façade panels further reinforce the message of value and durability (Fig. 3).

Stainless Steel Building Typology

To many, stainless steel is considered the ultimate opaque façade material. It combines superior durability, corrosion resistance, hardness, and a favorable strength-to-weight ratio with a unique aesthetic quality not found in other materials. There are numerous stainless steel finishes, textures and grades which can be selected by designers to establish the desired look and feel. As the preeminent metal fabricator Zahner has noted, "Today, stainless steel is synonymous with the modern age. Art conservationists consider it one of the so-called modern metals. Stainless steel's jewel-like

Fig. 3 200 Park street view and elevation (Jason O'Rear/Gensler)

Fig. 4 Chrysler Bldg (Postdif), Inland Steel (Inland Steel), Disney Hall (Carol Highsmith)

luster adorns building facades and roofs on many major structures around the world. Perhaps the most compelling reason to consider using stainless steel is the brightness of the metal. The shine of stainless steel is rich and intense. Unlike the warmth attributed to silver and pewter, stainless steel conveys a sense of modern achievement as a material that can stand the test of time" [1].

Three well known examples of stainless steel façades are the Chrysler Building in New York by William Van Alen from 1930, the Inland Steel Building in Chicago by Bruce Graham/SOM from 1957, and the Walt Disney Concert Hall in Los Angeles by Frank Gehry from 2003. Each architect used stainless steel for different effects, but the designs are connected by the way light interacts with the facades. The Chrysler Building showcased super-sized ornamental details. Inland Steel was a case study in modernist modularity and precision. The Walt Disney Concert Hall used the inherent structural capabilities of stainless steel panels to achieve a flowing form of expression. (Fig. 4).

Stainless Steel Material Properties and Finishes

Stainless Steel Austenitic grades (e.g., 304/316) typically used on façades provide excellent corrosion resistance via a self-healing chromium oxide layer with type 316 being preferred in urban exposures. Mechanical characteristics (elastic modulus ~ 200 GPa; density ~ 8000 kg/m^3) enable relatively thin-gauge cladding with high surface hardness. Finish families span mill (2B), bright annealed (BA), brushed/satin (e.g., No. 4), bead-blasted, and embossed/textured sheets. Texturing increases panel stiffness and mitigates oil-canning while tuning specular and diffuse reflections. Key trade-offs include potential glare from highly specular finishes, fingerprint visibility on very smooth surfaces, and cost premiums offset by reduced maintenance and long service life.

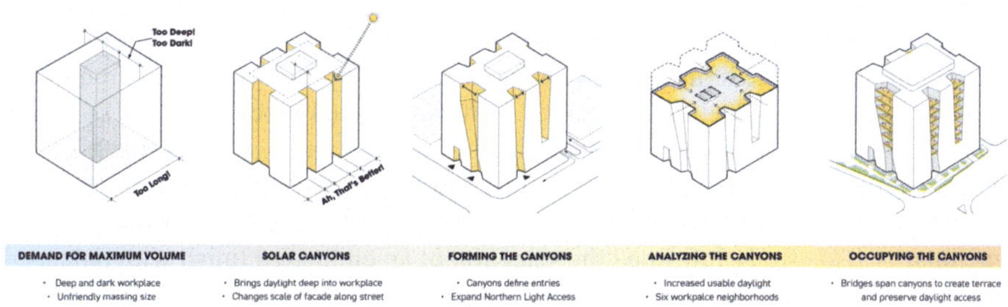

Fig. 5 200 Park design parti (Gensler)

2 Methodology

2.1 Façade Design Goals

The site for the building was very deep and almost square. This geometry resulted in a lease span of + 80' from the core to the perimeter. To break down the scale and provide daylight further into the floorplates, solar canyons were carved into the volumes. The solar canyons were shaped and scaled to respond to the urban context and solar orientations. Building entries were developed at key canyon locations and integrated with sidewalk gardens. Bridges across the canyons were added to provide an extensive amount of outdoor space. Opaque stainless steel clad panels were integrated into the façade panels to reduce solar heat gain. The width and configuration of the panels were optimized on each façade to account for solar orientation and provide a favorable window-to-wall ratio. Glass units within the unitized curtainwall panels were also angled away from direct sunlight. The collective result is a high performing façade. (Fig. 5).

2.2 Façade Study and Analysis Approach

Several important façade studies and areas of analysis were performed to assist the design team with façade material decisions and to understand sustainability impacts for the use of stainless steel panels. A few of these investigations are described below.

Solar and WWR Optimization Procedure: Parametric elevation studies mapped incident solar radiation by orientation; vertical opaque panel widths were iterated ($\approx$12–34″) to calibrate WWR on the south and west façades where solar exposure is greatest. Candidate WWR distributions were checked against energy performance calculations to validate reductions in cooling load while preserving daylight quality.

Glare and Finish Selection Procedure: Full-scale visual mock-ups (VMUs) were reviewed on site under correct solar orientation. A glare study (by Loisos + Ubbelohde) evaluated stainless steel panel reflectance characteristics and urban glare risk using site-specific solar data for downtown San Jose and comparative urban benchmarks. This study informed the selection of an embossed finish with balanced specular and diffuse behavior.

Embodied Carbon Study: In collaboration with the façade contractor, Permasteelisa North America (PNA), three unitized curtain wall variants were defined for GWP/m^2 comparison: Version 1 (As-Built) with stainless steel-clad aluminum fins and stainless steel sheet opaque panels; Version 2 with extruded aluminum fins (no cladding) and aluminum sheet opaque panels; and Version 3 with aluminum sheet cladding at fins and opaque panels. Results are reported in Sect. 3.3 of this paper.

2.3 *Façade Wall Types*

See Fig. 6.

WT-1 Main Unitized Curtain Wall

The majority wall type is designated as WT-1 Main Unitized Curtain Wall and is composed of custom extruded thermally broken aluminum mullions with projected frames and double insulated glass units which are angled in plan. The insulated glass units are structural silicone bonded to the custom frames on four sides. The frame depth with projection is 1'-4″ and the frame height with projection is 3 5/8″. The typical curtain wall unit is 14' high by a typical width of either 5'-8″ wide or 4'-4″ wide. Each unit is hung from the slab above and spans one floor. The blades or fins which form the projections are custom extruded aluminum shapes and are clad with 1 mm thick stainless steel sheet. The opaque horizontal spandrel and opaque vertical panels are composed of custom extruded aluminum frames clad with 1.5 mm thick stainless steel sheet. These panels are insulated with 50 mm of mineral wool insulation at the outer panels and 3″ of foil faced fiber insulation on the inboard side.

The WT-1 curtain wall panel overall *U*-value is 0.39 Btu/h*ft2*F with an SHGC of 0.267 Btu/h*ft2*F. WT-1 uses glass type GL-01 which is an AGC/Interpane IGU composed of all low iron extra clear substrates with a laminated outer lite. The outer laminated lite is 5/8″ thick and is composed of a 5/16″ (8 mm) low iron substrate, a 0.060″ (1.52 mm) clear acoustic PVB interlayer, and a 1/4″ (6 mm) low iron inner lite with an Interpane Stopray Vision 50 low emissivity coating on the #4 surface. A 5/8″ (16 mm) argon gas filled space separates a 1/4″ (6 mm) low iron fully tempered, heat soak tested inner lite. This overall composition yields a Visible Light Transmittance of 52%, a *U*-value Winter of 0.25 Btu/h*ft2*F, a SHGC of 0.27 and an Exterior Reflectance Value of 19%. This coating was specifically chosen for the balance between the correct visible light factor for an office space, the low exterior reflectivity for an urban environment, and the high energy performance values (Fig. 6).

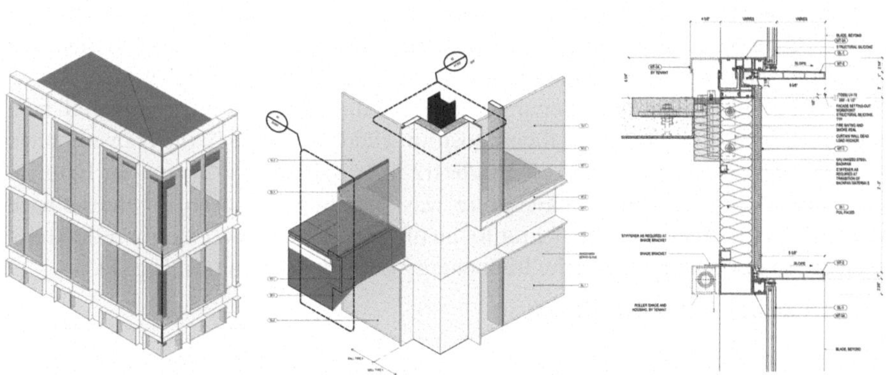

Fig. 6 200 Park WT-1 details (Gensler)

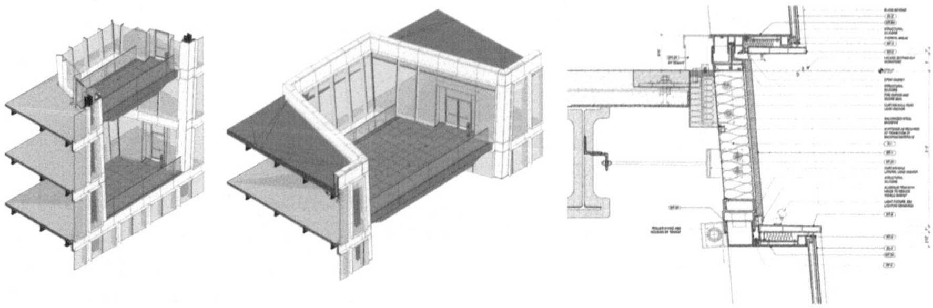

Fig. 7 200 Park WT-2 details (Gensler)

WT-2 Solar Canyon Unitized Curtain Wall

The Solar Canyons wall type is designated as WT-2 Solar Canyon Unitized Curtain Wall. This unitized system features inclined curtain wall units which intersect with outdoor bridge terraces. The system is composed of custom extruded thermally broken aluminum mullions with projected frames and double insulated glass units which are angled in plan and section. The insulated glass units are structural silicone bonded to the custom frames on four sides. The frame depth with projection is 1'-4″ and the frame height with projection is 3 5/8″. The typical curtain wall unit is 14' high by a typical width of either 5'-8″ wide or 4'-4″ wide. Each unit is hung from the slab above and spans one floor. The blades or fins which form the projections are custom extruded aluminum shapes and are clad with 1 mm thick stainless steel sheet. The opaque horizontal spandrel and opaque vertical panels are composed of custom extruded aluminum frames clad with 1.5 mm thick stainless steel sheet. These panels are insulated with 50 mm of mineral wool insulation at the outer panels and 3″ of foil faced fiber insulation on the inboard side.

The WT-2 curtain wall panel overall U-value is 0.39 Btu/h*ft2*F with an SHGC of 0.267 Btu/h*ft2*F. WT-2 uses glass type GL-02 which is an AGC/Interpane IGU composed of all low iron extra clear substrates with a laminated outer lite. The outer laminated lite is 5/8″ thick and is composed of a 5/16″ (8 mm) low iron substrate, a 0.060″ (1.52 mm) clear acoustic PVB interlayer, and a 1/4″ (6 mm) low iron inner lite with an Interpane Stopray Vision 50 low emissivity coating on the #4 surface. A 5/8″ (16 mm) argon gas filled space separates a 1/4″ (6 mm) low iron fully tempered, heat soak tested inner lite. This overall composition yields a Visible Light Transmittance of 52%, a U-value Winter of 0.25 Btu/h*ft2*F, a SHGC of 0.27 and an Exterior Reflectance Value of 19%. This coating was specifically chosen for the balance between the correct visible light factor for an office space, the low exterior reflectivity for an urban environment, and the high energy performance values (Fig. 7).

WT-3A Ground Floor Curtain Wall

The majority grade level wall type is designated as WT-3A Ground Floor Unitized Curtain Wall and is composed of custom extruded thermally broken aluminum mullions with projected frames and double insulated glass units which are angled in

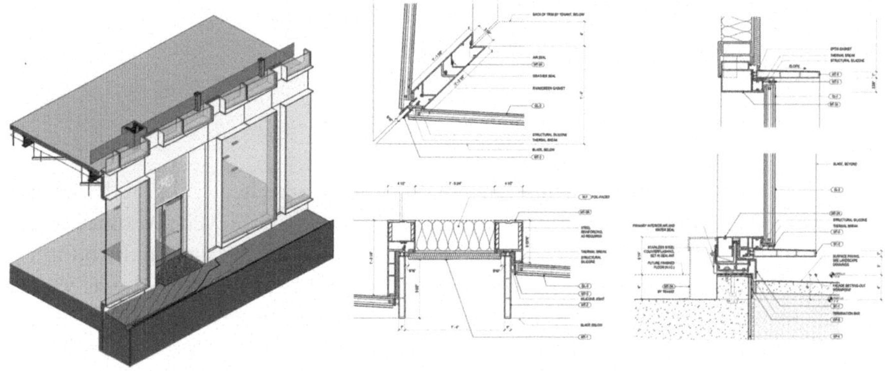

Fig. 8 200 Park WT-3 details (Gensler)

plan. The insulated glass units are structural silicone bonded to the custom frames on four sides. The frame depth with projection is 1’-4″ and the frame height with projection is 5 1/4″. The typical curtain wall unit is either 15’-8″ or 17’-8″ high by 10’ wide. Each unit is base loaded to a starter sill and curb at the ground floor slab and out-of-plane restrained through a head receptor connected from a hung unit above. Extruded aluminum bars within the extruded members provide additional support. The blades or fins which form the projections are custom extruded aluminum shapes and are clad with 1 mm thick stainless steel sheet. The opaque horizontal spandrel and opaque vertical panels are composed of custom extruded aluminum frames clad with 1.5 mm thick stainless steel sheet. These panels are insulated with 50 mm of mineral wool insulation at the outer panels and 3″ of foil faced fiber insulation on the inboard side.

The WT-3A curtain wall panel overall *U*-value is 0.39 Btu/h*ft2*F with an SHGC of 0.53 Btu/h*ft2*F. WT-3A uses glass type GL-03 which is an AGC/Interpane IGU composed of all low iron extra clear substrates with laminated outer and inner lites. The outer laminated lite is 11/16″ thick and is composed of a 5/16″ (8 mm) low iron substrate, a 0.060″ (1.52 mm) clear acoustic PVB interlayer, and a 5/16″ (8 mm) low iron inner lite with an Interpane I-Plus 1.1 low emissivity coating on the #4 surface. A 5/8″ (16 mm) argon gas filled space separates the inner laminated lite which is 1/2″ thick. This laminated inner lite is composed of a 1/6″ (6 mm) low iron substrate, a 0.060″ (1.52 mm) clear acoustic PVB interlayer, and a 1/4″ (6 mm) low iron inner lite. This overall composition yields a Visible Light Transmittance of 83%, a *U*-value Winter of 0.25 Btu/h*ft2*F, a SHGC of 0.55 and an Exterior Reflectance Value of 12%. This coating was specifically chosen for the balance between the correct visible light factor for retail spaces, the low exterior reflectivity for an urban environment and for retail spaces, and the high energy performance values (Fig. 8).

3 Working with Stainless Steel

3.1 Stainless Steel Panel Distribution

See Figs. 9, 10 and 11

One of the goals for the design team was to control solar heat gain on the façade based on each elevation's unique solar orientation. This was achieved through a series of parametric modelling exercises where solar intensity was mapped on the building elevations. The width of opaque vertical stainless steel clad wall panel was varied from 12″ to 32″ based on the need for that elevation to reduce the amount of vision

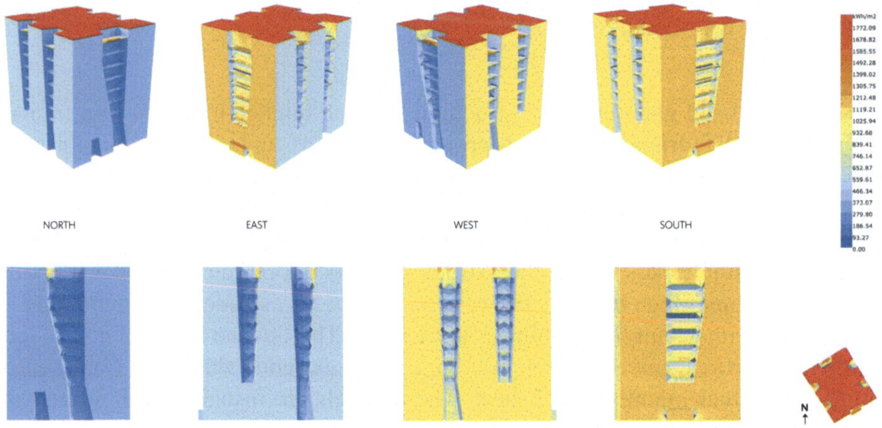

Fig. 9 200 Park solar analyses (Gensler)

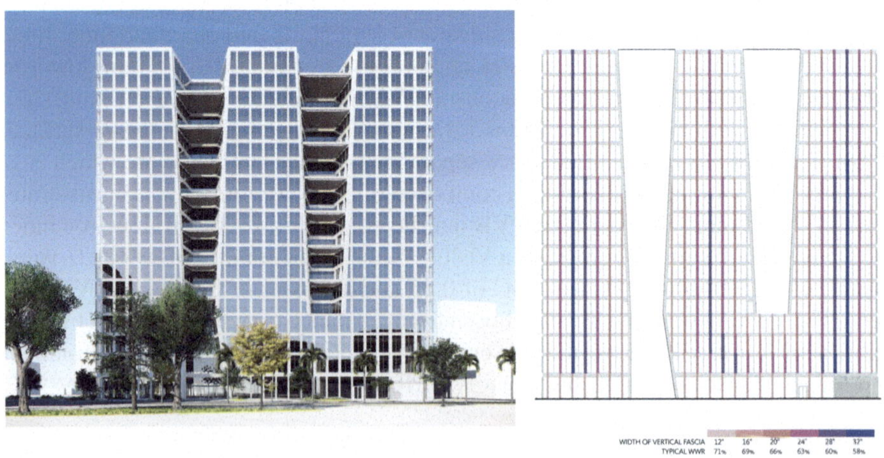

Fig. 10 200 Park WWR distribution (Gensler)

glass to control solar heat gain. The sum of the opaque panel area was then massaged to achieve a favorable window-to-wall ratio. For example, the south and west facades receive more intense solar radiation for longer periods of time. The façade design incorporated more of the stainless steel clad opaque panels on these elevations which had better thermal performance due to the high level of insulation compared to the vision glass units (Figs. 10 and 11).

3.2 Stainless Steel Panel Selection

Many factors went into the selection of the stainless steel panel product which involved reviewing historic precedents, numerous material samples, detailing considerations, durability, and sustainability benchmarks. From an aesthetic perspective, the design team felt a balance needed to be found regarding how reflective versus how diffuse the stainless steel finish would be. An overly reflective product would result in a "shiny" looking building. Besides aesthetic detriments to the image of the building, this could lead to unwanted solar glare issues in an urban environment impacting vehicles, airplanes, and pedestrians. Too dull of a finish and the building might appear dirty, patinated, or have an aged industrial feel. A favored option was found which combined enough reflectivity with diffuse properties which was an embossed stainless steel product. Embossing is produced by a fabricator sending the stainless steel coil into a rolling mill machine which applies a textured or raised pattern. The raised texture pattern results in both reflective surface areas for the top level as well and more diffused light zones for the side and lowered surfaces.

The stainless product used at 200 Park was manufactured by Outokumpu, a fabricator based in Helsinki, Finland. The design team chose stainless steel grade 316 alloy for durability and corrosion resistance given the urban environment. A specific embossed pattern was chosen after reviewing several options as part of a large-scale visual mockup or VMU. The VMU incorporated several embossed finish options and grain direction variations and was viewed on the project site with correct solar orientation. The selected embossed product was Outokumpu Deco Linen without Slubs M28 with a vertical grain direction. This pattern is a tight, repeating series of alternating dots and dashes. The embossment pattern itself is not visible unless observed from a close distance but the effect of both light reflection and diffusion is very dynamic in real world light conditions both during the day and at night. There is an additional warmth to the stainless steel texture during sunrise and sunset.

To better understand the potential for glare from the stainless steel panels, an unwanted solar glare study was conducted by the firm Loisos + Ubbelohde. This study reviewed the stainless steel panel properties, the solar glare from the VMU, and compared these findings against benchmark solar glare data in similar urban areas such as Los Angeles and San Francisco. Site specific solar data from downtown San Jose were additionally factored into the analysis (Figs. 12 and 13).

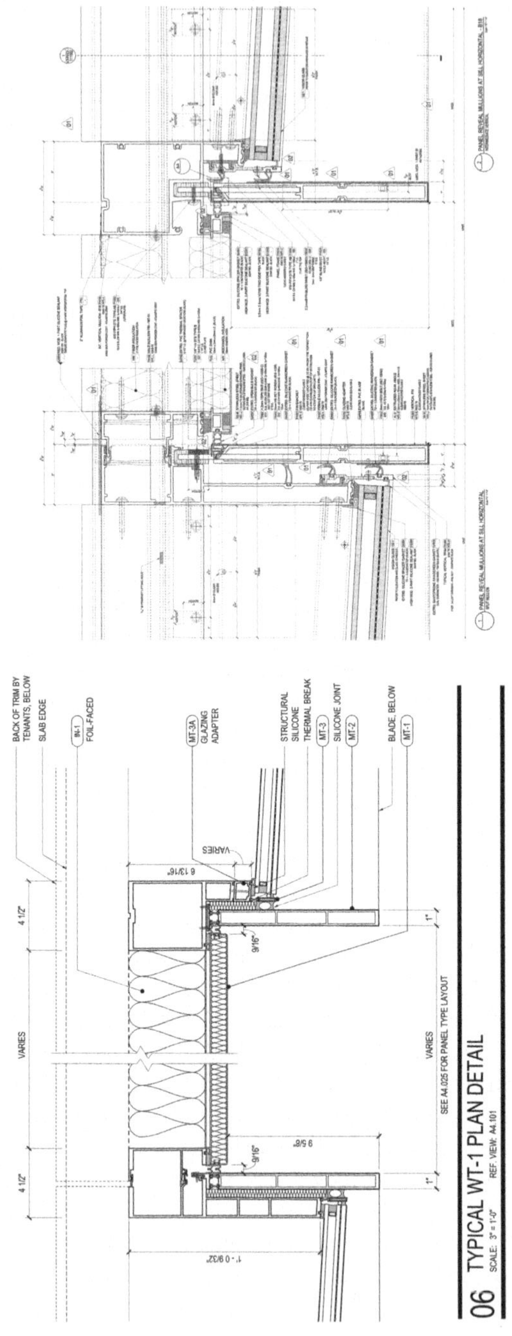

Fig. 11 WT-1 plan details architectural and PNA shop drawing (Gensler, PNA)

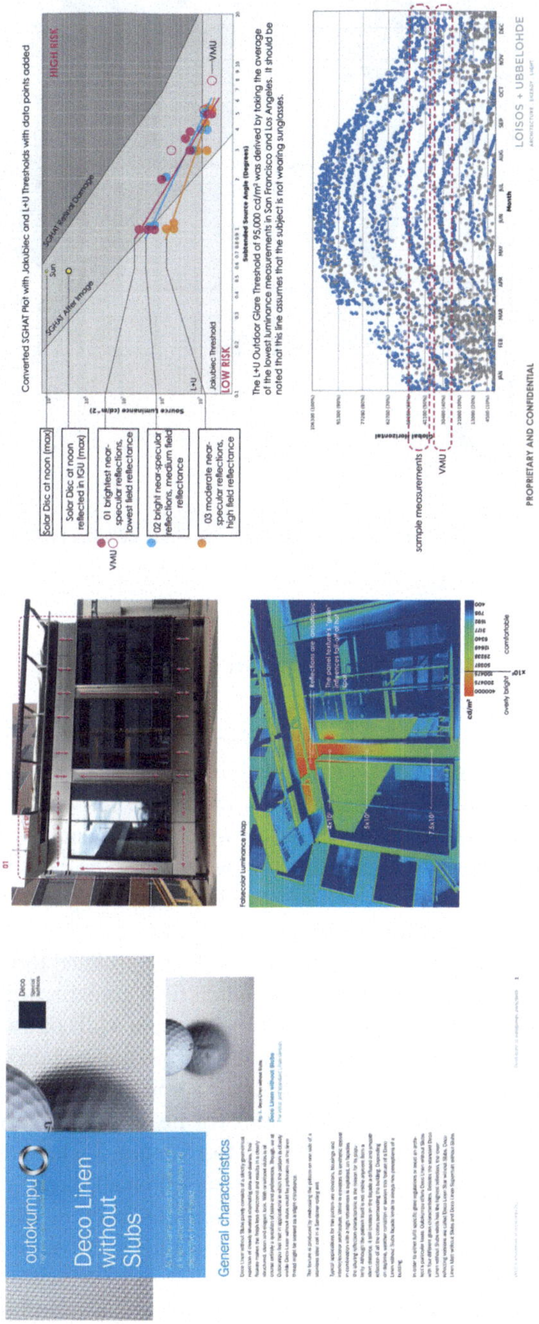

Fig. 12 Stainless data, VMU, Solar Study (Outokumpu, Gensler, Loisos + Ubbelohde)

Fig. 13 200 Park entrance view (Jason O'Rear/Gensler)

3.3 Curtain Wall Unit and Stainless Steel Embodied Carbon

One of the most pressing areas of study for façade design is that of embodied carbon calculations. While great advancements have been made for the calculation and reduction of operational carbon in buildings, embodied carbon calculations and savings strategies are still a work in progress for much of the enclosure design and manufacturing industry. This is partially due to how difficult the calculations can be given the variables which exist for a given unitized curtain wall panel. For example, the amount of recycled content for aluminum, steel, glass and other materials can vary depending on procurement sources and cycles. Energy related to shipping costs can also vary from when the project was bid to when final construction occurs.

The design team worked closely with the 200 Park curtain wall contractor, Permasteelisa North America (PNA), to study the amount of embodied carbon within typical curtain wall units. Three versions were analysed for comparison; Version 1 As Built condition with stainless steel clad aluminum fins (or projections), Version 2 where the fins are extruded aluminum with no cladding, and Version 3 where the fins would be clad with sheet aluminum in lieu of stainless steel. For all three versions, there is still an opaque metal panel (horizontal and vertical) factored in which is sperate from the fin cladding. The results indicate a GWP/m2 for each version as follows: (Table 1).

The comparison of Version 1 to Version 2 is the closest way to understand the carbon impact of cladding this curtain wall system with stainless steel both as opaque

Table 1 Version GWP calculations

Version 1	As Built	322.8	GWP/m^2
Version 2	Alum panel – No fin clad	309.4	GWP/m^2
Version 3	Alum panel + Alum fin clad	325.8	GWP/m^2
V1 versus V2	Delta between versions 1 and 2	13.4	GWP/m^2 (4.16%)

panels and extruded fin cladding. In the Version 2 scenario the extruded aluminum fins would only require a protective coating, not an aluminum or stainless steel cladding. It is worth noting that comparing sheet aluminum panels to sheet stainless panels not on the fins yields a 9.96% GWP/m^2 carbon savings (Fig. 14).

3.4 Discussion

The use of stainless steel façade panels at 200 Park can also be evaluated through the lens of long-term asset management and operational resilience. Beyond first cost and embodied carbon, owners increasingly evaluate facades based on durability, adaptability, and lifecycle predictability. Stainless steel's resistance to corrosion, UV degradation, and surface oxidation allows it to maintain appearance and performance over decades with minimal intervention. Unlike painted aluminum or coated steel systems that require periodic recoating, stainless steel panels rely on their inherent material properties rather than applied finishes. This reduces both maintenance frequency and the risk of aesthetic failure due to coating delamination, fading, or chalking.

From a performance standpoint, stainless steel façade cladding supports improved moisture management with flatter and more consistent panel surfaces which permits proper weeping and water shedding. The embossed pattern increases surface stiffness and helps reduce oil-canning while simultaneously allowing micro shape variations to be visually absorbed. This characteristic improves tolerance to fabrication and installation variables, which is critical in large-scale unitized systems where thousands of panels must visually align. In practice, this allowed the project team to maintain tighter quality control without imposing unrealistic fabrication tolerances which would have driven cost increases.

The integration of stainless steel panels into the façade design enabled a high degree of formal clarity. The verticality of the façade panels reinforces the canyon concept while visually elongating the building. The material's subtle reflectivity responds to changing daylight conditions, reinforcing the parametric intent behind the facade's solar-driven logic. This creates a direct relationship between performance analysis and architectural expression, where data-driven decisions become legible in the built form.

The design team leveraged stainless steel's recyclability to support circular economy goals. Stainless steel is one of the most recycled construction materials

Version 1 **AS BUILT**

Category	Global warming (TRACI) kg CO2e
Aluminum Frame	2086.7
Glazing	821.3
Aluminum Fin	552.2
SS Fin Clad	192.9
SS Panel	165.7
Factory	156.3
Silicone	47.5
PVC	46.8
Gasket	40.1
Other classifications	87.1
GWP total	4196.5
GWP/m2	**322.8**

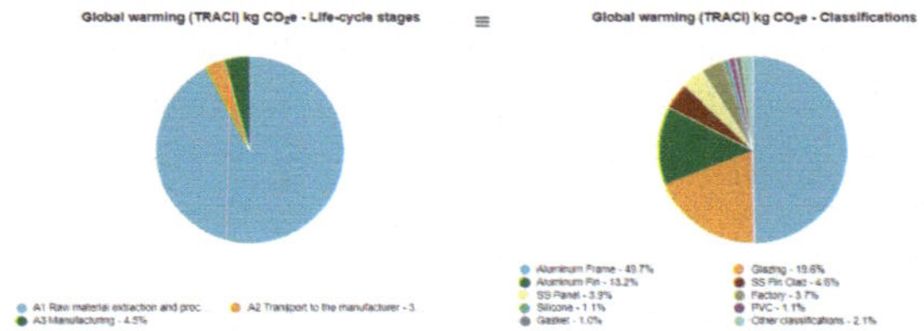

Version 2 **Aluminum Panel - No fin Clad**

Category	Global warming (TRACI) kg CO2e
Aluminum Frame	2086.72363
Glazing	821.3259903
Aluminum Fin	552.168784
Aluminum sheet	183.9791071
Factory	156.2643727
Silicone	47.50069486
PVC	46.78993859
Gasket	40.11279312
Insulation	30.32914496
Other classifications	56.74380451
GWP total	4021.93826
GWP/m2	**309.4**

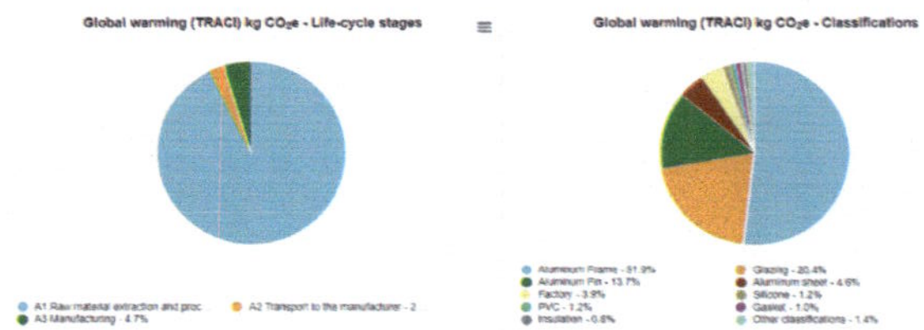

Version 3 **Aluminum Panel + Aluminum Fin Clad**

Category	Global warming (TRACI) kg CO2e
Aluminum Frame	2086.72363
Glazing	821.3259903
Aluminum Fin	552.168784
Aluminum sheet	397.8025139
Factory	156.2643727
Silicone	47.50069486
PVC	46.78993859
Gasket	40.11279312
Insulation	30.32914496
Other classifications	56.74380451
GWP total	4235.761667
GWP/m2	**325.8**

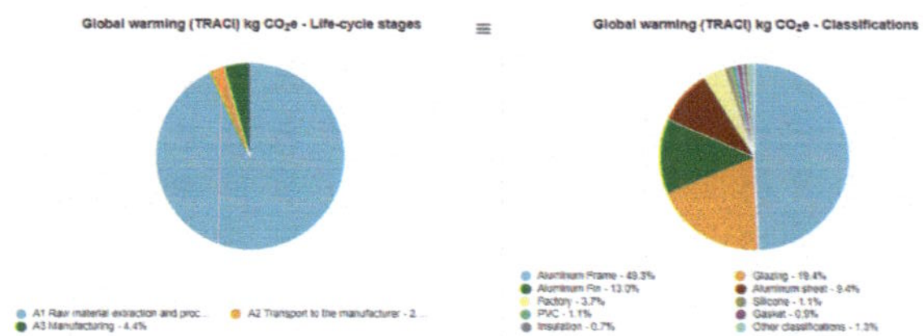

Version 1 SS Fin Clad	192.9	
Version 1 SS Panel	165.7	
Version 1 All SS sheet	358.6	
Version 3 Alum sheet on Fin only	213.8	
Version 3 Alum sheet not on Fin	184.0	
Version 3 All Alum sheet	397.8	
Fin Alum to SS delta	20.9	**9.78%**
Sheet Alum to SS delta	18.3	**9.96%**
All Alum to SS delta	39.2	**9.86%**

Version 1 AS Built	322.8	
Version 2 Aluminum Panel - No fin Clad	309.4	
Version 3 Aluminum Panel + Aluminum Fin Clad	325.8	
V1 vs V2 Delta Between Versions 1 and 2	13.4	**4.16%**

Fig. 14 200 Park embodied carbon table (PNA, Gensler)

globally, with typical recycled content levels exceeding 60% depending on alloy and sourcing. At end of life, the material retains high scrap value, which increases the likelihood that it will be recovered and reused rather than landfilled. This aligns with emerging ESG frameworks and tenant-driven sustainability metrics that go beyond operational energy use. In the context of future adaptability, the façade system was detailed to allow for panel replacement without dismantling adjacent units. This approach supports long-term flexibility if interior program needs change or if localized façade upgrades are required. Stainless steel's dimensional stability and resistance to creep support this modular logic, ensuring that replacement panels will remain visually consistent with original installations.

From a procurement perspective, early engagement with the fabricator and facade contractor was essential. Mockups, finish samples, and glare studies were not treated as isolated exercises but as integrated decision-making tools. This allowed the project team to align aesthetic intent, performance criteria, and construction feasibility. The success of the stainless steel facade at 200 Park is therefore as much a product of process as of material selection.

Additional design and technical issues addressed include thermal movement, attachment strategy, and tolerance management within the unitized system. Stainless steel panels experience different coefficients of thermal expansion than the aluminum subframes to which they are attached. To accommodate this, the panel attachment details were designed with slotted connections and isolating pads that allow controlled movement while maintaining panel alignment and weather tightness. This approach minimizes stress concentrations at fasteners and reduces the risk of long-term deformation, noise, or fastener fatigue under daily and seasonal temperature cycles.

Another critical aspect was the interface between stainless steel panels and adjacent glazing components. Careful coordination was required to maintain continuous air and water barriers while allowing for differential movement between opaque and transparent elements. Backpans, gaskets, and sealants were specified to ensure compatibility with stainless steel surfaces and to avoid galvanic corrosion at material transitions. Detailing also considered cleanability and access, allowing façade maintenance teams to wash and inspect panel joints without disrupting adjacent systems.

Wind loading and panel stiffness were evaluated in detail. The embossed stainless steel panels were tested for deflection limits under design wind pressures to ensure compliance with serviceability criteria and to avoid visual distortion. Panel thickness, emboss depth, and subframe spacing were coordinated to achieve a balance between structural performance and material efficiency. This optimization reduced unnecessary material use while maintaining a high-quality appearance.

Constructability and sequencing influenced design decisions. Stainless steel panels were prefabricated and installed as part of the unitized curtain wall modules, reducing on-site labor and improving quality control. The design team worked closely with the facade contractor to resolve tolerances between slabs, anchors, and curtain wall units, ensuring that stainless steel elements remained visually aligned across floor transitions. These technical considerations demonstrate how material selection, detailing, and construction logistics must be integrated to successfully deliver a high-performance stainless steel facade system.

The stainless steel panels at 200 Park demonstrate how advanced facade design can merge sustainability, performance, and architectural expression into a coherent system. Rather than treating opaque panels as secondary elements to glass, the project elevates them to active contributors to energy control, urban identity, and long-term value. This approach suggests a broader role for stainless steel in future high-performance office buildings, particularly in climates and markets where durability, adaptability, and visual distinction are critical drivers.

3.5 *Technology Readiness Level (TRL) and Market Status*

TRL Assessment: Embossed stainless steel sheet as cladding within unitized curtain wall systems is a mature, widely deployed technology for high-rise façades. The TRL level for this project is therefore understood to be TRL8 Actual System Completed and TRL 9 Actual System Proven in Use. The supply chain (global mills and regional fabricators) supports consistent coil-to-panel quality, PMU testing, and integration with standard aluminum unitized frames.

Market Status and Product Notes: Project-proven finishes include embossed and textured sheets which temper specular reflection; grade selection typically favors 316 in urban exposures. Visual mock-ups and glare studies are considered best practice prior to final selection. For this case, Outokumpu's Deco Linen M28 with vertical grain provided the target balance of diffusion and brightness.

Relevant Wall Types: For stainless-centric scope, the primary reference assemblies for this project are WT-1 (Main Unitized Curtain Wall with stainless-clad fins and stainless sheet opaque panels), WT-2 (Solar Canyon units with inclined geometry and the same stainless components) and WT-3A Ground Floor Curtain Wall (also with the same stainless components).

4 Conclusions

The selection of façade materials has traditionally been rooted in consideration of performance values such as air/water resistance, thermal conductivity, material structural properties, and connection details to name a few. A new set of parameters has emerged which force Architects and Engineers to understand the degree to which the materials may impact whole project carbon calculations in the form of embodied carbon. One of the major questions to answer is will the impact on embodied carbon for a given material reduce the operational carbon footprint of the building enough to justify the façade design. Factors such as durability, source material recycling, and manufacturing and transportation of the façade elements requires study. Such analyses are complicated but becoming more accessible as key partnerships have shown. Social, economic, and historic contexts must be considered when looking at which façade materials represent and enhance the overall architectural design.

Key Findings:

(1) Orientation-specific WWR strategies paired with stainless steel opaque panels effectively reduce solar heat gain while preserving daylight and urban image quality.
(2) An embossed 316 stainless steel panel finish can mitigate glare risk and visually absorb minor flatness variations.
(3) Embodied carbon impacts for stainless-clad variants can be within ~ 4–5% of unclad baselines at the unitized panel level, while opaque sheet comparisons favor stainless by ~ 10% depending on sourcing and thickness.
(4) Lifecycle durability and low maintenance strengthen stainless steel's value proposition in Silicon Valley's quality-driven market.

Recommendations and Guidelines:

Establish WWR targets by orientation studies early; require VMU with finish and grain direction trials; commission a project-specific glare assessment; coordinate slotted/isolated attachments for thermal movement; specify compatible sealants and backpans; and plan for replaceable panels within unitized modules to support future adaptability.

Future Work and Outlook:

Future research should deepen comparative embodied carbon datasets for stainless versus aluminum systems at component resolution (sheet gauges, recycled content, transport scenarios), expand simulation-to-field validation of glare behavior across sky conditions, and quantify maintenance lifecycle benefits in total cost of ownership models. Further parametric multi-objective optimization (energy, carbon, glare, cost, and image metrics) can formalize trade-offs and support repeatable decision frameworks for Silicon Valley and similar markets (Fig. 15).

Fig. 15 200 Park street view from above (Jason O'Rear/Gensler)

Acknowledgements Select Project Team
Owner and Client: Jay Paul Company
General Contractor: Level 10 Construction
Architect: Gensler
Façade Engineering Consultant: Eckersley O'Callaghan
Façade Contractor: Permasteelisa North America
Façade Access: CS Caulkins
Select Façade Engineering: Ficcadenti Waggoner and Castle
Waterproofing: RDH
Structural Engineer: Magnusson Klemencic Associates
Wind Engineering: RWDI
Mechanical, Electrical, Plumbing, Fire Protection Engineer: PAE, Meyers + Engineers
Landscape Architect: Field Operations
Civil Engineer: Kier + Wright
Lighting Design: HLB
Select Glare: Loisos + Ubbelohde

Reference

1. Zahner, Stainless steel guide part 1: design, fabrication & building applications. https://www.azahner.com/blog/stainless-steel-king-of-metals

Performance and Place: Innovations in Arctic Façade Design

Peter Dushenski

Abstract Designing façades for the Canadian Arctic means succeeding at the intersection of extreme environmental conditions and the unique cultural priorities that shape daily life in the North. In Iqaluit, where winter temperatures regularly approach − 50 °C or colder, and construction materials arrive only during a short two-month sealift window in late summer, building envelopes must sustain long-duration cold, intense winds, and tightly sequenced logistics while still accommodating the daylight quality and outward orientation integral to life in Nunavut. Earlier institutional buildings often responded with inward-looking and highly insulated forms, trading transparency for thermal stability and limiting connection to the surrounding landscape. Next-generation façade engineering offers new ways to address these constraints. At Nunavut Arctic College Nunatta Campus, an academic and cultural hub for students traveling from communities across the territory, a fibreglass-framed curtain wall system forms the basis of a glazing strategy that combines triple glazing with insulated translucent units filled with aerogel. This mix achieves a 27.8% glazing-to-mass ratio, with the aerogel sections providing centre-of-glass U-values of 0.32 W/m^{2}K (R-18) and delivering diffuse, full-spectrum light during the low-sun winter months. This approach stabilizes interior comfort, mitigates glare, and preserves meaningful views toward Koojesse Inlet, aligning spatial openness with occupant well-being. This chapter examines the climatic, cultural, and logistical forces shaping the project's façade, including the thermal behavior of materials in deep cold, condensation control, and the reliability afforded by domestic fabrication. Situating Nunatta Campus within a broader context of circumpolar architecture, we demonstrate how coordinated high-performance systems can create durable, daylight-rich environments grounded in northern values and the practical realities of remote construction.

Keywords Arctic architecture · façade design · Fibreglass framing · Daylighting · Embodied carbon · Cold-climate engineering · Indigenous design context · Aerogel glazing · Supply-chain resilience · Curtain wall systems

---

P. Dushenski (✉)
GlasCurtain inc., Edmonton, AB, Canada
e-mail: peter@glascurtain.ca

A. Mejorin et al. (eds.), *Performance-Based Façade Design*, Springer Tracts in Civil Engineering, https://doi.org/10.1007/978-3-032-28835-6_13

1 Introduction

Façade design in the Arctic must reconcile two demands long considered incompatible: exceptional thermal resilience and a parallel need for buildings that remain open to light, landscape, and cultural experience. In communities across Nunavut, prolonged winter darkness, sustained temperatures below − 30 °C, powerful winds, and the constraints of remote construction complicate every design decision for building envelopes [1–4]. Under these conditions, building science is not a secondary or even tertiary concern, but rather the primary technical boundary within which architecture operates.

Earlier generations of northern buildings responded to these pressures with highly opaque envelopes, modest glazing ratios, and compact forms that privileged protection from the elements over openness to nature. While technically robust, these strategies often produced interiors that felt disconnected from their surroundings and spatially constrained, especially for occupants of First Nations and Inuit heritage accustomed to open horizons and daily engagement with the land [5, 6]. Today, the environmental realities shaping technical architecture remain unchanged, but the materials available to counter them have evolved.

Fiberglass-framed curtain walls, improved thermal breaks, warm-edge spacers, high-performance triple glazing, and translucent aerogel units now offer avenues for increasing glazing ratios without compromising condensation safety, surface-temperature stability, or long-term durability [7–12]. These tools do not completely eliminate all risk—Arctic conditions still impose harsh constraints—but they introduce a broader design space for occupant experience and cultural expression.

The Nunavut Arctic College Nunatta Campus expansion exemplifies this next-generation opportunity. Located in Iqaluit, the territorial capital and a hub for education, governance, and community life, the building serves students who arrive from across Nunavut, many from small coastal or inland communities where the land provides orientation, memory, and continuity [19]. In this context, the quality of interior space, its light, openness, and ability to sustain a visual relationship with the surrounding environment carries weight far beyond mere aesthetics.

Teeple Architects and Cibinel Architects approached the project as both a technical and cultural undertaking. Their façade strategy interlaced clear triple glazing with translucent aerogel units, aligned glazing placement with view corridors and daylight characteristics, and leveraged the thermal stability of fibreglass framing to support a higher window-to-wall ratio than typically seen in Arctic institutional buildings [6, 13]. The result is an envelope that resists extreme cold while remaining spatially and visually connected to the landscape.

This chapter analyzes the scientific, architectural, and logistical considerations underlying such an approach. It begins by outlining the environmental conditions that govern envelope performance in circumpolar climates, then situates the project within the cultural and geographic context of Iqaluit and Nunavut Arctic College. The discussion then turns to the material properties of fibreglass-framed systems, triple glazing, and aerogel daylighting assemblies, alongside issues of condensation control and supply-chain reliability. The chapter culminates in a detailed examination of the Nunatta Campus façade and concludes with broader reflections on the implications of this work for high-performance architecture in the Canadian North.

2 Environmental Framework for Arctic Envelope Design

2.1 Climatic Extremes and Their Architectural Consequences

Designing façades for the eastern Arctic requires careful engagement with a climate that pushes material systems to their limits. Iqaluit experiences some of the coldest sustained winter temperatures in North America, with ambient conditions commonly between − 25 and − 35 °C from December through March, punctuated by extreme lows approaching − 50 °C [1–4]. These temperatures interact with high relative humidity, rapid temperature swings, and wind events that can reach 80–120 km/h, creating a façade environment defined by thermal shock, persistent heat loss, and complex loading patterns.

For the building envelope, these conditions translate into three principal technical challenges:

1. Maintaining surface temperatures above the interior dew point to prevent condensation and subsequent ice formation on glazing and framing surfaces [7–10].
2. Limiting thermal bridging, particularly at structural penetrations, fasteners, and frame-to-wall interfaces, where sub-zero exterior metals can rapidly draw heat from interior assemblies [9–11].
3. Preserving airtightness against prolonged negative pressures driven by wind and stack effect, which become magnified in taller or highly ventilated interior spaces [3, 4].

In the past, these demands discouraged highly transparent or even moderately glazed façades. Small punched openings were easier to detail, easier to heat, and less prone to interstitial failure. But such approaches had spatial consequences: interior rooms often felt inward-facing, artificially lit, and visually separated from the environment outside. In a more equatorial setting, this separation might be incidental. In the Arctic, where the horizon is a primary orienting feature and daylight is scarce for much of the year, the consequences are more pronounced in terms of everyday experience.

The physical climate therefore sets strict thresholds: façades must keep interior surfaces warm enough to avoid condensation under high indoor relative humidity; they must resist cold-induced contraction and pressure-driven leakage; and they must be buildable within the short seasonal window allowed by the sealift-based logistics of Iqaluit's construction economy [1–4]. Any façade strategy, transparent or opaque, must begin here.

Fig. 1 Comparison of conventional low-glazing buildings in Nunavut (left) and the higher glazing ratios achieved at Nunatta Campus (right). Right photograph by Julie Jira, reproduced with permission

Figure 1 illustrates this contrast showing earlier low-glazing typologies compared to the larger, more deliberate apertures employed at Nunatta Campus.

2.2 Seasonal Light Patterns and Their Implications for Interior Environments

Daylight in Iqaluit changes dramatically depending upon the time of year, ranging from roughly four hours in late December to continuous illumination in June and July [5]. For an educational building, this seasonal amplitude is a performance condition that profoundly shapes how the interior environment functions.

Winter daylight sits low on the horizon throughout the school day. It moves horizontally across snow-covered terrain, producing long shadows, strong ground-reflection, and rapidly shifting luminance as cloud cover and blowing snow evolve across the tundra landscape. These dynamics influence how much useful daylight reaches the façade, how it enters interior spaces, and how occupants experience visual comfort.

From a performance standpoint:

- Glare increases when solar elevation remains low for the entire day, making direct-beam penetration more difficult to control [5, 9].

- Circadian-supportive daylight is limited to concentrated pockets of diffuse brightness produced by snow reflectance or overcast sky conditions [5, 9].
- Visual comfort relies on reducing interior contrast ratios so occupants are not forced to adapt between bursts of intense brightness and deeper interior shadows during the shortest winter days [8–10].

These conditions affect both façade composition and interior spatial planning. Larger glazed areas risk admitting sharp contrast gradients unless their orientation, solar exposure, and adjacent reflectance conditions are thoughtfully and carefully calibrated. Conversely, insufficient daylight access can leave classrooms dependent on electrical lighting during all but the brightest winter hours.

The resulting design question is not simply how to admit light, but how to effectively curate it, ensuring that the interior luminous environment remains comfortable, legible, and predictable, especially during periods of relatively rapid transitions between darkness, twilight, and full daylight. For a building intended to accommodate full-day educational activity throughout the winter season, this stability becomes a fundamental component of envelope performance.

2.3 *Material Behavior Under Prolonged Cold Exposure*

In Arctic climates, materials are defined not only by nominal thermal performance values, but by their behaviour under sustained cold exposure. At temperatures near − 30 °C, exposure alters elasticity, moisture movement, and dimensional stability long before assemblies show visible signs of stress [1, 14, 15]. To understand these constraints, several material behaviours become key factors in prolonged Arctic cold:

- Metals contract in deep cold, stressing perimeter seals, gaskets, and structural connections [10, 16].
- Sealants stiffen, reducing their ability to accommodate movement.
- Plastics and composites can become brittle if not reinforced for sub-zero conditions [10, 14].
- Moisture accumulation at even minor thermal discontinuities can generate frost over weeks or months of uninterrupted cold [17].
- Wind-driven air leakage exacerbates these behaviours by forcing cold air into micro-openings that would be inconsequential in milder regions [15, 17].

For the Nunatta Campus, an institutional building with limited access to specialized maintenance trades, the façade must sustain reliable surface temperatures, tolerate cyclical movement, and resist vapour drives throughout long periods when repairs are neither practical nor feasible [1, 4, 14].

3 Architectural Intent and Façade Logic

3.1 Landscape, Form, and Orientation

The architectural strategy for the Nunatta Campus begins with a direct reading of its physical setting. The site sits between exposed bedrock rising to the inland ridge and open slope facing Koojesse Inlet, a transitional zone shaped as much by geology as winter wind patterns. Rather than treating these forces as environmental constraints to be mitigated, Teeple Architects and Cibinel Architects leveraged them as generative inputs. The building massing is positioned so that the building partially shelters itself behind the rock outcrop, reducing exposure to prevailing northwest winds while easing snow drifting at the primary entrance [6, 13].

This orientation also establishes a precise relationship with winter light. With the sun moving low across the southern sky, the south and southeast elevations become primary daylighting surfaces. The design aligns these façades with the low-angle illumination typical of December through February, ensuring that limited winter daylight interacts with the building in ways that enhance interior visibility and comfort. Meanwhile, the more opaque and insulated portions of the envelope face north and west, where wind-driven heat loss is highest. The resulting form is therefore less of an artificially imposed geometry than a considered composition shaped by ridgeline, solar path, wind vectors, and circulation routes through the campus.

The architects describe the building as emerging "from the dark bedrock of the hill behind as an ambiguously natural form" shaped in large part by the wind. This characterization reflects a deliberate design intention: to produce an envelope that reads as belonging to its setting while meeting the stringent performance demands of Arctic construction [6]. Rather than rely on a single continuous glazed surface or rigidly repetitive modules, the façade adopts a carefully calibrated distribution of materials that responds to the changing conditions around it. In this way, the building's relationship to the site is operational, not symbolic, and the form is a direct consequence of its distinct environmental framework.

3.2 Distribution of Vision and Daylighting Units

Within this framework, clear triple glazed units and translucent Solera aerogel panels are deployed with distinct purposes. Clear glazing is used in places where outward visibility deepens orientation and enhances learning in classrooms, study areas, circulation corridors, and social spaces that benefit from a visual relationship to the inlet, the slope, and the sky [6, 13]. These are the portions of the building where occupants spend extended periods of time, and where external reference points contribute to comfort and alertness during the short winter days (Fig. 2).

The aerogel units address a related but separate set of needs. Their diffuse visual light transmission softens the contrast created by low-angle winter sunlight and

Fig. 2 Exterior elevation of Nunatta Campus with calibrated glazing distribution. Photograph by Julie Jira, reproduced with permission

provides stable brightness for areas where even illumination is more constructive than direct exterior views. They are placed along sections of the façade adjacent to program areas requiring visual calm, glare control, or privacy; locations where light quality, rather than transparency, is the primary driver [6, 12, 13].

What makes this distribution possible in a climate like Iqaluit's is the thermal behaviour of the Thermaframe 7 fibreglass-framed curtain wall system. This performance allows both clear and translucent units to be used in greater quantities without introducing condensation risk or cold-surface discomfort. Conventional metal-framed systems would impose stricter limits on glazing distribution due to thermal bridging at the frame. Here, those constraints are reduced, allowing the façade composition to respond more directly to program and daylighting objectives [6–11].

Through this interwoven design strategy, this next-generation façade behaves as a calibrated surface, open where views enrich the interior experience, modulated where diffuse illumination is needed, and insulated where heat loss must be minimized. This leads to a window-to-wall ratio near 30%, a level rarely feasible in northern institutional construction while preserving thermal stability [6, 13].

3.3 Interior Spatial Strategy Supported by the Façade

The interior depends heavily on this façade behaviour. The campus is arranged around a central gathering space, with teaching and support areas radiating outward. Clerestory cuts bring southern light deep into the plan, while internal glass partitions allow daylight to move freely through classrooms and circulation spaces without sacrificing acoustic privacy. The goal is an interior that feels connected, legible, and open, a significant improvement from earlier northern educational buildings that often relied on enclosed rooms and artificial lighting for most of the day (Fig. 3).

The interlaced façade reinforces this spatial intent. Clear glazing is positioned where occupants need outward reference points and where natural light can reduce the dependence on electrical illumination during the brief winter daylight hours. The translucent aerogel units provide more stable luminance required for study zones, transitional spaces, and smaller rooms where direct sun would create glare or visual strain. By matching the light qualities of each façade element to the needs of the rooms behind them, the envelope shapes not only the exterior aesthetic, but arguably more importantly, the interior atmosphere of the building [5, 6, 9].

This spatial arrangement relies on a consistent thermal baseline, with interior glazing, clerestories, and open sightlines functioning without introducing localized discomfort. In this way, the façade and the interior planning operate as a single, holistic system.

Fig. 3 Interior view showing balanced daylight from triple glazing, fibreglass framing, and aerogel units. Photograph by Julie Jira, reproduced with permission

3.4 *Performance Integration and Envelope Stability*

The Thermaframe 7 fibreglass system produces a curtain wall assembly with modeled frame U-values in the range of 0.9–1.1 W/m^{2}K (R5–6), capable of withstanding continuous sub-zero exposure while keeping interior surface temperatures safely above dew-point conditions [6–13]. The fibreglass framing minimizes thermal bridging, protects IGU edge seals under cyclical loading, and preserves airtightness despite wind-driven pressure differentials that are magnified during winter storms.

This performance foundation enables greater transparency than is typical of Arctic institutional buildings, without compromising thermal stability across the envelope. Without it, the building would have faced the conventional constraints of northern construction: small punched openings, conservative glazing ratios, and interior environments separated from the land and sky.

In the Nunatta Campus, architecture and building science converge through a façade that responds to program, climate, and culture simultaneously. The façade operates as a coordinated system in which material, form, and performance are aligned with the climatic conditions of Iqaluit. It functions because each component works in concert within this context. The result is a building that performs reliably despite the environmental extremes of Nunavut, while offering an interior experience grounded in daylight, landscape, and long-term durability.

4 Material Systems and Performance Rationale

The façade at the Nunatta Campus is shaped not only by architectural intent but by the material systems capable of delivering high transparency while providing thermal resilience. In Iqaluit's climate, the rate at which materials conduct heat, expand or contract under cold, and interact with moisture determines the viability of any glazing-heavy envelope. The project team selected a combination of pultruded fibreglass framing, high-performance triple glazing, and aerogel daylighting units to meet these demands.

4.1 *Fiberglass Framing: Thermal Stability and Structural Behavior*

In a climate where extended cold can persist for months, the performance of the frame becomes as critical as the glazing it supports. The fibreglass-framed curtain wall system used on the Nunatta Campus enables stable thermal characteristics through prolonged cold exposure. Unlike metal frames, which readily transmit exterior temperatures to interior surfaces, GFRP framing exhibits much lower thermal conductivity, resulting in frame U-values of 0.9–1.1 W/m^{2}K (R5–6). This reduction

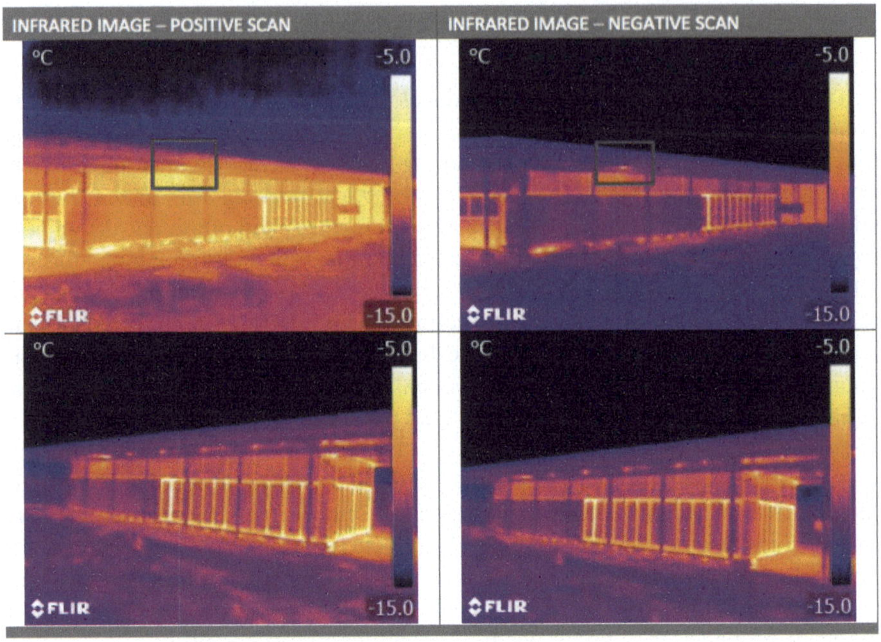

Fig. 4 Infrared thermographic views of the building perimeter showing both aluminum and fiberglass curtain wall framing under identical conditions. Photography by Entuitive

in conductive heat flow helps keep interior mullion surfaces warm enough to avoid condensation even during prolonged exposure to temperatures below − 30 °C [7–11] (Fig. 4).

This stability has direct implications for façade design. This expands the use of glazing, both in quantity and in locations more sensitive to thermal comfort than would be feasible with conventional aluminium. It also reduces the differential expansion between frame and IGU, relieving stress on seals during the rapid temperature swings common in Iqaluit. Over decades of service, this mitigates one of the primary failure pathways for Arctic glazing assemblies, where repeated contraction fatigues insulated glass unit (IGU) edge seals long before visible defects appear [10, 15, 16].

Fibreglass's rigidity further benefits the campus façade. With fewer thermally induced dimensional changes, the system has tighter tolerances, consistent gasket compression, and long-term airtightness, all essential in a region where wind-driven pressure differentials can be severe [3, 4]. These characteristics form the foundation upon which the building's expanded glazing strategy rests (Fig. 5).

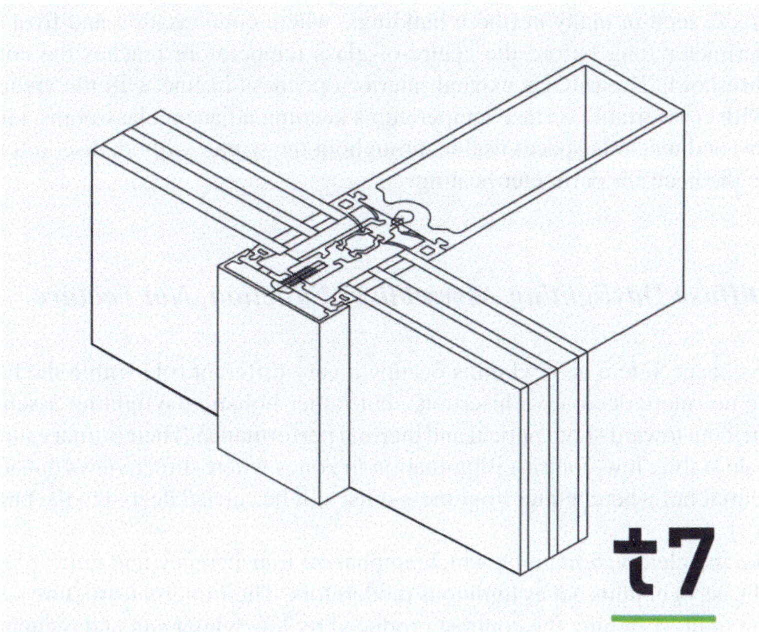

Fig. 5 Thermaframe 7 assembly schematic showing fibreglass framing and triple-glazed IGU integration. Diagram provided by GlasCurtain

4.2 *Triple Glazed IGUs: Edge Performance and Interior Surface Temperatures*

The clear-glazed portions of the façade rely on high-performance triple glazed insulating glass units engineered for sustained cold-weather operation. These assemblies achieve centre-of-glass U-values of 0.6–0.8 W/m^{2}K (R7–9), depending on coating configuration and inert gas fill (e.g. argon, krypton). Comprising three lites of glass separated by two sealed cavities with warm-edge non-metallic spacers, the units keep interior surface temperatures safely above dew-point thresholds during prolonged winter exposure. Glazing adjacent to occupied spaces remains usable without cold-surface discomfort or condensation risk. In Iqaluit, the exterior lite of an IGU can reach temperatures tens of degrees below ambient air during wind events. For the interior environment to remain comfortable, the inner pane must maintain a surface temperature above the dew point while resisting conductive losses through the spacer and edge assembly [8–11].

Warm-edge spacers and inert gas fills help achieve this condition, but their success depends on the thermal continuity provided by the surrounding frame. For Nunatta Campus, the pairing of triple glazing with fibreglass-framing results in a system where interior mullion intersections and glass edges remain warmer and more stable than comparable aluminium-framed assemblies. This prevents the cold "ring effect" seen in many northern buildings, where condensation and frost begin at the perimeter long before the centre-of-glass temperature reaches the condensation threshold. These IGUs expand interior openness in line with the architects' intent, with comfortable surface temperatures keeping adjacent classrooms, circulation areas, and teaching spaces usable throughout the winter without discomfort and reducing the need for perimeter heating.

4.3 Diffuse Daylighting Assemblies: Function, Not Feature

The translucent Solera aerogel units occupy a very different role within the façade. They are not mere decorative insertions, but rather holistic daylighting assemblies that contribute towards both optical and thermal performance. Their primary function is to create stable, low-contrast illumination in zones where direct views outside are less essential but where winter brightness must still be carried deep into the building [6, 12, 13].

Unlike the clear vision units, which emphasize transparency and outward orientation, the aerogel units act as luminous moderators. Their internal structure scatters incoming light, softening the contrast produced by low winter sun and reducing the need for supplemental lighting during the brief daylight window. Thermally, the aerogel units perform closer to insulated wall assemblies than conventional glazing. The silica-based aerogel core suppresses conductive and convective heat transfer through its nano-porous structure. The panels achieve thermal transmittance values of U-0.32 W/m^2K (R-18) while still transmitting diffuse daylight. This combination increases glazing coverage without a proportional increase in envelope heat loss.

In the context of the façade logic developed in Sect. 3, these units extend the building's capabilities rather than compensate for limitations. They support the level of glazing deployed across the south and east elevations, while the building "borrows" daylight and avoids the thermal penalties that would normally accompany additional glass.

4.4 Integrated Envelope Performance Under Arctic Conditions

The overall performance of the Nunatta Campus façade does not reside in any individual material but in the interaction between them. Fibreglass framing establishes a thermal and structural baseline that resists bridging and contraction under deep cold, while the triple glazed units deliver the optical clarity and interior surface temperatures needed for outward-facing program areas, and the translucent aerogel units complete the system by modulating brightness and contributing insulation where direct views are less critical.

Together, these components create an envelope that operates within the narrow tolerances dictated by the extreme climate: airtight under strong negative pressures, durable through intense freeze-thaw cycles, and warm enough at interior surfaces to prevent condensation even as indoor humidity fluctuates with occupancy patterns [6–13, 17]. This integration permits the campus to achieve a glazing ratio roughly three times higher than the northern institutional norm.

The result is not simply a high-performance wall but an envelope that realizes the architectural ambitions of the project: daylit interiors, openness to the landscape, and a spatial quality aligned with the cultural and environmental context of Iqaluit. In this setting, these qualities shape how the building is used and experienced on a daily basis, influencing how occupants orient themselves, reducing visual and thermal strain during long winter days, and maintaining continuity with the land in a built environment where that connection is often lost. The façade's technical capacity and its architectural intent are inseparable; each reinforces the other.

5 Construction Logistics, Manufacturing, and the Realities of Remote Delivery Constraints

5.1 Sealift Dependencies and Scheduling Realities

Every building in Iqaluit is shaped, in some way, by the logistics that make its construction possible. With neither road nor rail connection to the rest of Canada, the community depends upon a short annual sealift season during which building materials, equipment, and prefabricated assemblies arrive by barge [1–4]. This constraint introduces a timeline that has little flexibility: if a crate or component misses the seasonal shipment, the consequence is not just a modest delay but a full year added to procurement.

For façade design, this means every element, from the largest structural component to the smallest fastener, must be planned with lead time and sequencing in mind. The envelope is often one of the first major packages to be delivered and one of the last to be fully installed, placing additional pressure on reliable delivery and predictable performance.

Triple glazing, curtain wall framing, sealants, anchors, weather barriers, and specialized components must therefore arrive complete, correctly labelled, and ready for installation in a context without immediate access to replacement parts. Contingencies must be built into the design itself, not left to chance or on-site improvisation. In this environment, materials must be specified, coordinated, and procured from trusted partners to minimize risk and mitigate disruptive delays.

5.2 *Benefits of Domestic Manufacturing for Cold-Climate Assemblies*

Within this logistical structure defined, domestic manufacturing introduces a highly desirable measure of predictability. Materials and systems manufactured and fabricated in Canada enter this context with shorter shipping routes, more stable supply timelines, and fewer geopolitical uncertainties compared to transcontinental procurement chains [7–11]. While domestic sourcing does not eliminate the challenges of Arctic transport, it simplifies and derisks them: fewer border crossings, fewer opportunities for weather-related shipping disruptions, and clearer communication between designers, fabricators, and contractors.

For Arctic projects, this predictability carries considerable architectural upsides. When fabrication and delivery schedules are more reliable, design teams have greater freedom to specify assemblies that rely on precise performance tolerances. In contrast, imported systems, especially those dependent on multi-leg freight paths, risk variability in lead times that can affect not only installation sequencing but structural coordination and enclosure strategy.

Domestic fabrication also improves consistency in quality control. When frames and components are produced in the same climatic and regulatory context as their intended installation environment, material formulations, sealant specifications, and thermal assumptions are calibrated to Canadian building-science standards rather than generalized international norms. Over the lifespan of the building, this alignment reduces maintenance uncertainty, a significant advantage in communities where specialized trades are limited and service visits are expensive to mobilize [3, 4].

5.3 *Advantages of Stick-Built Systems in Arctic Context*

Panelized systems offer advantages in certain climates and applications, but in the Arctic, they carry risks related to weight, transportability, and on-site tolerances. Stick-built assemblies reduce shipping complexity by dividing components into more manageable units for sealift handling, and also allow incremental erection during periods of workable weather, providing installers with flexibility to address variable field conditions and substrate irregularities without requiring excessively large and cumbersome equipment [2, 6].

For Nunatta Campus, this method aligned with the design intent and the realities of northern construction logistics, resulting in a façade installation process that remained predictable within a climate where predictability is the exception rather than the norm.

6 Case Study: The Nunatta Campus Façade as an Integrated System

6.1 *Performance Outcomes in Occupied Conditions*

When the Nunatta Campus opened, one of the most immediate impacts reported by students and staff was the difference in interior light quality during winter days. Spaces that would traditionally feel enclosed or artificially lit during the short daylight window instead remain bright, with daylight distributed through both direct views and diffuse illumination. The façade's composition reduces glare from low sun angles, promotes stable interior temperatures, and limits cold downdrafts that often plague northern glazing [5, 6, 9, 12, 13] (Fig. 6).

Thermally, the façade has performed within predicted ranges. Interior mullion temperatures remain well above dew-point thresholds even during periods of deep cold, validating the material assumptions underpinning the façade design [7–11]. This reliability contributes to consistent interior conditions during extended cold periods.

The building's energy profile also benefits from the envelope's performance. Lower U-values (i.e. higher R-values) in the translucent daylighting units compensate for the expanded glazing ratio, allowing the architects to pursue their daylighting

Fig. 6 Winter night view of the Nunatta Campus, with interior illumination making the mix of clear vision glazing and aerogel-filled translucent units clearly visible. Photograph by Julie Jira, reproduced with permission

objectives without causing disproportionate heat loss. Combined with airtightness strategies coordinated across the wall assembly, the campus achieves predictable heating loads despite its openness.

6.2 Cultural and Spatial Experience Supported by the Envelope

In a setting where many students arrive from communities with wide horizons and close relationships to the land, outward visibility reinforces orientation and continuity with familiar spatial cues. Classrooms and circulation spaces framed by clear glazing provide wayfinding cues, views of changing atmospheric conditions, and visual continuity with the landforms surrounding Koojesse Inlet [6, 13]. These views contribute to a sense of orientation within an urban environment where built form otherwise dominates the immediate experience.

The translucent units complement this by maintaining brightness in areas that do not face outward, preventing the interior from becoming a sequence of bright pockets surrounded by dim corridors. The result is a campus where daylight reaches spaces that would otherwise depend entirely on artificial lighting for much of the winter.

The most significant spatial feature, the circular gathering space at the building's heart, benefits indirectly from the façade strategy. By allowing daylight to permeate classrooms and corridors, the interior can remain visually open without sacrificing acoustic or thermal control, reinforcing the architects' intent to create a socially central space that feels connected to the rest of the building.

6.3 Longevity and Maintenance Expectations

Northern buildings must be understood not only as completed works but as long-term systems expected to endure decades of challenging conditions. The façade strategy at Nunatta anticipates these expectations by minimizing materials prone to cold-induced fatigue and by reducing reliance on components that require frequent or specialized maintenance.

Fiberglass's resistance to corrosion and unwanted dimensional movement dramatically lowers the probability of frame deterioration over time. Triple glazing and high-performance edge assemblies reduce the frequency of IGU replacement cycles. Lightweight framing and simplified connections allow future maintenance crews, who may not have immediate access to specialized labour, to work with assemblies that are predictable, stable, and forgiving of extended periods between service intervals [1, 4, 14].

In a community where repair logistics are as constrained as construction logistics, this resilience becomes part of the building's long-term value. The façade continues to perform without requiring levels of maintenance that are difficult to sustain locally.

6.4 Lessons for Future Arctic Institutional Projects

The Nunatta Campus offers a model for Arctic building envelopes that integrate environmental performance, cultural priorities, and logistical feasibility.

Several key principles emerge:

- Material systems must be selected with both thermal demands and local maintenance capacity in mind. A façade that performs well but requires frequent specialist intervention is misaligned with Arctic conditions.
- Design freedom is created through performance stability. The ability to increase glazing ratios without increasing risk depends on systems that provide consistent thermal performance under prolonged cold.
- Daylighting and thermal design are inseparable in circumpolar regions. The quality, duration, and angle of winter light must inform material choice, glazing distribution, and interior organization.
- Supply chain reliability is a performance criterion. The durability of a façade is tied to its logistical feasibility as much as its technical specification.
- Façade strategies must acknowledge the lived experience of occupants. In a region where light, landscape, and climate hold deep cultural significance, envelope design shapes not only thermal performance but the daily experience of learning and gathering.

Taken together, these insights suggest a pathway for future Arctic buildings: one where the envelope is not a dividing line between inside and outside but a calibrated interface that responds to climate, addresses community needs, and remains feasible within the operational realities of northern construction.

7 Conclusions

Façade design in the Canadian Arctic requires a type of precision that extends far beyond meeting prescriptive performance thresholds. In Iqaluit, where materials must arrive by barge, where winter temperatures persist for months without relief, and where daylight changes in both quantity and character across the year, the building envelope becomes a mediator between climate, culture, and long-term feasibility. The Nunatta Campus demonstrates how these demands can be held together in a single architectural strategy, one that leverages building envelope performance as a constructive constraint while enabling spatial openness, cultural continuity, and reliable operation.

What distinguishes the project is the way its elements operate as a coordinated system. The façade integrates thermal performance, daylighting, and spatial intent in response to the climatic and cultural conditions of Iqaluit. The massing responds to wind and topography, while the logistical realities of constructing in Iqaluit shape the sequencing, material selection, and long-term maintenance expectations. Each decision reinforces the others, producing a façade that remains legible as a technical, spatial, and cultural system.

For future Arctic institutional buildings, several lessons emerge. Increasing transparency is achievable when glazing is treated as part of a coordinated envelope assembly rather than an isolated element; domestic manufacturing reduces logistical uncertainty, enabling design teams to commit to envelope strategies that rely on predictable performance; and in northern educational settings, where students often arrive from communities with close relationships to landscape, envelope design must consider how light, view, and thermal comfort influence daily experience. These insights point toward an expanded conception of northern resilience, one that includes material durability as well as energy performance, but also the social and cultural dimensions of building use.

As Arctic communities continue to grow in the twenty-first century, and public infrastructure expands alongside them, future architectural possibilities in this region will depend increasingly on systems that are robust, adaptable, and grounded in the realities of remote construction. The Nunatta Campus illustrates how the integration of material performance, logistics, and cultural context can produce a façade that performs reliably while positively shaping the lived experience of northern life. In doing so, it signals a shift from merely surviving the environment to working with it, using design to reinforce both the technical and human dimensions of circumpolar architecture.

References

1. Environment and Climate Change Canada (2019) Canadian climate normals 1981–2010: Iqaluit, NU. Government of Canada
2. Government of Nunavut (2020) Nunavut climate data and trends. Department of Environment
3. National Research Council Canada (2018) Building in the north: guidelines for cold climate construction. NRC
4. Proskiw G, Parekh A (2017) Challenges of building envelope design in extreme cold regions. In: Canadian building science insight
5. Figueiro M, Rea M (2016) Light, circadian rhythms, and health in high-latitude environments. Light Res Technol
6. Teeple Architects, Cibinel Architects (2019) Nunatta Campus expansion: design narrative and technical documentation. Project submission
7. RDH Building Science (2019) Thermal bridging and frame performance in cold climate curtain wall systems. RDH Technical Report
8. Passive House Institute (2020) Criteria for cold climate fenestration systems. Darmstadt
9. Gustavsen A, Grynning S, Arasteh D (2011) High-performance windows for cold climates. Energ Buildings

10. Onysko D et al (2017) Condensation and frost accumulation in cold-climate fenestration. Cold Reg Sci Technol
11. Cascadia Windows & Doors (2020) Fiberglass framing and thermal performance in sub-zero applications. Technical Bulletin
12. McCormick C (2020) Built on air: window manufacturers and architects capitalizing on the remarkable properties of aerogel. Glass Canada Magazine
13. The Plan Magazine (2019) Nunavut Arctic college expansion—project feature. The Plan Award Publication
14. Lstiburek J (2010) Moisture control for high-performance enclosures. ASHRAE J
15. Künzel H (1995) Simultaneous heat and moisture transport in building components. Fraunhofer Institute for Building Physics Report
16. Bomberg M, Kumaran K (2015) Long-term moisture performance of building envelopes in cold climates. J Build Phys
17. Wolf A (2018) Sealant performance limitations at low temperatures. Façade Engineering Review
18. Straube J, Burnett E (2005) Building science for building enclosures. Building Science Press
19. Aporta C (2011) Inuit orientation, wayfinding, and environmental knowledge. Polar Geogr
20. Bennett J (2018) Settlement patterns and infrastructure development in Nunavut communities. Arctic Institute of North America Report

From Concept to Reality: The Facade of Cirrus and Cascade, Chicago

Srdjan Avram and Jon Gately

Abstract Located at the edge of Lake Michigan within Chicago's Lakeshore East master plan, the Cirrus and Cascade development embodies a performance-based design process that integrates aesthetics, constructability, and sustainability from concept through completion. The design reconciles high-density urban living with a strong connection to the lakefront, using three tower volumes that cascade toward the water and maximize panoramic views while respecting neighborhood sightlines. The facade strategy—anchored in the serrated geometry of the towers—balances architectural expression with rigorous performance criteria. Through iterative collaboration among architect, consultants, client, contractor, and manufacturer, the system evolved into a rationalized, modular, unitized wall assembly reducing unique profiles while ensuring precision, repeatability, and economy. Over fifty glass types were reviewed to optimize transparency, color, solar control, acoustic attenuation, and bird-friendly attributes, achieving USGBC LEED Gold certification. Comprehensive environmental analyses informed each design phase: wind tunnel testing for comfort and structural loads; acoustic modeling for traffic-induced noise; and bird-friendly studies shaped by dialogue with the American Bird Conservancy. A full-scale performance mock-up underwent over thirty sequential tests—thermal cycling, structural load, and dynamic water penetration—to validate facade resilience. This synthesis of design intent and technical rigor demonstrates how performance-based facade design transforms architectural vision into tested, measurable reality—delivering a responsive, sustainable envelope suited to Chicago's climate and urban context.

Keywords Performance-based facade design · Multi-criteria design · Facade rationalization · Glazing selection · Energy performance · Predicted EUI (pEUI) · Wind tunnel testing · Pedestrian wind comfort · Acoustic performance · Bird-friendly design · Facade testing · Chicago · Lake Michigan

S. Avram (✉) · J. Gately
bKL Architecture, LLC, Chicago, Illinois, USA
e-mail: savram@bklarch.com

A. Mejorin et al. (eds.), *Performance-Based Façade Design*, Springer Tracts in Civil Engineering, https://doi.org/10.1007/978-3-032-28835-6_14

Nomenclature and Abbreviations

BIM	Building information model
BPS	Building performance simulation
PRM	Performance rating method
IEQ	Indoor environmental quality
IGU	Insulating glass unit
Low-E	Low-emissivity coating
U-factor/U-value	Overall heat transfer coefficient
SHGC	Solar heat gain coefficient
VLT	Visible light transmittance
WWR	Window-to-wall ratio
STC	Sound transmission class
OITC	Outdoor–indoor transmission class
NC	Noise criteria
GEM	Gust equivalent mean
PMU	Performance mock-up
QA/QC	Quality assurance/quality control
LEED	Leadership in energy and environmental design
ASCE 7	Minimum design loads and associated criteria for buildings and other structures
ASTM	ASTM international
AAMA/FGIA	Fenestration standards organization
EUI	Energy use intensity

1 Introduction: Performance-Based Design in a Lakefront High-Rise Context

Over recent decades, facade design has shifted from predominantly compositional curtain-wall concepts toward integrated environmental systems in which the envelope is required to satisfy structural, thermal, hygrothermal, acoustic and visual performance requirements simultaneously. This shift has been driven by heightened awareness of environmental impact, progressively stricter energy codes, the increasing use of computer-based simulation tools, and more systematic laboratory testing, all of which emphasize the facade as a quantifiable performance system rather than solely a visual artifact.

Within this context, contemporary practice increasingly treats design intent, analytical modelling and physical testing as components of a single iterative process rather than independent activities. Architectural decisions are therefore evaluated against measurable performance criteria, while performance targets remain explicitly linked to the spatial and formal ambitions of the project.

Cirrus and Cascade, two residential towers with a shared park podium in Chicago's Lakeshore East neighborhood, are used in this chapter as an applied case study of performance-based facade design in a dense, waterfront high-rise context. Rather than being generated from a fixed formal concept and subsequently checked for compliance, the tower massing and facade configuration were developed through an iterative process in which architectural, urban and environmental objectives were evaluated concurrently by the architect, facade consultant, engineers, manufacturer, client and contractor.

The project occupies a transitional zone between downtown, the Chicago River, and Lake Michigan, within a master plan that concentrates tall buildings while preserving view corridors and public access to the waterfront. The immediate surroundings include multi-level roadways, a tall service podium from neighboring buildings, and proximity to Lake Shore Drive. As a result, the site experiences complex microclimatic conditions: strong lake winds, substantial temperature variation between seasons, and elevated traffic noise levels.

The design brief for the facade, developed collaboratively by the architect, client, and consultant team, embraced a wide range of performance considerations: energy and comfort, wind and structural behavior, acoustic quality, bird-friendly design, durability, constructability, and the ability to verify performance through full-scale testing. In parallel, the project pursued LEED Gold sstainability certification, relying on the envelope as a primary instrument for reducing energy demand and improving both indoor and outdoor environmental quality.

This chapter documents how the facade of Cirrus and Cascade progressed from initial concept to completed construction within this performance-based framework. Drawing on project experience, design documentation, simulation outputs and laboratory and field test results, the chapter provides a structured account of the decision-making process and summarizes the main lessons that are potentially transferable to other high-rise residential projects.

1.1 Conceptual Framework and Positioning (Practice-Led)

This chapter uses "performance-based facade design" in a practice-led sense: the facade is treated as a multi-criteria system whose performance is defined by explicit objectives (energy performance, wind loading and pedestrian comfort, acoustic comfort, bird-friendly design, durability, and constructability) and verified through staged evidence rather than through prescriptive rules alone. The approach documented here links architect-led design decisions (geometry, module rationalization, and glazing/system selection) to analytical studies that quantify trade-offs and to a verification program comprising full-scale performance mock-up testing, field testing, and quality-assurance observations across fabrication and installation. The example of Cirrus and Cascade is therefore presented as a case study that contributes a transferable process-level framework for reducing facade performance risk in a cold, windy, lakefront urban context. Where relevant, contextual background is drawn from established standards and guidance (ASTM/AAMA/FGIA; ASCE 7; LEED; American Bird Conservancy) and from selected peer-reviewed work on multi-criteria facade decision-making; however, the primary evidentiary basis for the chapter's conclusions is the project's documented analyses and verification outcomes.

The chapter is organized as follows: Sect. 2 summarizes the urban-form drivers that establish facade boundary conditions; Sect. 3 defines the performance brief and evaluation framework; Sect. 4 describes system rationalization and key interfaces; Sect. 5 documents the glazing and material selection workflow; Sect. 6 summarizes supporting analytical studies (energy, wind, acoustics); Sect. 7 documents the digital workflow and decision loop; Sect. 8 presents full-scale and in situ verification testing; and Sects. 9, 10 and 11 synthesize transferable lessons, and conclusions.

2 Urban Form as a Performance Instrument

The urban configuration of the Cirrus and Cascade towers was treated as an explicit performance variable, with tower placement and massing adjusted iteratively in response to quantified environmental drivers (wind exposure, views, solar access) and programmatic requirements. In this sense, tower form, serrated perimeter and podium configuration constitute the first layer of the performance-based facade strategy rather than a purely compositional decision.

2.1 Lakeshore East and the Cascading Towers

The Lakeshore East neighborhood, developed over former boat slips and rail yards, has become a dense ensemble of residential and mixed-use towers surrounding a central park and extending toward Lake Michigan. It is framed by the Chicago River to the north, Millennium Park to the south and the historical Chicago "Loop" to the west, and Lake Shore Drive and the lakefront to the east. Within this context, Cirrus and Cascade occupy a prominent waterfront parcel planned for three point towers rising from a continuous, landscaped podium. Two towers were realized in the first phase, with a third planned for a later phase (see Fig. 1).

Within the Lakeshore East master plan, Cirrus and Cascade were required to accommodate high residential density while maintaining long-distance views and public access routes to the lakefront for both new and existing residents. These planning constraints imposed explicit performance objectives on tower placement and massing related to protected view corridors and avoiding a continuous "wall" condition along the lakefront (see Fig. 2).

Within these parameters, tower position and height become performance variables as much as compositional ones. Sightline studies and three-dimensional massing analyses were used to confirm that the new towers would not unduly block key view corridors from existing residential buildings and from the park. The cascading

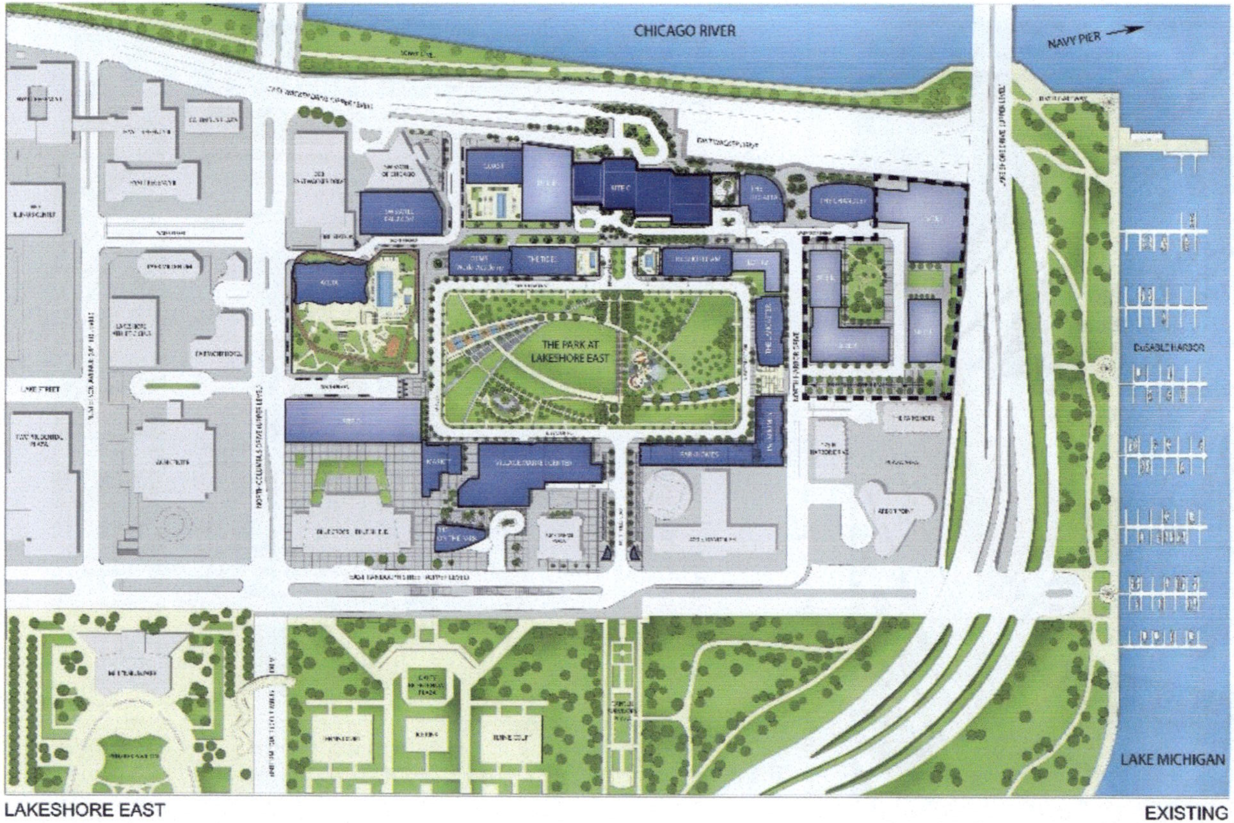

Fig. 1 Lakeshore East masterplan and project site within downtown Chicago. Aerial plan diagram showing the Lakeshore East neighborhood, adjacent buildings, lakefront access, and the "Project Site" highlighted within the Lakeshore-East masterplan

Fig. 2 Before image illustrating the pre-development site conditions (multi-level roadways, neighboring service podiums)

arrangement—stepping the towers down toward the lake—responds directly to these studies, allowing more distant buildings to see "over and between" new volumes (see Figs. 3a–c, and 4).

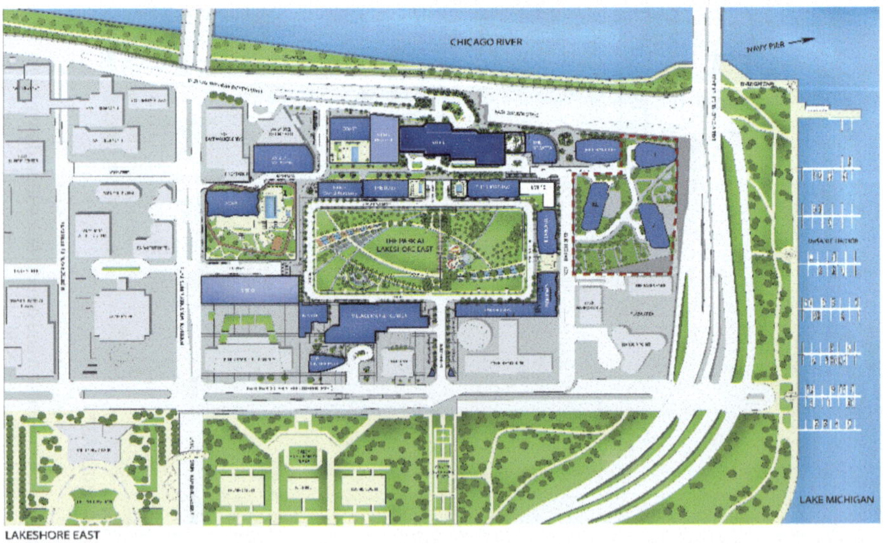

Fig. 3 **a** Three-point tower concept in a cascading landscaped podium setting. **b** Positioning and arrangement of towers maximizing access to views and natural light. View from Lake Michigan at Chicago River gate-locks. **c** View of development from Lake Shore Drive

Fig. 3 (continued)

Fig. 3 (continued)

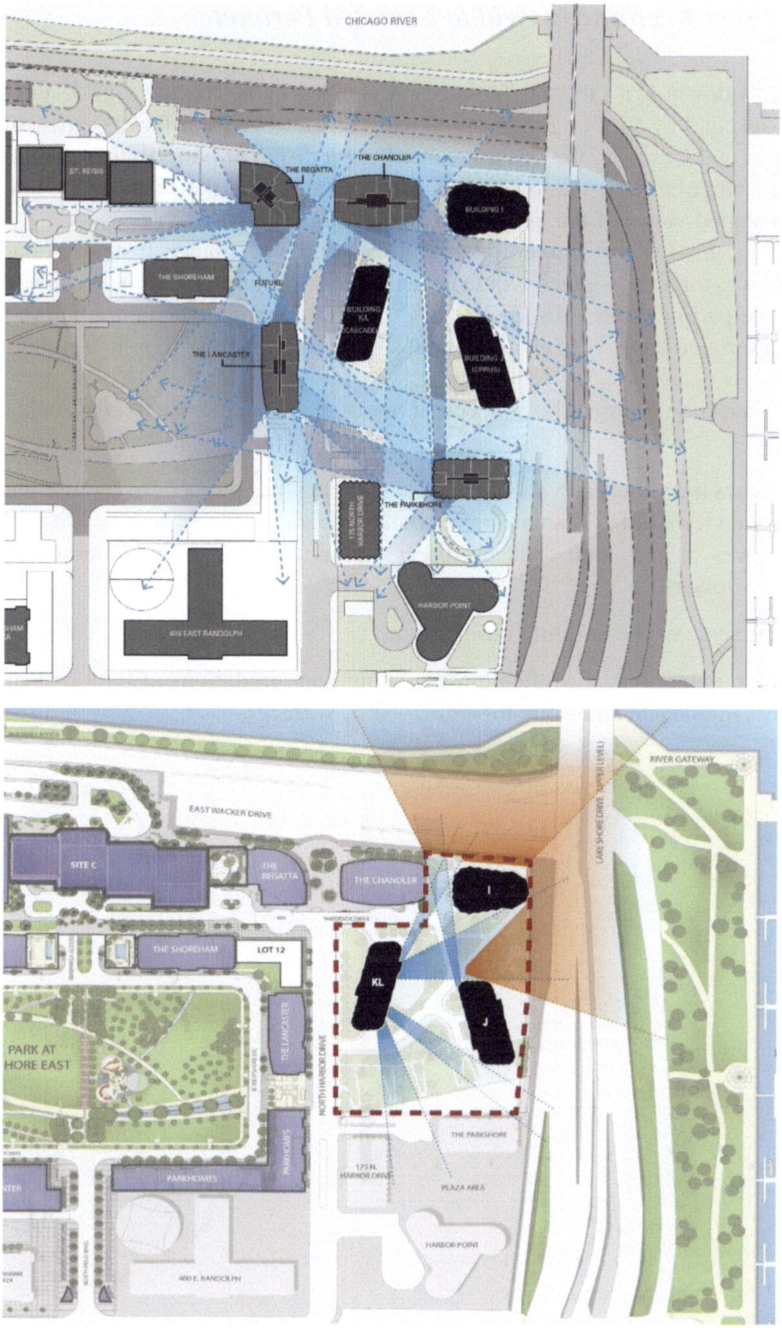

Fig. 4 **a, b** Maximizing views in a dense urban context. Massing and analytical diagrams illustrating how tower rotation, spacing, and non-orthogonal positioning preserve view corridors for existing neighbors

2.2 From Rectilinear Grid to Serrated Perimeter

Chicago's downtown fabric is famously organized by a rectilinear grid, and many towers respond by adopting orthogonal plans, even when their facades are sculpted. At Cirrus and Cascade, the design team chose a different path. Rather than simply rotate rectangular bars, the towers are also shaped so that their residential units fan toward angled views of the lake and river (see Figs. 4a, b and 5).

In plan, this leads to curved tower outlines at the perimeter. Instead of building those curves directly—which would have resulted in countless unique facade geometries—the perimeter is realized as a series of short, straight segments. These facets create a serrated edge, with each segment slightly rotated relative to its neighbor (see Fig. 5).

This serration became the conceptual backbone of the facade. At the scale of the unit, it:

- Increases the variety of viewing angles, allowing living spaces to capture multiple oblique views rather than a single frontal aspect.
- Improves privacy and reduces direct "glass-to-glass" facing conditions with nearby towers.

At the scale of the building envelope, it:

- Breaks up large planar reflections, creating finer gradations of light and shadow across the facade and mitigating mirror-like surfaces.

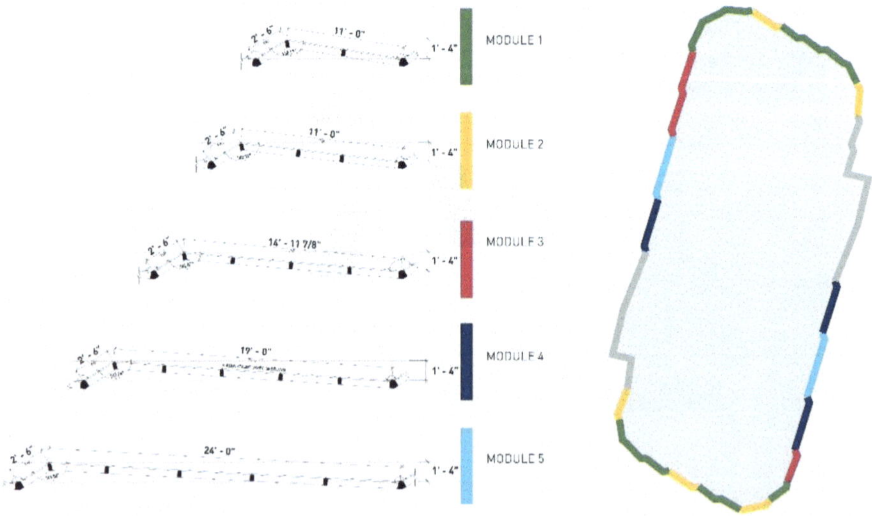

Fig. 5 Serrated facade geometry and plan modules 1–5. Plan diagrams illustrating the serrated perimeter, five primary plan modules, and their role in rationalizing the unitized window-wall system and reducing unique angle profiles

- Disrupts wind flows that would otherwise peel around sharp corners, helping distribute pressure more evenly and contributing to improved comfort around the base.

The serration was developed in close dialogue with the facade system. Rather than sculpt an arbitrary form and then ask the facade to follow, the design treats geometry, modularity, and performance as intertwined factors: any viable perimeter shape had to support repetition, manufacturability, and testing.

2.3 The Landscaped Podium

At the base of the towers, a five-story podium steps and terraces across the site and functions as the primary interface between the high-rise development and the park and streetscape. From a distance it reads as a continuous horizontal datum unifying the future three-tower composition, while at pedestrian level it is articulated into a sequence of planted terraces, outdoor rooms and glazed common spaces with distinct microclimatic and programmatic roles (see Figs. 6a–d and 21a–d).

Materially, the podium deliberately contrasts with the lighter tower envelopes. Stone, concrete, and metal panels combine with recessed glazing and shading elements to give the base a calmer, more tactile presence. This allows the podium to register at the scale of the park and streets, even while functionally connecting the towers above.

From a performance perspective, the podium massing and detailing play several roles. Setbacks and planters help to slow and redirect deflected wind and wind downwash flows from the towers, creating calmer microclimates at entries and terraces. Overhangs and recessed glazing support solar control and reduce overheating risk in public spaces. Vegetated surfaces and planted areas contribute to stormwater management and localized reduction of surface temperatures, supporting the broader environmental strategy of the development.

Collectively, these urban-form decisions established boundary conditions for envelope performance and constructability; they therefore serve as the starting point for the explicit facade performance brief.

3 Performance-Based Brief: Setting Goals and Targets

Past experience with tall buildings has shown that performance goals are most effective when they are articulated early, clearly, and in language shared by all disciplines. For Cirrus and Cascade, the design team translated the client's aspirations and local regulatory context into a set of performance themes that guided concept development long before detailed numerical values or testing pressures were fixed.

a

b

Fig. 6 **a** Form, materiality, and podium-tower relationship. Glassy towers rising from an "earthy" 5-story landscaped base. *Source* Tom Rossiter Photography. **b** Form, materiality, and podium-tower relationship. Cascading urban park podium, connecting the neighborhood to the lakefront via multi-level pedestrian and bicycle paths. *Source* Tom Rossiter Photography. **c** Tall point towers stepping down toward Lake Michigan. *Source* Tom Rossiter Photography. **d** Vertical accent spandrel strips emphasize the towers' slenderness. *Source* Tom Rossiter Photography

c

Fig. 6 (continued)

d

Fig. 6 (continued)

The facade was expected to:

- Support a high-performance energy strategy, aligned with LEED Gold certification ambitions. This meant limiting unwanted heat loss and gain, managing solar exposure by orientation, and coordinating glazing selection with thermal comfort and peak-load management objectives.
- Provide robust weather protection in Chicago's climate, which includes frequent low temperatures, driven rain, and high wind speeds along the lakefront. The facade needed to resist:
  - air infiltration per ASTM E283/E283M standards
  - water infiltration per ASTM E331 and AAMA 501.1 standards, under elevated pressures
  - structural load performance per ASTM E330/E330M
  - manage condensation risks in a way compatible with residential comfort expectations.
- Deliver acoustic comfort for residents living adjacent to substantial traffic flows, particularly along Lake Shore Drive, without over-building areas that are naturally quieter.
- Integrate bird-friendly strategies appropriate to a major migratory pathway, minimizing collision risk while maintaining transparency and long views—a growing expectation in contemporary facade practice
- Maintain durability and serviceability over decades, with details that tolerate movement, thermal cycling, and realistic maintenance practices without losing performance

These expectations were framed qualitatively in early design discussions and subsequently tied to specific target values, tests and acceptance criteria. This sequencing allowed the design team to explore forms and materials through a performance lens without prematurely constraining the architecture to a rigid checklist.

3.1 Methodology and Multi-criteria Design Framework

The facade design process followed a structured sequence of definition, exploration and verification steps.

- First, the project team collaboratively established a set of facade performance objectives spanning energy use, thermal and visual comfort, acoustic performance, wind and pedestrian comfort, bird safety, durability, constructability and capital cost.
- Second, design alternatives were generated and screened using a combination of option matrices, design workshop-based evaluations and targeted simulations, with trade-offs between criteria made explicit at each stage.

- Third, a reduced set of solutions was refined and validated through more detailed analytical modelling and full-scale testing (performance mock-up and field tests), closing the loop between predicted and measured performance.

With the performance objectives and evaluation approach defined, the next step in design process defined how the facade perimeter was translated into a rationalized, testable unitized facade system.

4 From Serration to System: Modularity and Interfaces

The serrated facade geometry was rationalized into a finite set of repeatable modules so that the geometric complexity of the tower perimeter could be reconciled with the fabrication and testing constraints of a unitized system. This modularization ensured that key performance variables such as air and water tightness, structural capacity and tolerance management could be verified on a limited number of representative conditions at mock-up scale prior to full production.

4.1 Rationalizing Geometry

The key challenge in turning the serrated geometry along the curved perimeter into a long term, reliable exterior wall system was controlling its complexity. Without discipline, each facet would become a custom condition, undermining the benefits of unitized construction.

To avoid that, the design team developed a limited family of plan modules, each representing a standard angle and bay width. These modules were used repeatedly around the perimeter, creating a rhythmic, saw-tooth pattern that still reads as a continuous curve at urban scale. The slight differences between modules, which include slight changes in angle or glass width, are almost imperceptible from inside the units but allow the facade to approximate the desired curvature (see Fig. 5).

Each unitized panel is composed of a thermally broken aluminum frame, an insulated glass unit or spandrel assembly, and provisions for anchoring, movement, and drainage. Transoms and mullions are aligned with interior planning grids where possible, so that framing lines reinforce the unit layouts instead of fighting them.

By constraining the design to five primary module types and a small number of special conditions, the project achieves:

- Manufacturability and quality control in the factory
- Repeatable detailing in critical performance areas such as stack joints and slab edges
- A manageable scope for laboratory testing, since representative modules and corners can be assembled into a single performance mock-up

In short, the serration, instead of being a source of uncontrolled variation, becomes a carefully tuned pattern that underpins the entire facade strategy.

4.2 *Interfaces with Doors and Podium Systems*

Where facade systems meet other building components, the risk of performance breakdown increases. On Cirrus and Cascade, particular attention was paid to:

- Balcony doors and operable elements, which needed to match the surrounding wall's air, water, and acoustic behavior while also satisfying accessibility and user-experience requirements. Thresholds had to be low and robust; seals and gaskets had to remain continuous without compromising operability.
- Slab edges and soffits, where structural tolerances, fire-stopping, and enclosure continuity must all be satisfied simultaneously. In a unitized system, these lines become critical "hinges" that accommodate movement while maintaining air and water barriers.
- Transitions between tower systems and podium facades, where differences in materials, wall depth, and structural support have the potential to create discontinuities in waterproofing, insulation, and vapor control if left uncoordinated.

These interfaces were modeled three-dimensionally, reviewed in coordination sessions, and incorporated into visual and performance mock-ups (see Fig. 14a, b). The guiding principle was simple: conditions critical to performance in the finished building should be made visible and testable at mock-up stage, allowing corrections in a controlled environment rather than on the completed structure.

5 Glass and Material Selection: Balancing Multiple Performances

Once the module family and critical interfaces were resolved, the primary lever for balancing comfort, energy performance, and ecological responsiveness became material selection—particularly the glazing system. Glass selection for the project was treated as a multi-criteria optimization task in which optical properties, energy performance, acoustic attenuation, and bird-friendly characteristics were considered simultaneously. Through iterative review of more than fifty candidate glass types, the final family of insulating glass units was configured to provide differentiated performance by orientation and exposure while maintaining a controlled overall appearance.

5.1 Criteria and Process

The glass and material palette shapes not only the building's appearance but also energy demand, indoor comfort, and ecological impact. Recognizing this, the team adopted a structured selection process supported by parametric sensitivity analysis (see Fig. 7), physical samples, and visual mock-ups.

More than fifty glass types and combinations were considered. Candidate products were evaluated against a matrix of criteria, including:

- color neutrality and compatibility between vision and spandrel conditions
- exterior reflectance and the way the facade would read from the park and lake
- interior reflectance reduction for elevated tenant experience
- visible light transmittance, solar heat gain characteristics, and overall thermal performance
- acoustic performance in laminated configurations for noise-exposed facades
- suitability for bird-friendly treatments, including compatibility with ceramic frit patterns and other surface modifications
- fabrication quality, including flatness and distortion behavior

Glazing parametric study (SHGC trade-off): A project glazing study evaluated facade energy sensitivity to SHGC and found that overall energy performance diminished below approximately SHGC 0.30, with the most favorable range between SHGC 0.30 and 0.40. When applying representative energy prices, SHGC 0.32 was identified as approximately cost-neutral, while values closer to SHGC 0.30 were advised to reduce cooling equipment demand. The study was intentionally parametric—used to rank options and characterize sensitivity rather than to forecast absolute consumption (see Fig. 7).

The decision-making process was iterative. Early samples helped bracket acceptable visual and performance ranges, while subsequent rounds refined color tone, reflectance levels, and the relationship between tower facades and podium materials. Ultimately, a compact family of vision and spandrel types was selected, deployed differently across the two towers to produce subtle variation while preserving a unified language (see Figs. 8 and 9).

5.2 Acoustic and Orientation-Based Zoning

Acoustic criteria and orientation-based energy considerations were layered onto this glass palette. Facades facing higher traffic noise were assigned laminated units with interlayers tuned to improve attenuation at critical frequencies, while more sheltered facades used simpler configurations where appropriate. In some zones, glass compositions were further adjusted to align with unit types—for example, bedrooms receiving additional acoustic attenuation where their primary exposure is toward the ever-busy Lake Shore Drive (see Fig. 16).

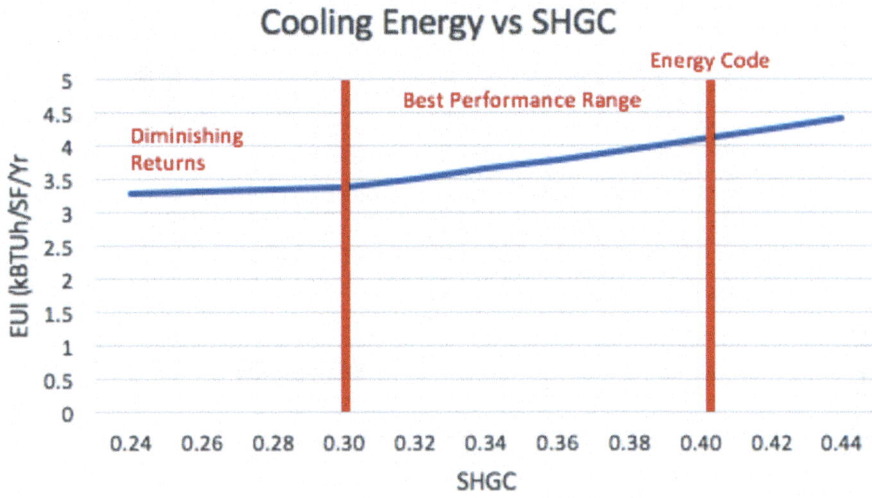

SHGC Impact on Cooling

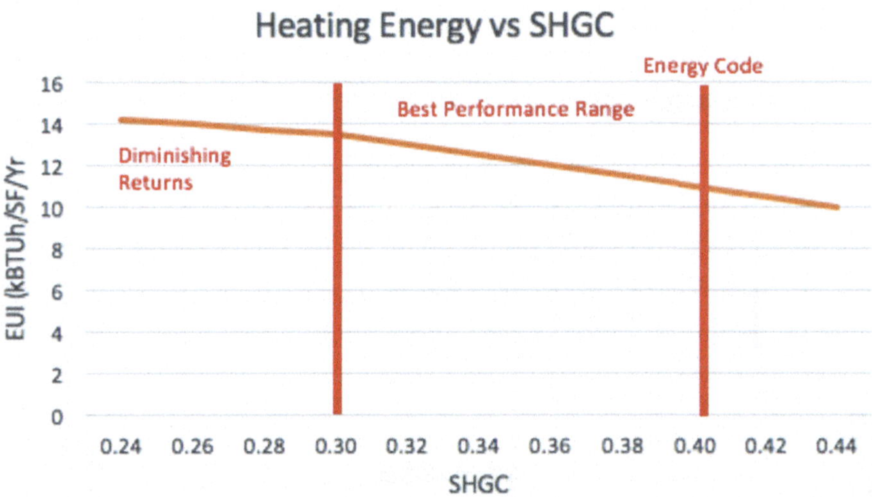

SHGC Impact on Heating

Fig. 7 Solar heat-gain coefficient (SHGC) sensitivity: parametric study of SHGC impact on heating and cooling energy and cost neutrality. *Source* Glazing Study Report—"Glazing and Building Energy Performance" (Cyclone Energy Group, 22 March 2018)

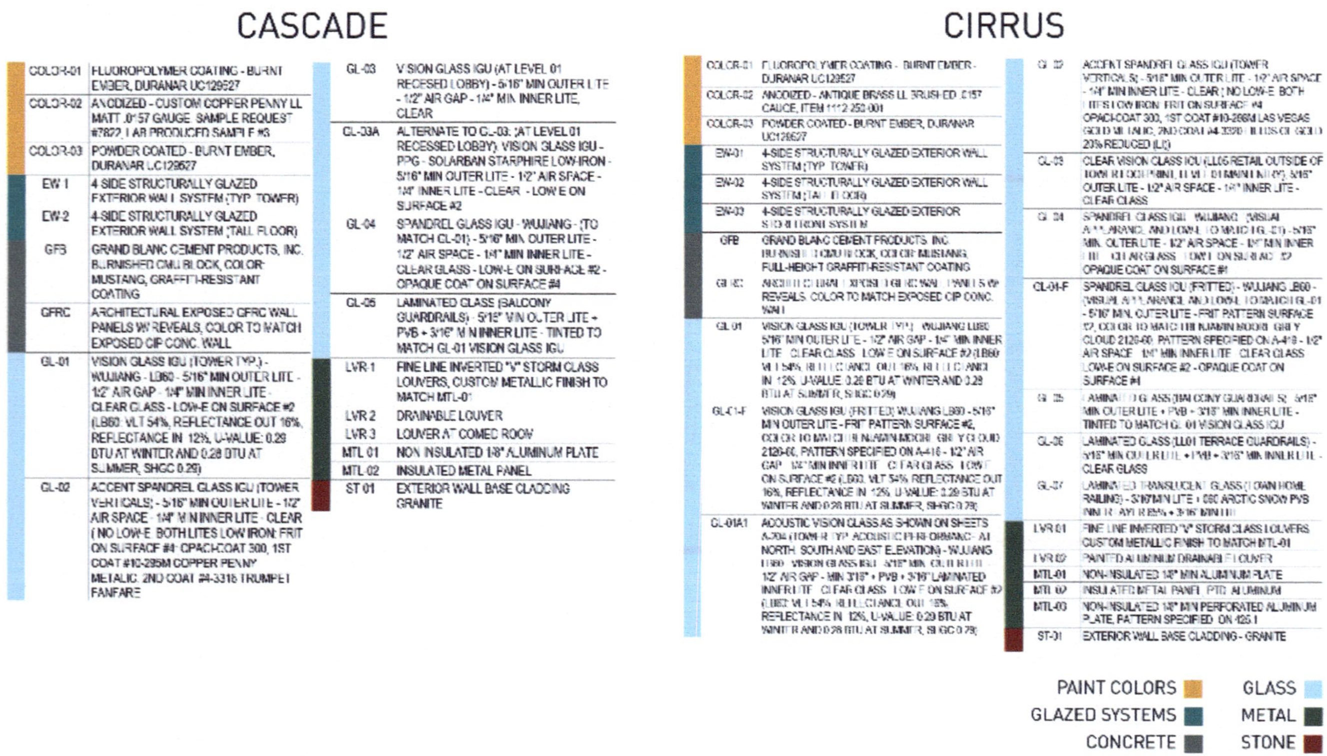

Fig. 8 Facade material selection board: summarizing parallel evaluation of glass transparency, reflectivity, color, low-E coatings, spandrel matching, energy and acoustic performance, bird-friendly characteristics, flatness, and viewing distortion

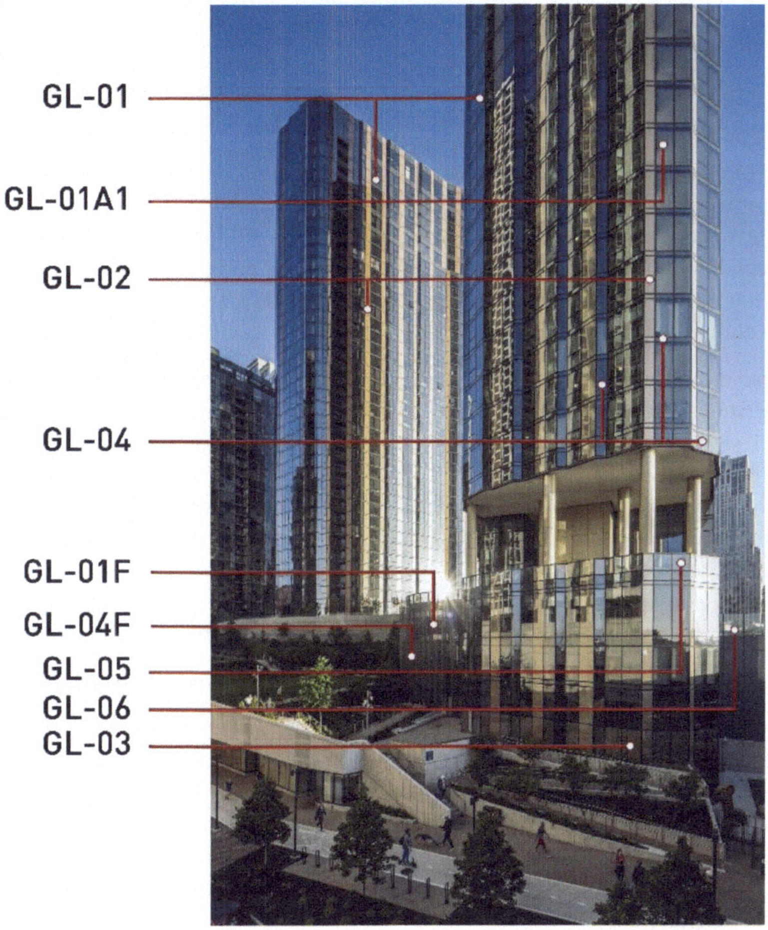

Fig. 9 Glass type selection and placement per performance criteria requirements

Orientation also influenced the balance of glass to opaque wall areas and the choice of coatings. South-facing facades sought to admit winter sun while controlling summer heat gain; east and west facades had to address low-angle sun and glare potential; and north facades could accept more glass with less solar penalty. The serrated geometry adds a local twist to these global orientations, as each facet encounters light and reflections at a slightly different angle.

5.3 Bird-Friendly Measures

Bird-friendly considerations cut across visual, energy, and constructability constraints. In Cirrus and Cascade, strategies included:

- fine-tuning reflectance and shading near vegetated podium areas to avoid creating large "mirrored landscapes"
- using fritted or patterned glass within the most collision-prone height bands, especially adjacent to planted terraces and interior greenery
- relying on the serrated geometry to reduce broad, uninterrupted sky or canopy reflections

These measures are part of a broader shift in facade practice toward ecological responsiveness. While the project cannot yet claim long-term collision data, it embeds a set of strategies that reflect current understanding of bird-safe design in urban environments (see Fig. 10a, b).

The selected module geometry and material palette were then subjected to targeted analytical studies to quantify trade-offs and reduce uncertainty before full-scale verification testing.

6 Analytical Support: Simulations, Studies and Tracking

A performance-based facade is supported by analysis, not as a purely numerical exercise, but as a way of testing and refining intuition. For Cirrus and Cascade, the analytical backbone combined whole-building energy modelling with wind engineering studies and acoustic modelling to quantify trade-offs, refine details, and define representative conditions and acceptance criteria for subsequent laboratory and field verification.

6.1 Energy Performance

Early comparison studies based on Total Building Performance compliance approach for annual energy simulations demonstrated viability of lower than baseline energy use for varying combinations of glazing percentage, window assembly U-value and opaque surface R-value parameters. It illustrated impacts and pointed to recommendation for glass thermal transmittance performance in the range of U-0.30 to 0.35 (see Table 1).

Energy simulations were used to test envelope configurations against the project's LEED and comfort goals. The design team examined how adjustments in glass type, framing strategy, opaque wall areas, and shading affected:

- heat loss and gain across seasons

a

b

Fig. 10 **a** Frit pattern sized and spaced to make glass wall barrier perceivable to birds. **b** Glass frit pattern strategically placed on building for bird-friendly design. *Source* Tom Rossiter Photography. **c** Energy modeling PRM summary (baseline vs. proposed): Cascade. *Source* Cyclone Energy Group [10] PRM compliance report; outputs generated in IES Virtual Environment (VE) v2019. **d** Energy modeling PRM summary (baseline vs. proposed): Cirrus. *Source* Cyclone Energy Group [9] PRM compliance report; outputs generated in IES Virtual Environment (VE) v2019. **e** American Institute of Architects (AIA) (2022) AIA 2030 DDx Project Stats export: Cascade Building (reporting year shown: 2018; updated 10 Dec 2022). **f** American Institute of Architects (AIA) (2022) AIA 2030 DDx Project Stats export: Cirrus Building (reporting year shown: 2018; updated 10 Dec 2022)

c

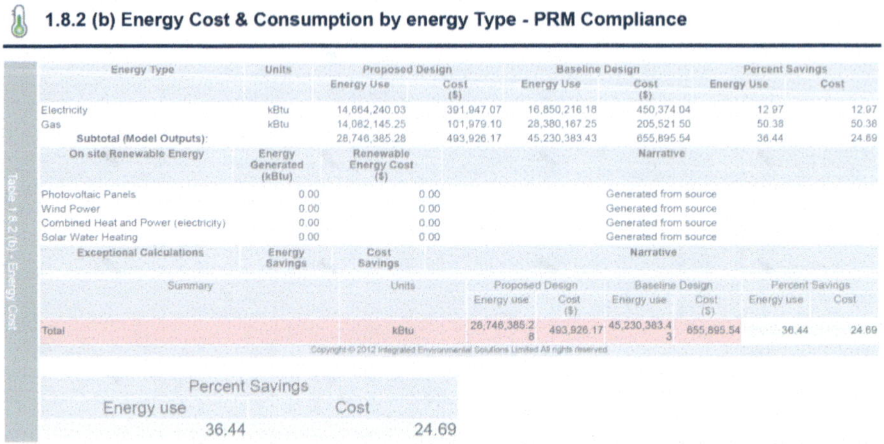

1.8.2 (b) Energy Cost & Consumption by energy Type - PRM Compliance

Table 1.8.2 (b) - Energy Cost

Energy Type	Units	Proposed Design		Baseline Design		Percent Savings	
		Energy Use	Cost ($)	Energy Use	Cost ($)	Energy Use	Cost
Electricity	kBtu	14,664,240.03	391,947.07	16,850,216.18	450,374.04	12.97	12.97
Gas	kBtu	14,082,145.25	101,979.10	28,380,167.25	205,521.50	50.38	50.38
Subtotal (Model Outputs):		28,746,385.28	493,926.17	45,230,383.43	655,895.54	36.44	24.69

On site Renewable Energy	Energy Generated (kBtu)	Renewable Energy Cost ($)	Narrative
Photovoltaic Panels	0.00	0.00	Generated from source
Wind Power	0.00	0.00	Generated from source
Combined Heat and Power (electricity)	0.00	0.00	Generated from source
Solar Water Heating	0.00	0.00	Generated from source

Exceptional Calculations	Energy Savings	Cost Savings	Narrative

Summary	Units	Proposed Design		Baseline Design		Percent Savings	
		Energy use	Cost ($)	Energy use	Cost ($)	Energy use	Cost
Total	kBtu	28,746,385.28	493,926.17	45,230,383.43	655,895.54	36.44	24.69

Copyright © 2012 Integrated Environmental Solutions Limited All rights reserved

Percent Savings	
Energy use	Cost
36.44	24.69

d

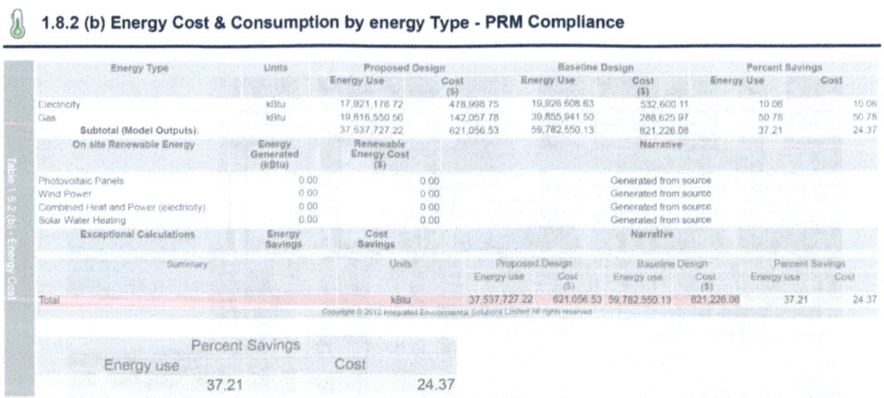

1.8.2 (b) Energy Cost & Consumption by energy Type - PRM Compliance

Table 1.8.2 (b) - Energy Cost

Energy Type	Units	Proposed Design		Baseline Design		Percent Savings	
		Energy Use	Cost ($)	Energy Use	Cost ($)	Energy Use	Cost
Electricity	kBtu	17,921,176.72	478,998.75	19,926,608.63	532,600.11	10.06	10.06
Gas	kBtu	19,616,550.50	142,057.78	39,855,941.50	288,625.97	50.78	50.78
Subtotal (Model Outputs):		37,537,727.22	621,056.53	59,782,550.13	821,226.08	37.21	24.37

On site Renewable Energy	Energy Generated (kBtu)	Renewable Energy Cost ($)	Narrative
Photovoltaic Panels	0.00	0.00	Generated from source
Wind Power	0.00	0.00	Generated from source
Combined Heat and Power (electricity)	0.00	0.00	Generated from source
Solar Water Heating	0.00	0.00	Generated from source

Exceptional Calculations	Energy Savings	Cost Savings	Narrative

Summary	Units	Proposed Design		Baseline Design		Percent Savings	
		Energy use	Cost ($)	Energy use	Cost ($)	Energy use	Cost
Total	kBtu	37,537,727.22	621,056.53	59,782,550.13	821,226.08	37.21	24.37

Copyright © 2012 Integrated Environmental Solutions Limited All rights reserved

Percent Savings	
Energy use	Cost
37.21	24.37

Fig. 10 (continued)

e

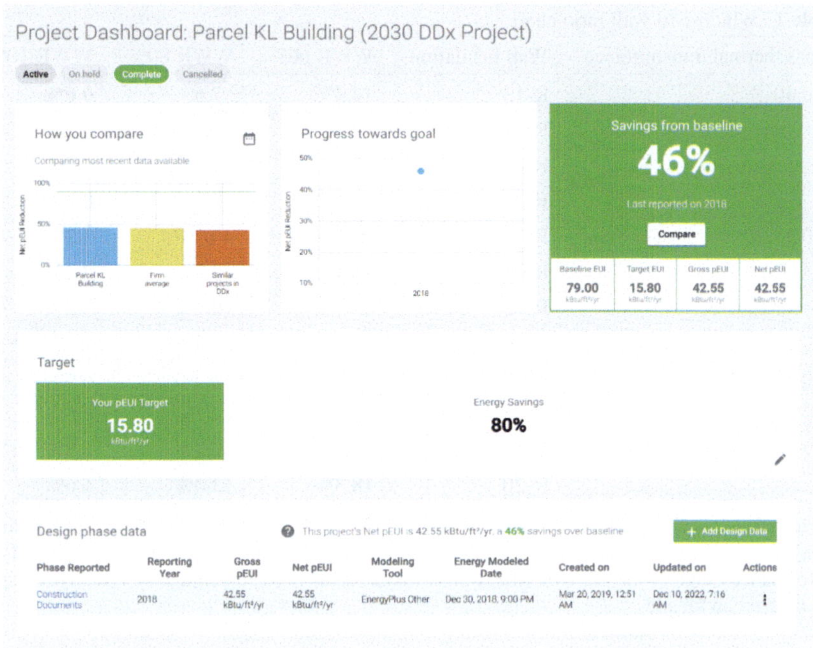

f

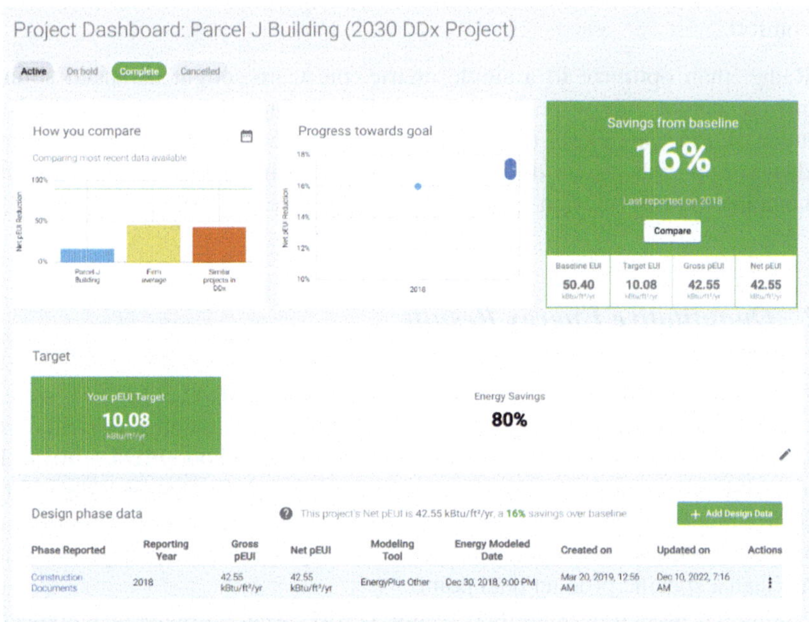

Fig. 10 (continued)

Table 1 Window-to-wall ratio chart

Glass thermal transmittance	Wall insulation	WWR 60%	WWR 80%	WWR 100%
U-0.40	R-13	14.4%	7.6%	**0.0%**
	R-15	14.6%	7.7%	0.1%
	R-18	14.9%	7.8%	0.1%
	R-20	15.0%	7.8%	0.2%
U-0.35	R-13	16.3%	10.2%	3.2%
	R-15	16.4%	10.2%	3.4%
	R-18	16.7%	10.4%	3.5%
	R-20	16.8%	10.5%	3.4%
U-0.30	R-13	18.0%	12.6%	6.5%
	R-15	18.2%	12.7%	6.5%
	R-18	18.4%	12.8%	6.5%
	R-20	**18.5%**	13.0%	6.6%

Cyclone Energy Group [8] Glazing and Building Energy Performance. Technical memorandum, 22 March 2018. Bold values point to the extremes of allowable limits in the upper and lower performance range of the glass thermal transmittance relative to percentage of visually transparent panels used on facade

- peak cooling and heating loads, with implications for mechanical sizing and comfort

Rather than optimize to a single metric, the team sought balanced solutions: improved annual energy performance without compromising the project's core promise of generous views to the lake and park. The modelling results formed a feedback loop with façade design, supporting adjustments in glazing percentages and helping to justify design choices to the client (see Fig. 10c, d).

6.2 *Quantitative Energy Results*

Initially, whole-building energy simulations based on ASHRAE 90.1-2007 were used to evaluate facade and glazing options in a comparative manner (baseline versus proposed), consistent with the Performance Rating Method (PRM). In the PRM models, the proposed fenestration was represented with an area-weighted U-factor of approximately 0.33 Btu/h ft^2 °F, SHGC 0.29, and VLT 0.44 as reported in the compliance outputs; these values reflect modelling inputs for comparative analysis rather than a specific product data point.

Additional modeling quantified results (PRM) were subsequently updated per ASHRAE 90.1-2010 Appendix G (Performance Rating Method). These updated compliance reports were used for the final LEED certification and indicate total annual energy-use savings of approximately 37.21% for Cirrus Building and 36.44%

for Cascade Building, with corresponding total energy cost savings of approximately 24.37% (Cirrus) and 24.69% (Cascade), relative to the respective baseline models (see Fig. 10c, d).

Though not captured in updated reports, the actual value of the typical glass unit panel of the system had an area-weighted U-factor 0.24 Btu/h ft^2 °F which improved energy performance of the façade component by 27% compared to the baseline assumption metric used in the energy model, while actual SHGC 0.30 is within the recommended range.

6.2.1 AIA 2030 Commitment Performance Tracking (DDx)

In parallel with LEED PRM compliance modelling, the project's early design-stage energy performance was also reported through the AIA 2030 Commitment, using the AIA Design Data Exchange (DDx) platform to benchmark predicted energy use intensity (pEUI) against a building-type baseline and the program's reduction targets. This reporting is not a substitute for LEED PRM; rather, it provides an additional, industry-standard mechanism for office-wide performance tracking and year-over-year comparison across US projects.

For the reporting year shown in the DDx project dashboards (2018), both Cirrus and Cascade buildings were entered with a net pEUI of 42.55 kBtu/ft^2 yr. The DDx baseline and target values differ by building type and assumptions: Cirrus (condominium building) lists a baseline EUI of 50.40 kBtu/ft^2 yr and a target EUI of 10.08 kBtu/ft^2 yr, corresponding to a reported 16% savings from baseline; Cascade (rental apartment building) lists a baseline EUI of 79.00 kBtu/ft^2 yr and a target EUI of 15.80 kBtu/ft^2 yr, corresponding to a reported 46% savings from baseline. These DDx entries should be interpreted as practice-level tracking snapshots (inputs and normalization rules differ from PRM), but they reinforce the project's broader objective of quantifying and communicating energy performance outcomes within established professional frameworks. In this case, it is evident how the market driven demands for different residential dwelling unit space needs impact the energy use intensity in different building typologies (see Fig. 10e, f).

6.3 *Wind, Structural Behavior and Pedestrian Comfort*

Wind tunnel testing addressed both facade cladding pressures and pedestrian comfort, providing zone-specific pressures and directional sensitivity for the serrated perimeter and podium. The cladding-load outputs informed mullion/anchor design, required minimum glass thickness, and movement-joint strategy; the pedestrian-wind outputs informed placement and shaping of outdoor program.

6.3.1 Cladding Wind-Load Results

Directional boundary-layer wind tunnel testing produced recommended cladding design wind pressures for localized facade zones. RWDI reported a largest recommended negative cladding pressure of −70 psf (occurring on the south elevation of Cascade Tower) with the majority of negative pressures in the − 30 to − 50 psf range. The largest recommended positive cladding pressure was + 50 psf (reported at parapets on select elevations), with the majority of positive pressures in the + 25 to + 40 psf range. Internal pressure allowances were also documented (e.g., main facades +6/−12 psf; roofs/soffits +6/−6 psf) and were combined with external pressures to derive differential design pressures. Pressures were provided without load/safety factors and intended to be applied in the same manner as code-derived cladding loads (see Fig. 11a–c and Tables 2, 3).

6.3.2 Quantitative Wind and Pedestrian Comfort Results

Pedestrian wind criteria and safety: Pedestrian comfort was evaluated using RWDI's Gust Equivalent Mean (GEM) criteria, which relate combined mean and gust effects

Fig. 11 **a** Wind tunnel model photograph. *Source* RWDI Cladding Wind Load Study (24 Oct 2018). **b** Directional distribution of local wind speeds. *Source* RWDI Cladding Wind Load Study (24 Oct 2018). **c** Wind Loads blocking diagram. *Source* RWDI Cladding Wind Load Study (24 Oct 2018)

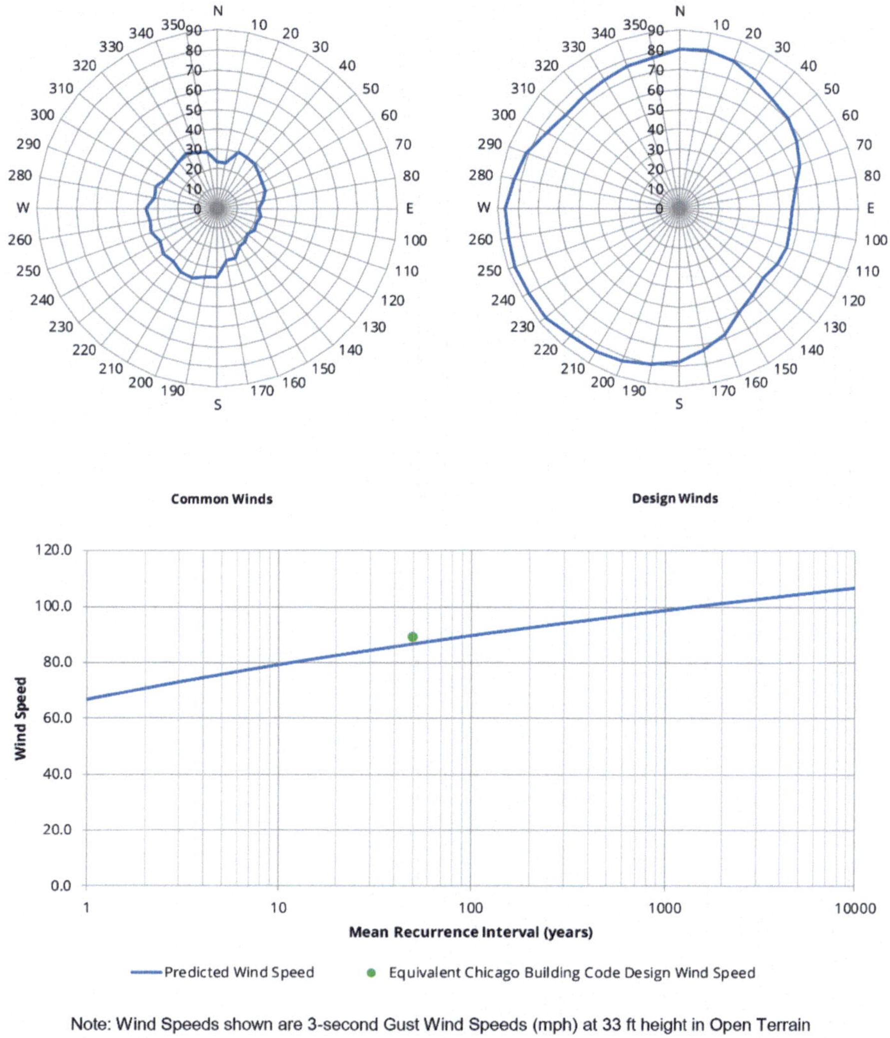

Fig. 11 (continued)

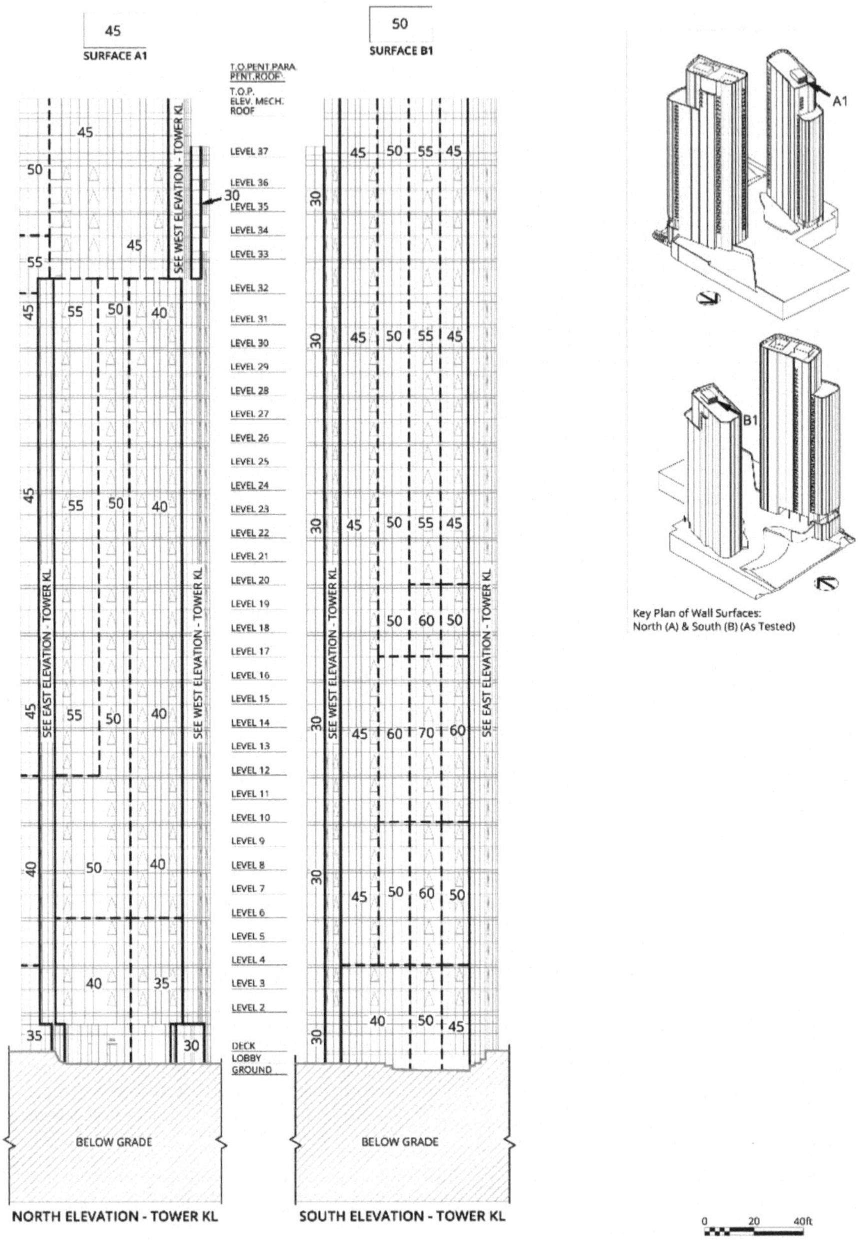

Fig. 11 (continued)

Table 2 Recommended internal pressures

Scenario	Internal pressures	
	Positive (psf)	Negative (psf)
Main facades	+ 6	−12
Roofs and soffits	+ 6	−6

Cladding wind loads summary. *Source* RWDI Cladding Wind Load Study (24 Oct 2018)

Table 3 Summary of cladding design wind pressures

Element	Negative pressures			Positive pressures		
	Peak value (psf)	Location	Range of majority (psf)	Peak value (psf)	Location	Range of majority (psf)
Main facades	−70	Fig. 4	−30 to −50	+ 50	Figs. 14 a and b	+ 25 to + 40
Roofs and soffits	−55	Fig. 11	−30 to −40	+ 35	Fig. 14	+ 20 to + 25

Cladding wind loads summary. *Source* RWDI Cladding Wind Load Study (24 Oct 2018)

to activity-based comfort thresholds. For reference, RWDI's comfort categories define approximate GEM thresholds of < 6 mph (sitting), < 8 mph (standing), < 10 mph (strolling), and < 12 mph (walking), with conditions > 12 mph categorized as uncomfortable for typical pedestrian activities (see Fig. 12a). A separate safety criterion flags gust speeds exceeding approximately 56 mph as potentially hazardous to balance and footing, based on a low annual exceedance threshold.

Quantitative wind and pedestrian comfort results: Boundary-layer wind tunnel testing compared existing and proposed massing configurations and evaluated comfort at key outdoor locations (entries, terraces, and primary pedestrian routes). Iterative adjustments to podium shaping, canopy placement, and landscape/wind-control features were used to reduce exceedances of the intended comfort categories at programmed gathering areas (see Fig. 12b, c).

6.4 *Acoustic Modeling*

Acoustic modelling supported facade and glazing zoning in response to spatially varying traffic noise exposure, particularly from Lake Shore Drive. The project's design-development acoustic memorandum combined measured exterior sound levels with CadnaA propagation modelling to estimate facade incident levels at multiple elevations and orientations, and then aligned glazing transmission targets to interior comfort criteria.

Acoustic criteria and zoning: The report classified facades as "noise impacted" where exterior sound pressure levels exceed approximately Ldn 65 dBA. For

a

RWDI Pedestrian Wind Criteria

Comfort Category	GEM Speed (mph)	Description
Sitting	≤ 6	Calm or light breezes desired for outdoor restaurants and seating areas where one can read a paper without having it blown away
Standing	≤ 8	Gentle breezes suitable for main building entrances, bus stops, and other places where pedestrians may linger
Strolling	≤ 10	Moderate winds that would be appropriate for window shopping and strolling along a downtown street, plaza or park
Walking	≤ 12	Relatively high speeds that can be tolerated if one's objective is to walk, run or cycle without lingering
Uncomfortable	> 12	Strong winds of this magnitude are considered a nuisance for all pedestrian activities, and wind mitigation is typically recommended

Notes:

(1) GEM speed = max (mean speed, gust speed/1.85);

(2) GEM speeds listed above are based on a seasonal exceedance of 20% of the time between 6:00 and 23:00. Nightly hours between 0:00 and 5:00 are excluded from the wind analysis for comfort since limited usage of outdoor spaces is anticipated; and,

(3) Instead of standard four seasons, two periods of summer (May to October) and winter (November to April) are adopted in the wind analysis, because in a colder climate such as that found in Chicago, there are distinct differences in pedestrian outdoor behaviors between these time periods.

b

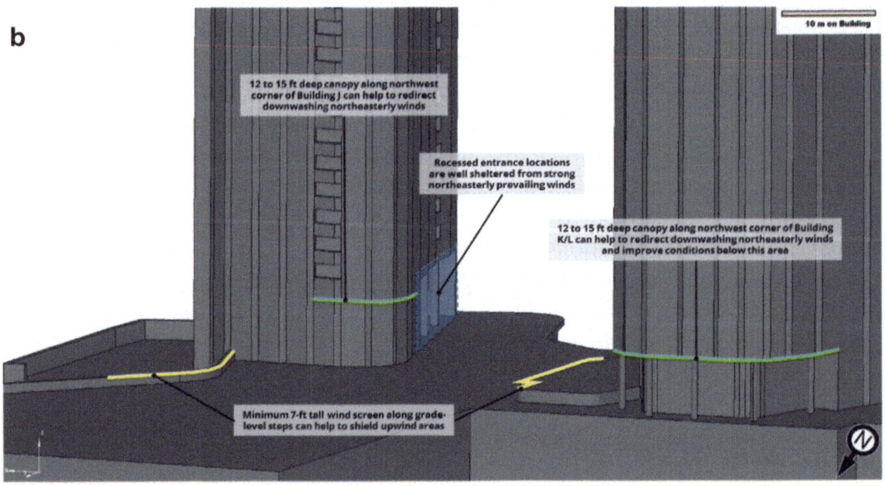

Fig. 12 **a** Pedestrian wind comfort and safety: a RWDI GEM criteria (comfort thresholds and safety gust criterion), *Source* RWDI Pedestrian Wind Study (19 Mar 2018). **b** Pedestrian wind comfort and safety: conceptual wind control measures.*Source* RWDI Pedestrian Wind Study (19 Mar 2018). **c** Pedestrian wind comfort condition plan for the adopted configuration. *Source* RWDI Pedestrian Wind Study (19 Mar 2018)

Fig. 12 (continued)

impacted facades, it recommended exterior window-system performance of approximately STC/OITC 38/33 (noting that OITC better reflects traffic-noise frequency content), while for non-impacted facades it recommended approximately STC/OITC 34/28. These targets were paired with facade sound-level mapping and informed zoning of laminated versus standard IGU constructions (see Fig. 13).

This targeted approach allowed the project to concentrate resources where they had the greatest impact on resident comfort. It also reinforced the broader performance-based attitude: not every surface requires the same solution, and performance follows exposure and use.

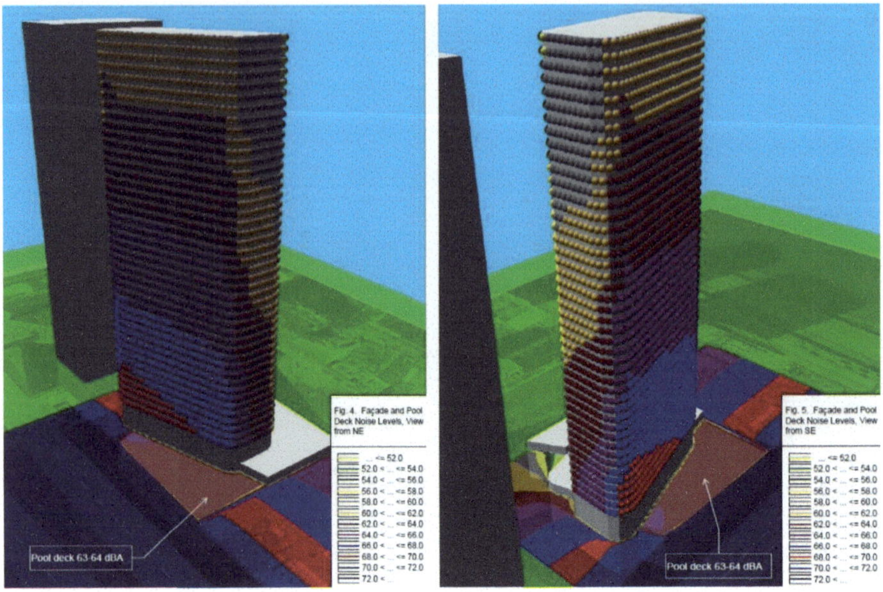

Fig. 13 Acoustic site analysis and facade sound-level mapping (CadnA): **a** exterior sound level summary and modelling approach; **b** facade-elevation sound level predictions informing glazing zoning. *Source* Acoustic review memorandum [13]

Quantitative acoustic outcomes: The result of the acoustic workflow was not a single "acoustic glass" selection, but an exposure-based zoning strategy that matched window-system performance to predicted incident noise levels.

The actual value of the used acoustic glass panels had an STC/OITC of 44/35, yielding + 6 STC and + 2 OITC compared to the recommendation of the acoustic analysis. Typical panels used had an STC/OITC of 35/31, yielding + 1 STC and + 3 OITC compared to the baseline recommendation.

7 Digital Workflow and Tools

A digital workflow coordinated facade geometry, consultant analysis, and fabrication-ready detailing. Model-derived geometry was used to generate analysis abstractions (e.g., orientation-based glazing groupings, representative detail families, and zone maps), and analysis outputs were translated back into a reduced module set and a targeted verification plan.

Fig. 14 **a**, **b** Visual and performance mock-up, weather performance testing

7.1 Modelling, Analytics, and Simulation Tools Used on the Project

- IES Virtual Environment (VE) v2019—whole-building PRM energy modelling and comparative facade studies (Cyclone Energy Group).
- RWDI boundary-layer wind tunnel testing—cladding wind loads and pedestrian wind comfort/safety assessment.
- CadnaA—acoustic propagation modelling and facade incident-level mapping to support glazing zoning.
- BIM authoring model (Revit-based) geometry and interface coordination.

7.2 Quality Assurance Loop: Closing the Verification-to-Detailing Loop

A key operational characteristic of the Cirrus and Cascade delivery process was the explicit QA loop linking predictive analysis, full-scale verification, and production documentation. Rather than treating simulation outputs and mock-up testing as isolated checkpoints, the team maintained a traceable chain from observed issue to root-cause hypothesis to detail revision to re-test/verification to release of updated shop drawings and installation guidance. This loop was applied at multiple scales: (A) macro-scale massing and podium adjustments informed by pedestrian wind testing; (B) system-scale module and interface refinement informed by PMU outcomes; and (C) production-scale corrective actions informed by factory observations and field testing. Documenting these closed-loop steps turned performance intentions into verifiable acceptance criteria and reduced the risk that late-stage changes or workmanship variability would erode the intended envelope performance.

Together, these analyses provided the evidence base for the staged verification program, in which representative assemblies were tested at full scale and validated in the field.

8 From Drawings to Reality: Mock-Ups and Testing

To verify that analytical predictions and design intent were realized in constructed work, the project implemented a staged verification program combining a full-scale performance mock-up (PMU) tested under laboratory conditions with in situ field testing of installed units. Findings from each stage were used to refine details and installation practices, extending performance-based decision-making into construction.

8.1 Visual and Performance Mock-Ups (VMU and PMU)

The performance mock-up assembled representative sections of the facade, including typical serrated modules, inside/outside corner conditions, operable elements, balcony doors, and interface zones. Built under controlled conditions at a testing laboratory, it was subjected to repeat cycles of air infiltration, static and dynamic water penetration, structural loading, and thermal cycling, while also confirming functional operability and accessibility load requirements for doors and vents (see Fig. 14a, b).

Photographs of the full-scale visual and performance mock-up under air and water static and dynamic testing, structural loading, and thermal cycling, (vision modules, stack joints, balcony interfaces).

Mock-up test matrix and criteria: The full-scale performance mock-up (PMU) incorporated representative portions of the serrated tower facade, corner conditions, balcony doors, and slab-edge interfaces. The PMU was subjected to a test matrix aligned with ASTM E283/E283M (air infiltration), ASTM E331 (static water penetration), AAMA 501.1 (dynamic water penetration), and ASTM E330/E330M (structural loading), along with AAMA 501.5 (thermal cycling) intended to evaluate sealant and gasket robustness. Both air, and water static and dynamic infiltration tests were repeated upon passed uniform load deflection test and upon passed thermal cycling test. Air infiltration tests were conducted at 6.24 psf positive pressure differential, while water infiltration tests were conducted at a positive static pressure differential of 12.0 psf with a water rate of minimum 5 gallons per square foot per hour.

PMU issue resolution and lessons: PMU testing and subsequent issue-tracking documented interface and installation sensitivities (e.g., seal continuity at transitions, drainage path robustness, and tolerance stack-up effects). The PMU 'Issues and Next Steps' log provided a traceable loop from observation to root-cause hypothesis to detail modification to re-test/verification (see Fig. 15).

The aim was not simply to pass tests, but to observe how the assembly behaved: where water traveled, how seals and gaskets responded to repeated assembly movement upon subjected to simulated loads, and how thermal gradients affected condensation and differential expansion. Where tests revealed vulnerabilities—such as leakage at complex junctions or unexpected flexing under load—the team adjusted details, repeated tests as necessary, and updated the production drawings.

This iterative process used the mock-up as an investigational prototype whose failures were as informative as its successes, benefitting the project in the long-term.

Fig. 15 PMU initial dynamic water test, infiltration at door gasket

The thermal insulations should be compacted well Non-compact areas	Photo #14 

Apply masking tape without cleaning up the excess structural sealant while preparing the weather caulking Masking tape Excess sealant	Photo #29 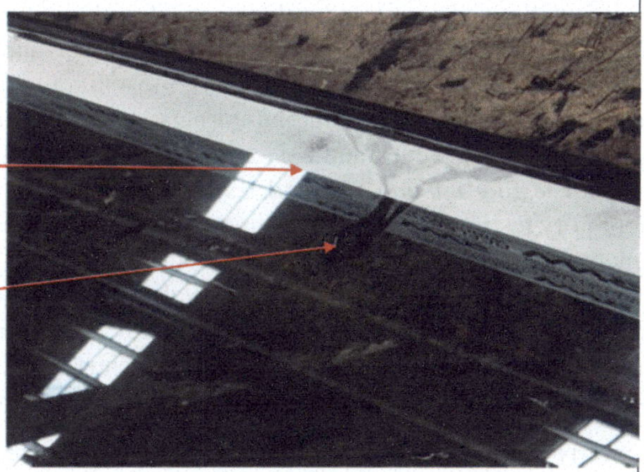

Fig. 16 **a**, **b** QAQC procedures undertaken at the manufacturing facility. *Source* CDC Factory production facility Quality Assurance independent observation report, May 2020

8.2 Quality Assurance—Factory Observations and QAQC Checklists

A performance-based facade delivery model depends on a closed quality-assurance loop that links factory process controls, installation verification, and corrective action based on measured outcomes.

Independent factory observations documented process controls and QA checkpoints during unitized-panel production have been performed (e.g., handling/storage of gaskets, insulation placement and compaction, glazing setting block placement, sealant mixing/application, and completeness of production end QC checklists). Observations were logged as actionable items with recommended corrective steps and follow-up verification (see Fig. 16a, b). Laminated glass panels were inspected, setting roll-wave distortion range limit acceptance criteria aligned with project's visual comfort goals.

8.3 Quality Control—Field Testing

Once installation began, randomly selected facade areas were tested in situ at different phases of construction using portable chambers and calibrated spray systems. These field tests focused on locations most likely to experience elevated loads or detailing challenges, including corners, transitions between tower and podium systems, operable vents, and balcony door zones.

Field testing campaign. Field testing served several purposes. In parallel with laboratory testing, an on-site quality assurance program included field air and water penetration tests on a defined sample of installed units. Test locations were selected to cover different orientations, floor levels, and interface conditions (typical bays, corners, podium transitions). Field tests were carried out per AAMA 501.2 requirements and project QA procedures; where deficiencies were identified, remedial work was executed and follow-up tests were performed to confirm closure:

- verify that the as-installed facade areas reproduced the performance demonstrated in the laboratory
- provide early warning of workmanship or interface issues that might not be apparent by visual inspection
- reinforce that air- and water-tightness are measurable outcomes supported by repeatable field procedures

Where field tests revealed deficiencies, these were treated as inputs to a structured corrective process. Local remediation measures were implemented and re-tested, and the underlying causes and solutions were communicated to installation crews so that similar issues could be prevented elsewhere on the facade. In practice, this approach extended the performance-based design methodology into the construction phase, rather than treating construction as a purely execution-focused activity.

9 Discussion: Reflections and Lessons for Performance-Based Facade Practice

The combined modelling and testing outcomes provide an empirical basis for discussing which aspects of the process are broadly transferable and which are specific to Chicago's lakefront context. The facade of Cirrus and Cascade is the product of a specific site, client, and team, and should not be treated as a universal template. Even so, several lessons from the project resonate with current discourse on performance-based facade design:

Geometry is a performance parameter. The decision to move away from a purely rectilinear perimeter toward a serrated edge shaped not only the building's appearance, but its view quality, wind behavior, and reflectivity. Treating geometry as an active performance variable—rather than a neutral container for performance—can unlock solutions that address multiple objectives simultaneously.

Modularity enables quality in complex facades. The disciplined development of a limited module family allowed a visually complex perimeter to be built with a high degree of control. Standardized modules simplify shop drawing production, fabrication, and testing, and they reduce the likelihood that isolated "one-off" conditions will undermine overall performance.

Material selection is a multi-dimensional negotiation. The glass selection process balanced often-conflicting criteria: visual clarity versus solar control, acoustic needs versus weight and cost, bird-friendly treatments versus transparency. A structured, documented process grounded in samples, mock-ups, and explicit trade-offs helps maintain coherence and build trust among stakeholders.

Testing is part of design, not an afterthought. The PMU and field tests informed the design as much as they confirmed it. Problems discovered under controlled testing conditions were resolved before they could become operational failures. Embedding testing as a design phase can improve both performance and confidence in the final result.

Ecology is part of facade performance. Bird-friendly measures are no longer peripheral. They intersect directly with decisions about geometry, glass reflectivity, and planting. The Cirrus and Cascade facade reflects this integration, showing how a performance-based approach can incorporate ecological considerations alongside energy and comfort.

9.1 Professional Decarbonization Context (2030–2050)

The AIA 2030 Commitment was established in 2009 as a voluntary, profession-led reporting program through which signatory firms track predicted energy use intensity (pEUI) across their portfolios via the Design Data Exchange (DDx) [17]. The program uses stepped reduction targets and has stated that its portfolio target increases from initial 80% to a 90% pEUI reduction in 2025 and to 100% (carbon–neutral operations) by 2030 (AIA 2023). Beyond 2030, the same logic of progressively tightening operational- and embodied-carbon performance aligns with sector-wide roadmaps, including Architecture 2030's "Zero CO_2 Emissions by 2040" scenario for the entire built environment [18] and the UIA/Architecture 2030 "2050 Imperative" calling for elimination of CO_2 emissions in the built environment by 2050 [19]. In this chapter, the DDx submission was presented as a 2018 snapshot of practice-level accountability mechanism complementary to PRM compliance modelling, and as a bridge between near-term project optimization and long-horizon decarbonization targets.

10 Future Work, Technology Readiness Level, and Market Status

Looking ahead, the most valuable next step for performance-based facade designs may lay in structured post-occupancy evaluation. Monitoring energy use and collecting occupant comfort feedback would close the loop between prediction and reality. The lessons learned would then inform future facade design, feeding back into both practice and research.

Technology readiness and market status: Most elements of the facade system and its associated testing regime are consistent with mature industry practice for high-rise residential buildings, and can therefore be considered to have a high technology readiness level (TRL ~ 8–9 rating). The more innovative aspects of the work lie in the degree of integration between multi-criteria simulation, iterative design decision-making and ecological considerations such as bird-friendly design. These integrated workflows are increasingly adopted on large projects but are not yet universal, suggesting a somewhat lower effective readiness level and highlighting opportunities for future research on standardization, automation and broader market uptake.

11 Conclusion

The facade of Cirrus and Cascade demonstrates how a performance-based mindset can guide a project all the way from early massing to final installation while maintaining a clear architectural narrative. The three cascading towers and their serrated glass envelopes are not merely formal gestures—they are the visible expression of a process linking urban context, system modularity, analytical studies, careful material selection, and rigorous testing into a single coherent strategy (see Fig. 17a–d).

Fig. 17 **a–d** Completed facade views of Cirrus and Cascade. Exterior photographs showing the towers from the lakefront and park, illustrating the serrated glass façades, vertical accent spandrels, landscaped podium, and overall integration within the Chicago skyline. *Source* **a**, **b** Tom Rossiter Photography; **c** Full Bars Media Inc; **d** YoChicago.com

Fig. 17 (continued)

In Chicago's demanding lakefront environment, this project demonstrates that high-rise residential facades can simultaneously offer expansive views, robust environmental performance, and responsiveness to ecological concerns. For architects, engineers, and researchers engaged in facade design, the case is best understood as a description of a reproducible working process rather than a prescriptive formula: facades are treated as hypotheses to be tested against clearly defined performance questions, and the resulting evidence can then be used to refine subsequent projects and associated design methods.

Acknowledgements Project Team
Developers: Lendlease Corporation, and Magellan Development Group
Architect: bKL Architecture LLC
Landscape Architects: Claude Cormier + Associés, Inc., and Confluence, Inc.
Structural Engineer: Thornton Tomasetti, Inc
MEP/FP Engineer Consultant: Salas O'Brien Engineers, Inc.
Sustainability Consultant: Salas O'Brien Engineers, Inc./Cyclone Energy Group
Energy Modeling Consultant: Cyclone Energy Group
Civil Engineer: Mackie Consultants, LLC
Wind Consultant: RWDI, Inc.
Acoustic Consultant: Shiner Acoustics, LLC
Facade Consultant: Curtainwall Design Consulting, Inc.
Contractor: Lendlease Construction, LLC
Facade Manufacturer: Reflection Window + Wall, LLC

References

1. American Bird Conservancy (2015) Bird-friendly building design, 2nd edn. American Bird Conservancy, The Plains, VA
2. AAMA/FGIA (2017) AAMA 501.1-17: standard test method for water penetration of windows, curtain walls and doors using dynamic pressure
3. AAMA/FGIA (2015) AAMA 501.2-15: quality assurance and diagnostic water leakage field check of installed storefronts, curtain walls, and sloped glazing systems
4. ASTM International (2019) ASTM E283/E283M-19: standard test method for determining rate of air leakage through exterior windows, skylights, curtain walls, and doors under specified pressure differences across the specimen
5. ASTM International (2016) ASTM E331-00(2016): standard test method for water penetration of exterior windows, skylights, doors, and curtain walls by uniform static air pressure difference
6. ASTM International (2021) ASTM E330/E330M-14(2021): standard test method for structural performance of exterior windows, doors, skylights, and curtain walls by uniform static air pressure difference
7. CDC (2020) Lakeshore East Building KL—CDC factory observation report (Quality Assurance independent observation report), 12 May 2020
8. Cyclone Energy Group (2018) Glazing and building energy performance. Technical memorandum, 22 March 2018
9. Cyclone Energy Group (2020) Lake Shore East Building J—LEED v3.0 certification submission: performance rating method (ASHRAE 90.1–2010 Appendix G) compliance report. Energy modelling outputs generated in IES Virtual Environment (VE) v2019

10. Cyclone Energy Group (2020) Lake Shore East Building KL—LEED v3.0 certification submission: Performance Rating Method (ASHRAE 90.1-2010 Appendix G) compliance report. Energy modelling outputs generated in IES Virtual Environment (VE) v2019
11. RWDI (2018) Lake Shore East Buildings K/L and J, Chicago, IL—cladding wind-load report, 24 Oct 2018
12. RWDI (2018) Lake Shore East Buildings K/L and J, Chicago, IL—pedestrian wind study, 19 Mar 2018
13. Shiner Acoustics, LLC (2018) Acoustic memorandum and CadnaA modelling outputs, 25 April 2018
14. U.S. Green Building Council (2009) LEED 2009 for new construction and major renovations (LEED v3). USGBC, Washington, DC
15. American Institute of Architects (AIA) (n.d.) The 2030 commitment. AIA, Washington, DC. https://www.aia.org/resources/6077661-the-2030-commitment
16. American Institute of Architects (AIA) (n.d.) Design Data Exchange (DDx) platform. AIA, Washington, DC. https://www.the2030commitment.org/
17. American Institute of Architects (AIA) (2024) AIA 2030 commitment by the numbers. Resource Center, 24 September 2024. https://www.aia.org/resource-center/aia-2030-commitment-numbers
18. Architecture 2030 (n.d.) Accelerating to zero by 2040! https://www.architecture2030.org/accelerating-to-zero-by-2040/
19. Architecture 2030 (2014) Architects to phase out carbon by 2050 (2050 Imperative). https://www.architecture2030.org/architects-to-phase-out-carbon-by-2050/

Designing the Sky Slide in Dubai

Agnes Koltay and Omar Oian

Abstract This paper presents a performance-driven façade engineering study for the design and validation of a fully glazed external slide installed between Levels 53 and 52 of a high-rise tower in Dubai. The system comprises a dry glass chute formed by structurally bonded laminated glass panels, representing a non-standard structural typology that is not explicitly addressed within existing international design codes or façade engineering guidelines. The research adopts a multi-layered analytical framework combining experimental testing, numerical simulation, and probabilistic risk assessment. Full-scale and component-level mock-up testing was undertaken to establish and verify baseline structural behavior, serviceability limits, and failure modes under representative loading conditions. Global and local finite element models were developed to simulate stress distribution, interlayer performance, adhesive behavior, and load transfer mechanisms across critical interfaces. These models were calibrated against physical test data to improve predictive reliability and reduce uncertainty in key design assumptions. Given the absence of prescriptive guidance for post-breakage behavior in such systems, a risk-based assessment methodology was implemented to evaluate residual load-bearing capacity and redundancy following glass fracture. This enabled the definition of project-specific acceptance criteria aligned with performance-based design principles. The findings demonstrate how a hybrid validation strategy—bridging empirical testing and advanced simulation—can be used to substantiate innovative glass structures where code-based approaches are insufficient. The paper contributes to the ongoing development of performance-based design methodologies for complex façade systems and proposes a transferable framework for assessing unconventional glass assemblies in high-risk applications.

Keywords Structural glass · Bonded glass assemblies · Finite element analysis · Post-breakage behavior · Risk-based engineering · Performance based design

A. Koltay (✉)
Koltay Facades, Dubai, UAE
e-mail: agnes.koltay@koltayfacades.com

O. Oian
Koltay Facades, Singapore, Singapore
e-mail: omar.oian@koltayfacades.com

A. Mejorin et al. (eds.), *Performance-Based Façade Design*, Springer Tracts in Civil Engineering, https://doi.org/10.1007/978-3-032-28835-6_15

1 Introduction

Structural glass applications in façades have expanded significantly in recent decades. However, the use of glass as a structural component presents a major challenge for structural engineers, as it requires a deep understanding of the glass material itself together with all the other materials that make it usable as structural elements. Glass, with its fragile and unforgiving nature, used in an obviously load bearing arrangement can result in breathtaking creations. Unlike conventional opaque materials like steel and aluminum, structural glass provides transparency, lightness and clarity, but must still satisfy strict safety, strength, and durability requirements. When applied to the design of a glass slide, these challenges intensify, demanding careful consideration of material behavior, redundancy, and user safety. Visual expectations often associate glass structures with qualities such as delicacy, slenderness, transparency and apparent fragility, whereas designs incorporating high levels of redundancy and safety tend to result in more robust, bulky, and visually heavier structural expressions, potentially compromising elegance. The result of considerate design and industry collaboration is a structure that seamlessly combines engineering precision with extensive R&D, transforming transparency into an active architectural and structural element (Fig. 1).

Fig. 1 Glass slide. Sky Views, Dubai

2 Project Overview

The Sky Slide is located between Levels 53 and 52M on the Burj Khalifa-facing elevation of Sky Views, a twin-tower high-rise development in Dubai. It comprises of two cantilevered glazed platforms connected by a straight, fully enclosed glass chute, forming a continuous transparent descent path that is both a thrill attraction and an architectural statement. The geometry, transparency, and alignment of the chute were key drivers of the design, as uninterrupted sightlines, controlled clearances, and a consistent slide gradient were essential to the user experience. As the superstructure had been completed prior to the introduction of the slide, the design was governed by retrofit constraints rather than pure greenfield structural integration. The engineering challenge was therefore centered on enabling the Sky Slide to behave as a coherent, self-supporting assembly, capable of accommodating user loads, dynamic effects, and operational requirements while maintaining tight tolerances along the length of the glazed chute. Inter-storey movements, construction tolerances, and restrictions on load transfer influenced the detailing of the Sky Slide connections, requiring movement-compatible supports that ensured alignment and serviceability of the platforms and chute without compromising the performance or clarity of the slide experience.

3 Theoretical Framework

Building on the experiential and technical ambitions of the Sky Slide, its design development was informed by a limited number of precedents worldwide. At the time of the Sky Slide design stage, the one and only main reference project offering a comparable user experience was the OUE Skyspace [1]. Installed at the U.S. Bank Tower in Los Angeles, this glass slide formed part of the OUE Skyspace observation experience and represented a rare example of a transparent, externally mounted slide in a high-rise environment. Commissioned in 2016, the OUE Sky Slide consisted of a glass tube attached between levels 70 and 69, enabling users to traverse the building façade through controlled motion within a fully glazed enclosure.

The OUE Skyspace highlighted several key challenges inherent to such attractions. These included the behavior of laminated structural glass under extreme conditions, requiring verification against combined static and dynamic actions such as user loading, impact, wind effects, and thermal gradients.

The need for redundancy and post-breakage capacity was fundamental, necessitating multi-layer lamination strategies and alternative load paths to ensure stability and safety even in the event of glass fracture.

Human-induced effects, including unpredictable occupant movement and psychological responses to transparency and height, introduced non-standard loading scenarios as well as serviceability considerations tied to vibration, deflection, and comfort.

Critical attention was also required at glass-to-structure interfaces, where connection detailing needed to accommodate construction tolerances, thermal movements, and differential deflections without inducing excessive stress concentrations in the glass.

Beyond structural performance, the precedent underscored the importance of managing the perception of safety alongside actual safety, balancing minimal visual obstruction with user reassurance to ensure public acceptance.

Finally, the absence of prescriptive codes for such installations necessitated a performance-based design methodology, supported by advanced numerical analysis, testing, and careful consideration of durability and maintenance, including long-term exposure to UV radiation, cleaning regimes, and surface wear.

4 Methodology

4.1 General Approach—Definition of Final Shape

After practicality considerations, we proposed to create two cantilevering glass boxes connected with a straight slide. This option brought the following benefits:

- Extended slide experience. While the slide itself takes less than 2 s, launching from a glass box and arriving onto a glass floor suspended at height outside of the building outline adds to the experience.
- More thrills on arrival. Instead of facing the interiors at arrival, the slider faces a glass wall and "about to slide through".
- Less risk of injuries. Sliding on glass is not like a waterslide. Bends bring risk of deviating from the nominal slide path and the slider hitting the walls or being dragged along the walls, resulting in discomfort and injury. A straight slide reduces this risk.
- Less components. A straight slide allows for the use of single pieces of flat glass, while a curved slide would necessitate more complex structure and may result in less transparency.
- Manufacturing lead times and duration of installation are reduced. Maintenance and potential replacement can be easier.

In terms of primary structure, the structural engineer implemented two steel trusses installed between the floor slabs, providing a discrete and stiff load receiving framework independent of the façade elements. The glazed platform enclosures were subsequently fixed to these trusses using point supports, establishing a direct and clear load path. The glass chute was then supported by dedicated steel brackets connected to the same truss system, concentrating vertical and horizontal actions into a limited number of predefined interfaces. This approach allowed accommodation of differential movements while maintaining structural clarity, minimizing secondary load

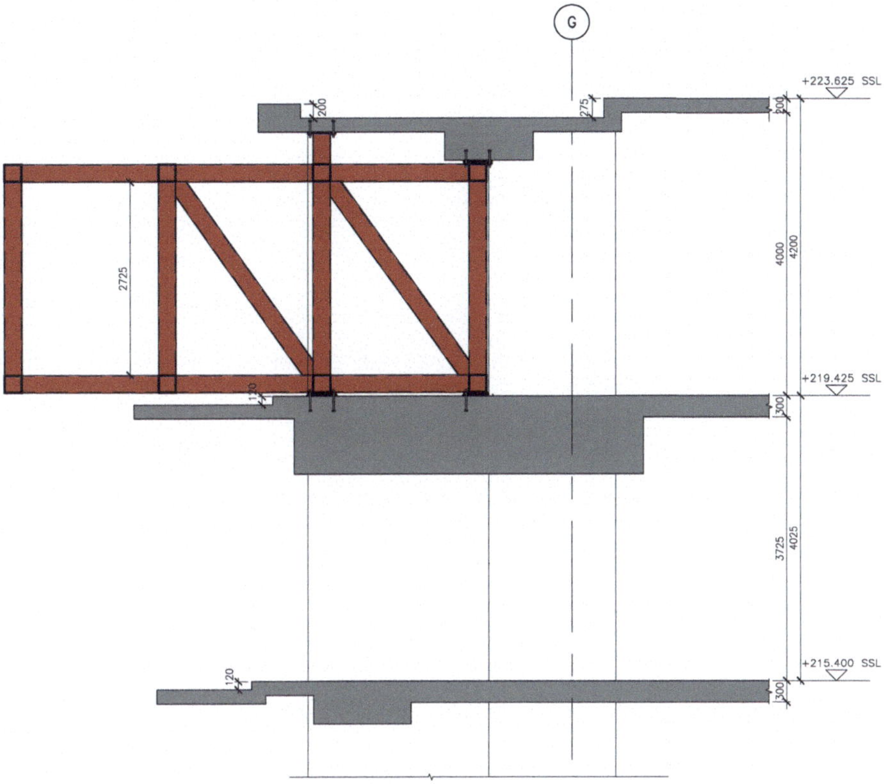

Fig. 2 Support structure diagram of the launch platform

effects on the glass components, and preserving the performance objectives of both the existing structure and the inserted system (Figs. 2 and 3).

The final geometry of the glass slide was not driven primarily by novelty or material expression, but by the requirement to achieve predictable user behavior and compliance with defined safety and performance objectives within a non-standard application. Although the project employs unusual materials and is located externally at high elevation, the fundamental functional requirement remained that of a controlled, gravity-driven sliding device with clearly defined transition, sliding, and run-out zones.

No single published standard directly addresses a fully enclosed, dry glass slide intended for adult users and integrated into a high-rise façade. Consequently, the geometric definition of the slide relied on selective interpretation of existing playground and amusement-ride standards, recognizing their limitations while extracting provisions directly relevant to user safety, posture control, inclination, clearance, and entanglement avoidance. In particular, standards developed for playground equipment and water slides—although not fully applicable—provide empirically derived geometric thresholds that inform safe sliding behavior.

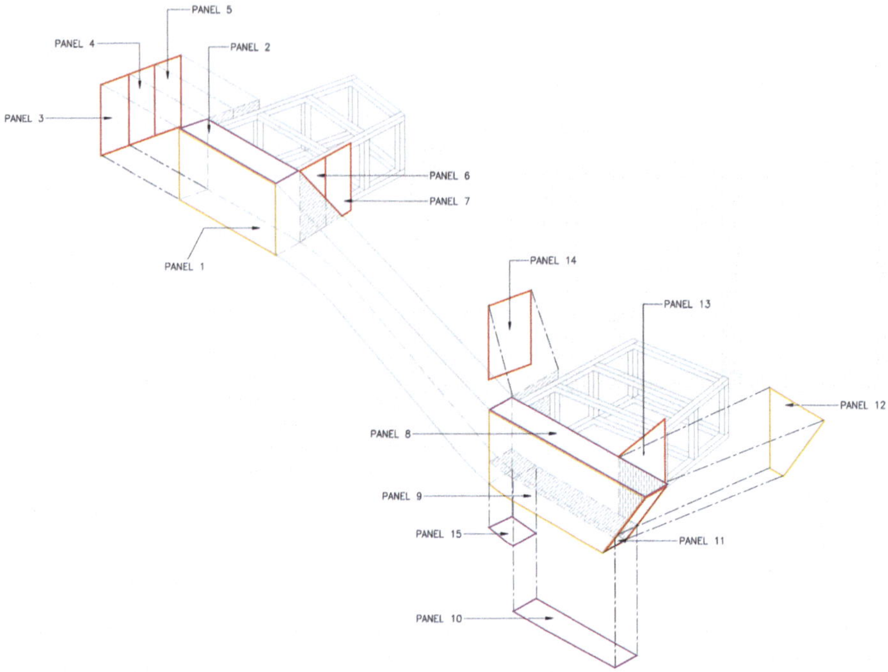

Fig. 3 Glass slide launch platform and arrival platform glazing panels

The following standards were reviewed as reference documents:

- ASTM F1487-01: Standard Consumer Safety Performance Specification for Playground Equipment for Public Use
- ASTM F2291-11: Standard Practice for Design of Amusement Rides and Devices
- ASTM F2376-06: Standard Practice for Classification, Design, Manufacture, Construction, and Operation of Water Slide Systems
- BS EN 1176-1:2008: Playground equipment and surfacing—General safety requirements and test methods.

Given that the slide is fully enclosed, dry, and intended for controlled single-user operation, several provisions within these standards were not directly applicable. However, ASTM F1487 and BS EN 1176 were identified as the most relevant sources for defining geometric limits governing slope, curvature, clearance dimensions, and transition behavior. These parameters directly influence user posture, speed development, and exit conditions and were therefore adopted as performance-defining constraints rather than prescriptive requirements.

From ASTM F1487, criteria governing maximum inclination, continuity of the sliding surface, and avoidance of gaps or abrupt geometric discontinuities were used to define the overall slope envelope and surface smoothness of the chute. Provisions relating to entrance configuration ensured reliable channeling of the user into a seated position, reducing variability in loading, contact forces, and kinematics. Requirements related to exit geometry informed the design of rounded terminations and controlled run-out behavior to mitigate impact risk.

Similarly, BS EN 1176 clauses relating to maximum and average slope, radiused transitions between changes in inclination, and minimum tunnel clearances were used to define acceptable curvature and cross-sectional geometry. These provisions were particularly relevant given the rigid nature of glass and the absence of compliance that would otherwise absorb local geometric irregularities. Radiused transitions were therefore treated as a key design variable, influencing both user comfort and stress concentrations in the glass panels.

By synthesizing the above criteria, the slide geometry was developed as a straight chute with controlled inclination, continuous surface transitions, and explicitly defined run-out length. This configuration minimized lateral acceleration, reduced the risk of unintended user contact with side panels, and allowed the use of flat laminated glass units with well-defined load paths. The resulting geometry therefore represents a performance-driven compromise between user safety, structural clarity, constructability, and material behavior, rather than direct compliance with any single codified slide typology.

The initial analysis of the design of the slide can be seen in Fig. 4.

The transition zone, chute inclination, and run-off length were subsequently verified against the defined safety criteria governing deceleration, occupant reach, and entrapment risk. Of particular importance was the definition of a minimum run-off distance equal to at least 0.3 times the overall slide length, which was treated as a governing parameter for controlling stopping behavior and reducing residual kinetic energy at exit. This zone was combined with an explicit entrapment-clearance

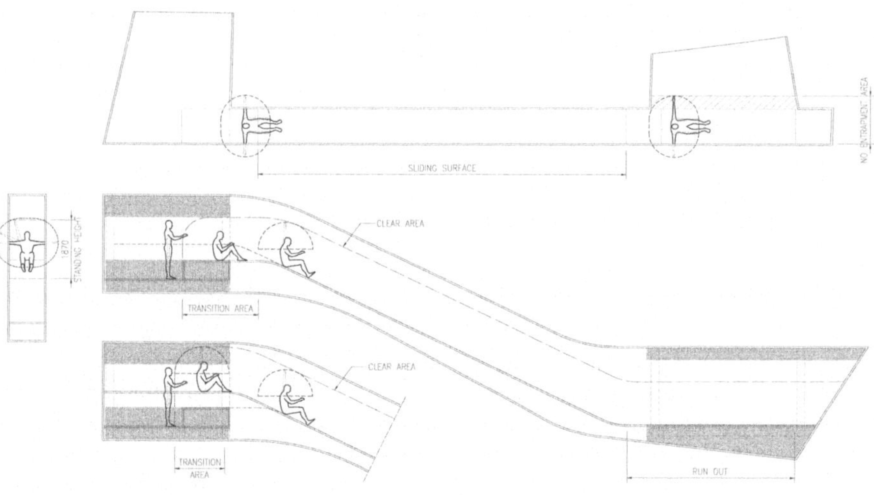

Fig. 4 Selection of inclination and runoff

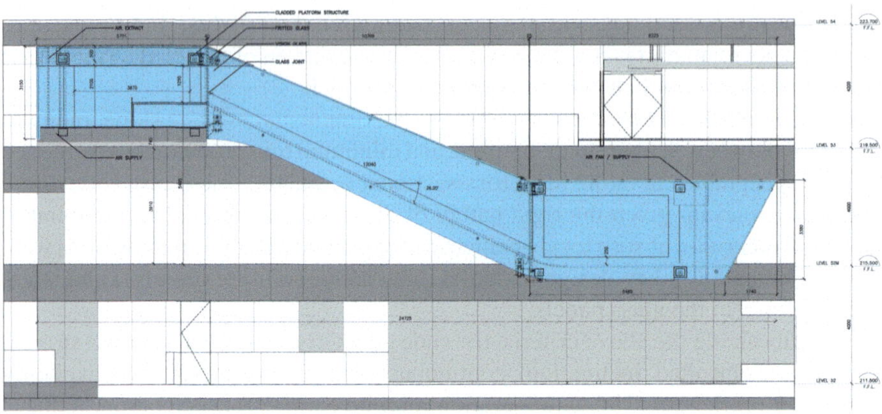

Fig. 5 Glass slide elevation

envelope derived from the reach of a 95th-percentile adult male, ensuring conservative accommodation of user interaction beyond average anthropometric assumptions (Fig. 5).

Within this combined clearance envelope, detailing constraints were imposed to eliminate secondary hazards. Any point fixings, protrusions, or local geometric discontinuities were prohibited unless fully flush or countersunk, thereby preventing snagging, impact, or finger entrapment during both user exit and emergency egress scenarios. The same clearance logic was extended beyond the user path to the landing platform, recognizing that operator personnel may occupy this area during operation. As a result, the run-off and landing zones were treated not merely as terminal elements of the slide geometry, but as shared safety-critical regions serving both users and operational staff. This approach ensured geometric continuity between functional slide performance and broader risk-management considerations associated with operation and maintenance.

4.2 Safety Approach—Definition of the Factor of Safety

In conventional structural design, factors of safety (FoS) are typically adopted directly from national or international standards such as ASTM, AISC, or EN codes. However, the glass slide incorporates components, geometries, and load conditions that are only partially addressed—or not addressed at all—by existing regulations. These limitations are not unique to this project, but are characteristic of transparent, human-occupied glass structures subjected to dynamic use. As a result, a uniform application of code-defined safety factors would neither reflect the relative criticality of individual components nor adequately capture the consequences of potential failure modes. A differentiated, performance-based approach to FoS selection was therefore adopted.

It is important to distinguish between project-specific constraints, which influenced the numerical values ultimately adopted, and generalizable design principles, which underpin the methodology and are applicable to similar glass attractions. The governing principle, applicable beyond this specific project, was that safety margins should be proportional to both the risk associated with component failure and the practical implications of repair, replacement, or loss of use. Project-specific considerations, such as accessibility of elements at height, replacement logistics, and anticipated operational downtime, informed the calibration of these margins but did not alter the underlying risk-based framework.

Components identified as safety critical, defined as those whose failure could directly compromise user safety, overall stability, or result in disproportionate interruption of the attraction, were designed with higher factors of safety, in some cases exceeding values typically prescribed by available standards. Less critical components, where failure would not result in immediate safety risk or where redundancy was demonstrably present, were designed in accordance with conventional values derived from established codes.

Risk classification considered not only the probability and consequences of failure, but also operational factors such as accessibility, replacement complexity, and interruption to use of the attraction. While the weighting of these factors is necessarily project-dependent, the process itself, explicitly linking FoS selection to risk, consequence, and recoverability, is generalizable. This framework allowed safety margins to be deliberately concentrated where they provided the greatest benefit, rather than applied uniformly across the system, avoiding unnecessary conservatism in non-critical elements while enhancing robustness where it mattered most.

Based on this approach, the following target factors of safety were adopted for the glass chute assembly under different design conditions.

Factors of Safety for Testing Conditions

- Glass fixings: FoS = 4.0
- Glass floor panels: FoS = 4.0
- Entire assembly (global behavior): FoS = 1.0.

Testing conditions were defined to reflect short-duration, controlled load scenarios intended to verify behavior rather than to establish long-term design capacity. While the reduced global factor is specific to the testing regime of this project, the underlying principle, maintaining elevated safety margins for local, safety-critical components even during controlled testing, is transferable to similar glass systems.

Factors of Safety for Design under Nominal Operating Conditions

- Glass fixings: FoS = 5.0
- Glass floor panels: FoS = 5.0
- Entire assembly (global behavior): FoS = 2.0.

Under nominal operating conditions, high safety factors were assigned to primary glass elements and their attachments due to their brittle behavior, limited warning prior to failure, and central role in user safety. This reflects a general design principle for structural glass subjected to public occupancy. The lower, but still conservative, factor applied to the global assembly reflects the presence of defined load paths, redistribution capacity within the supporting structure, and inherent redundancy at system level.

Factors of Safety for Post-breakage Design Scenarios

- Glass fixings: FoS = 3.0
- Glass floor panels: FoS = 3.0
- Entire assembly (global behavior): FoS = 1.0.

Post-breakage scenarios were assessed to verify residual load-bearing capacity and robustness rather than continued serviceability. Reduced safety factors were therefore adopted, consistent with the objective of preventing disproportionate collapse and ensuring controlled behavior following glass fracture. Although the numerical values are project-specific, the distinction between serviceability-driven and robustness-driven design objectives is directly applicable to other laminated glass installations.

Factors of Safety for Steel Components

- Steel elements in direct contact with glass panels: FoS = 4.0, based on yield strength
- Steel elements not in direct contact with glass panels: FoS = 1.0, with load factors applied in accordance with relevant EN standards.

Elevated safety factors were applied to steel elements in direct contact with glass to minimize the risk of stress concentrations, unintended restraint, or local damage that could compromise glass performance and ensure overall system redundancy. This reflects a general principle of interface-sensitive design in glass structures. Steel components isolated from glass elements were designed using conventional load factors, consistent with their ductile behavior, predictable failure modes, and lower sensitivity to localized over-stress (Fig. 6).

Since the viewing platforms do not represent special structural components in terms of geometry or use, their glass panels and steel bracketing were designed using available construction standards: ASTM E1300 and ASTM E2751 for the glass, and EN 1993 for the steel. This further illustrates the selective application of enhanced safety measures only where component behavior, risk exposure, and uncertainty justified deviation from prescriptive code-based design.

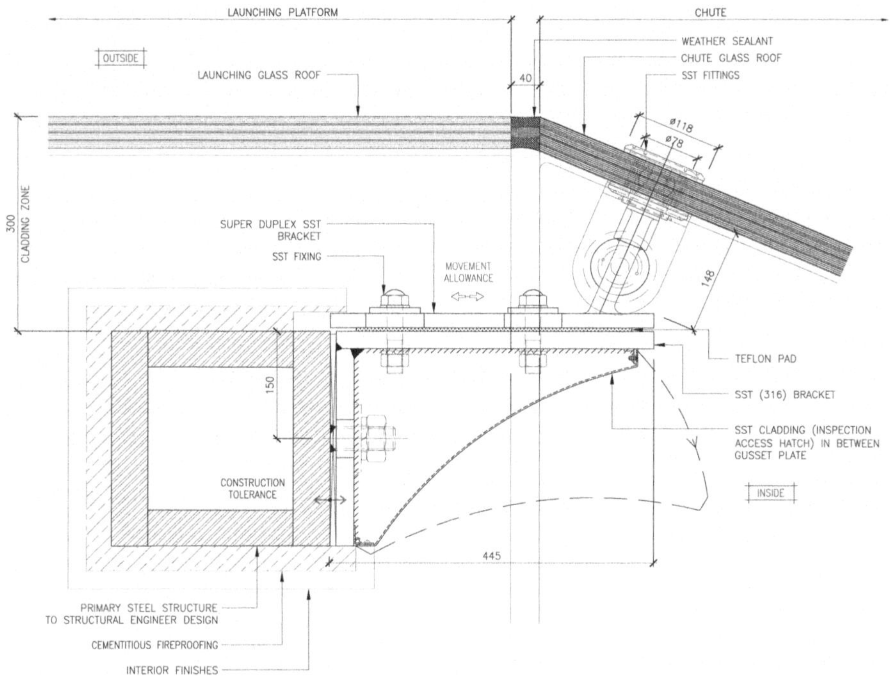

Fig. 6 Factor of safety for glass chute and relative connections to the main structure (red: FoS = 1.0, blue: FoS = 4.0)

5 Materials Applications

5.1 *Use of Structural Glass*

Among the principles that guide the structural performance-based design of glass, the two of major importance are:

1. Certainty of the Load Path
2. Redundancies.

These are well established guiding criteria as per Patterson [2]. Below is a description of how these aspects were taken into account in the final design.

Load Path

The continuous route that forces take from their point of application through structural elements to the primary concrete structure of the base building is referred as the load path. The load path is decided during conceptual phase and revalidated throughout the design.

The glass chute is conceived as a closed structural assembly composed of four primary laminated glass fins: two horizontal elements forming the floor and roof, and two vertical elements forming the inner and outer side walls. The load bearing behavior of the chute is governed by the directional stiffness of these fins, with primary actions intentionally resisted along the strong axes of the individual glass elements. This configuration allows gravity, user induced, and environmental loads to be transferred efficiently through the glass geometry rather than relying on secondary framing or discrete fittings.

Structural silicone is employed to bond the individual glass components into a unified chute assembly, enabling composite action and continuity along its length. In this configuration, the chute acts as a single structural element spanning between supports at Levels 53 and 52M, rather than as a collection of independently supported panels. To support this behavior, the laminated side walls incorporate stepped glass build ups that form integral shelves, providing direct bearing support to the floor panel and reducing reliance on point fixings. This geometry establishes a direct and legible vertical load path while minimizing eccentric load transfer into the glass.

Connection design was governed by the dual requirement of maintaining a clear load path while accommodating post installation movements of the primary structure. These movements include inter storey drift, differential thermal expansion, and construction tolerances. To prevent secondary stress accumulation within the glass, the chute supports were intentionally configured as nominally simply supported interfaces, avoiding moment resisting connections wherever possible. Dead load is supported at the lower interfaces, while controlled translational releases are provided to accommodate relative movements between the slide structure and the supporting platforms.

In the adopted support arrangement, individual connection points are assigned distinct restraint directions to avoid over constraint of the system. Vertical reactions are resisted where required for stability, while horizontal and out of plane restraints are selectively introduced only where necessary to ensure global equilibrium. This distributed restraint strategy ensures static determinacy at system level, limits unintended load redistribution, and reduces sensitivity to fabrication and installation tolerances.

The principal benefit of this approach is a structurally transparent system in which actions are transferred predominantly into the glass elements along their strongest axes, the number of fittings is minimized, and failure sensitive load paths are avoided.

A similar load path philosophy was applied to the fixing strategy of the launching and landing platforms. In these locations, the primary design challenge arises from the coexistence of mechanical fixings—which behave as relatively rigid supports—and structural silicone sealant—which exhibits significantly lower stiffness and nonlinear behavior. If not explicitly accounted for, this stiffness mismatch can result in unintended load sharing, with mechanical fixings attracting a disproportionate share of load or, conversely, local overstressing of the sealant.

To address this, a stiffness sensitivity study was conducted at an early design stage using solid finite element analysis. The study demonstrated elevated sealant stresses in the vicinity of rigid mechanical fixings, confirming the need for controlled interaction between support mechanisms. As a result, major point connections (routels) were detailed with translational releases in directions where the structural sealant was intended to act as the primary load carrying element. This strategy ensured predictable load distribution, prevented sealant overstress, and preserved the intended hierarchy between bonded and mechanical support mechanisms.

Redundancies

Glass exhibits brittle behavior characterized by limited tensile strength and the absence of plastic deformation, resulting in minimal warning prior to failure. As a consequence, structures relying on glass as a primary load-bearing material require explicit consideration of failure scenarios and the potential for progressive collapse following local damage. In the absence of inherent ductility, robustness cannot be achieved through material behavior alone and must instead be embedded at system level through deliberate redundancy.

Redundancy is an intentional inclusion of backup components or additional capacity, that are not required for primary functioning, but designed to ensure a continued, unaffected safe performance in case of failure of a primary component.

The design strategy for the glass chute therefore incorporates alternative load paths capable of maintaining structural stability following glass breakage. Redundancy was treated as a primary performance objective, ensuring that local failure of a glass component would not trigger disproportionate or cascading failure of the overall system. This is particularly critical given the public nature of the installation and the inability to rely on user exclusion immediately following damage.

Steel components were selectively introduced to fulfil this robustness function. Due to their high strength, predictable elastic-plastic behavior, and capacity to undergo large deformations without loss of load-carrying capability, steel elements provide a reliable secondary load-bearing mechanism following glass failure. Their application was deliberately limited in size and number to preserve the intended visual transparency while ensuring that residual structural capacity is retained under post-breakage conditions.

The resulting configuration allows primary service loads to be carried predominantly by the glass under normal operation, while steel redundancies activate under damage scenarios to stabilize the system and prevent collapse. This hierarchical load-sharing strategy ensures compatibility between architectural intent and structural safety requirements and forms a key component of the overall performance-based design approach adopted for the chute.

The angles connecting the vertical glass fins to the roof panel are designed to mechanically link these elements in the event of complete failure of the structural silicone sealant, thereby preventing separation of the chute enclosure. Similarly, a mechanical pin located at the bottom shelf is detailed to provide full support to the floor panel in the event of interlayer failure, ensuring that the primary walking and sliding surface remains supported even under extreme damage conditions.

5.2 Structural Silicone Sealant

Structural silicone is an established material for load-bearing glass-to-glass and glass-to-metal connections in façade engineering, supported by long-term use in structural glazing applications [3]. In the context of the glass chute, its adoption was driven by the requirement to realize a continuous structural system with a clearly defined load path, minimal mechanical interruption, and predictable deformation behavior under both serviceability and accidental conditions.

The primary functional motivation for employing structural silicone was to enable the chute to behave as a unified glass assembly, in which forces are transferred continuously along bonded interfaces rather than concentrated at discrete point fixings. This uninterrupted load transfer mechanism reduces local stress concentrations, improves load distribution along the length of the chute, and supports the intended composite structural action between floor, roof, and side panels. In addition, minimizing mechanical penetrations within the chute enclosure eliminates surface discontinuities that could otherwise introduce snagging hazards or local variations in friction, both of which are undesirable in a dry sliding application.

From a performance perspective, the use of bonded connections also facilitates controlled relative movement between glass elements. Unlike rigid mechanical fixings, structural silicone exhibits a deformable response that can accommodate differential thermal movements, construction tolerances, and limited structural displacements without inducing peak stresses in the glass. This behavior is particularly relevant for an externally exposed structure subject to temperature variation and building movement.

The adoption of structural silicone was, however, accompanied by recognized challenges. The stress distribution within bonded joints is not fully addressed by existing standards, necessitating detailed numerical analysis and explicit verification of load scenarios beyond simplified hand calculations. In addition, bonded structural systems require long-term inspection and maintenance strategies to ensure continued performance throughout their service life. These considerations were treated as integral aspects of the design process rather than secondary disadvantages.

5.3 Stainless Steel

While the selection of the glass components was governed primarily by structural performance and architectural intent, the choice of materials for fixings and redundancy elements was driven by a broader set of criteria, including environmental exposure, durability, required safety margins, and geometric efficiency. Given the external location of the slide and its exposure to a highly aggressive coastal and urban environment, corrosion resistance emerged as a governing performance requirement for all metallic components.

From the outset, carbon steel and protected alloy steels were excluded due to the long-term durability risks associated with coating deterioration and maintenance dependency. Stainless steel was therefore identified as the only viable material class capable of delivering the required corrosion resistance while maintaining predictable mechanical behavior over the service life of the installation. Within this material class, the design was further constrained by the elevated factors of safety applied to safety-critical components and the requirement to minimize the size and number of fixings in order to preserve visual transparency and avoid local stress concentrations in the glass. These constraints favored high-strength stainless steel grades with superior corrosion resistance, allowing reduced cross-sectional dimensions without compromising reliability.

Based on a review of available grades, technical literature, and consultation with stainless-steel manufacturers, a combination of Duplex and Super Duplex stainless steel was selected for the chute fixings and redundancy elements. Both materials offer a favorable balance between strength and ductility, but super duplex stainless steel provides enhanced corrosion resistance due to its higher chromium and molybdenum content. Corrosion performance was assessed using the Pitting Resistance Equivalent Number (PREN), a widely adopted indicator of resistance to localized corrosion in chloride-rich environments. Typical duplex stainless steel grades (EN 1.4462) exhibit PREN values of approximately 35, whereas super duplex grades (EN 1.4410) achieve PREN values exceeding 40, indicating significantly improved resistance to pitting and crevice corrosion.

The selection strategy therefore differentiated between components based on exposure severity and functional criticality. Super duplex stainless steel was adopted for elements subject to direct environmental exposure and those acting as primary redundancy mechanisms, where corrosion-induced degradation could directly compromise post-breakage performance. Duplex stainless steel was used for less exposed components where durability demands remained high but ultimate robustness requirements were lower. This approach allowed material performance to be matched to functional demand, avoiding unnecessary conservatism while maintaining long-term reliability.

The corrosion performance assumptions adopted in the design are supported by experimental exposure data from long-term atmospheric testing programmes in Middle-Eastern environments, including four-year field exposure studies comparing duplex and super duplex stainless steel grades [4]. These results corroborate the suitability of super duplex stainless steel for highly exposed, safety-critical applications in aggressive climates.

6 Approach—Detailed Components Design

6.1 Design Criteria and Loads

The basis of design is defined through the applied external load cases and corresponding deflection criteria adopted for the analysis. Strength limits for glass, structural sealant, and steel components are defined by the component-specific factors of safety established in the preceding sections, reflecting their differing material behavior and performance objectives.

Wind Load

- According to the Wind Tunnel Test report for the Sky Views towers, the maximum unfactored wind pressures on the façade at the slide locations are:
- Wind suction: 2.0 kPa;
- Wind pressure: 1.5 kPa.

As the wind tunnel testing for the main tower was completed prior to the introduction of the glass slide, the resulting pressure coefficients did not account for the local aerodynamic effects induced by the slide geometry itself. In particular, potential amplification of suction and pressure due to geometric protrusion, enclosure effects, and local flow separation at the chute location were outside the scope of the original study. To address this limitation, supplemental input was obtained from the wind engineering consultant in the form of project-specific peak net pressure values applicable to the glass slide elements. These values were subsequently adopted as governing wind actions for the detailed design and verification of the slide structure.

Inward facing vertical panels:

- Wind suction: 2.0 kPa;
- Wind pressure: 1.5 kPa.

All other panels:

- Wind suction: 3.0 kPa;
- Wind pressure: 2.0 kPa.

Live Loads

As existing standards do not fully cover all relevant load cases for the glass slide, a performance-based load definition was adopted. This approach combines conventional values taken from ASCE 7 with non-standard loads derived from comparable structural systems and conservative engineering assumptions tailored to the specific behavior of the installation.

Glass Chute

- Point Live Load (PLL) of 1.8 kN at any location;
- Uniformly Distributed Live Load (UDLL) of 5.0 kPa on the floor;

- Uniformly Distributed Live Load (UDLL) of 1.5 kPa on the roof;
- Impact Load (IL) on the roof and side glass 450 J impact load.

Platforms

- Horizontal Point Live Load (PLL) of 1.5 kN at 1.1 m from FFL;
- Horizontal Linearly Distributed Live Load (LDLL) of 1.5 kN/m at 1.1 m from FFL;
- Horizontal Uniformly Distributed Live Load (UDLL) of 1.5 kPa up to 1.1 m from FFL;
- Uniformly Distributed Live Load (UDLL) of 5.0 kPa on the floor;
- Uniformly Distributed Live Load (UDLL) of 1.5 kPa on the roof;
- Impact Load (IL) on the roof and side glass 450 J impact load.

Other Loads

All the other loads such as Seismic, Thermal etc. were accounted for during the design. However, their combinations and the magnitude of such loads when not relevant for the final design.

Glass Deflection Limits

The design considered the following glass deflection allowances:

- Glass center deflection min (L/60; 25 mm)
- Glass floor deflection min (L/240) under DL + LL
- Glass floor deflection min (L/360) under LL.

6.2 Glass Chute—The Global Model

Preliminary hand calculations based on available standards—such as AS 1288 for individual glass fin checks—were undertaken to establish an initial understanding of component-level behavior and to support early-stage sizing. While adequate for verifying isolated elements under simplified assumptions, these checks were not sufficient to characterize the global behavior of the slide system. The interaction between multiple glass panels, bonded interfaces, steel components, and viscoelastic interlayers results in coupled stiffness and load-sharing effects that cannot be captured through component-based calculations alone.

A more comprehensive analysis was therefore required to evaluate force redistribution, deformation compatibility, and stress development across the integrated system. To this end, a global finite-element model was developed using Strand7 [5, 6], enabling simultaneous assessment of structural movements and stresses across all primary components under combined loading scenarios. This approach allowed the influence of material interaction and connection behavior to be explicitly incorporated into the design verification process.

The size of the chute and the hybrid nature of the structural system introduced significant modelling challenges. The combination of multiple material types, large geometric dimensions, and non-linear behavior led to substantial computational demands. In addition, the iterative nature of the design process required that analysis runtimes be balanced against the need for timely design feedback, as changes triggered by analytical findings could necessitate substantial redesign effort at component and detailing level.

The overall complexity of the numerical modelling was driven by several interrelated factors:

- Non-linear material behavior of glass and sealant, including sealant deformation characteristics and temperature-dependent interlayer response
- Mesh definition, required to balance numerical accuracy against computational feasibility over a large structural domain
- Representation of structural sealant, which necessitated simplification strategies to capture stiffness effects without excessive model refinement
- Viscoelastic behavior of the laminated interlayers, requiring load-duration-dependent material properties
- Number and diversity of load cases, including serviceability, ultimate, and post-breakage scenarios.

These constraints informed the adopted modelling strategy, in which global system behavior was prioritized while local effects were examined selectively through targeted refinement. The resulting finite-element framework provided a rational basis for performance-based assessment of the chute under realistic operating and accidental conditions, complementing the limitations of prescriptive checks and enabling informed engineering decision-making.

Mesh

The reliability of finite-element analysis is dependent on the suitability of the adopted mesh in capturing both global structural behavior and local stress development. As standard modelling practice, mesh refinement is required in regions of high stress gradients or geometric discontinuity, while coarser discretization may be employed elsewhere to limit computational demand. In the present case, the objective of the global model was not to capture detailed local stress concentrations, but to accurately represent system-level stiffness, load redistribution, and deformation compatibility between glass, bonded interfaces, and supporting steel components.

The chute was modelled predominantly using eight-node solid (brick) elements to represent the glass panels, interlayers, and primary connection zones. From a numerical perspective, optimal accuracy is typically achieved when element aspect ratios approach unity. However, the geometric scale of the chute, combined with imposed limits on the total number of nodes (approximately 250,000), made uniform low-aspect-ratio meshing computationally impractical. As a result, portions of the model necessarily incorporated elements with high aspect ratios, in some cases exceeding 100.

Given this limitation, the modelling strategy prioritized consistency and verification over theoretical mesh optimality. The adopted meshing approach was reviewed in consultation with the finite-element software developer, and preliminary analyses were scrutinized to assess sensitivity to mesh quality. No numerical instability or non-physical behavior was identified, supporting the suitability of the approach for the intended global assessment.

To improve control over mesh distribution and element quality, glass panels, interlayers, and primary connections were generated by extrusion from a surface mesh rather than through full three-dimensional volumetric meshing. This technique allowed refinement of element size and internal angles while maintaining manageable computational requirements. Components that were not critical to the global response—such as supporting beams and bearing plates—were represented using beam or plate elements, reflecting their structural role without unnecessary mesh refinement.

The resulting global model was therefore capable of capturing the relative stiffness contributions of all primary components while remaining computationally efficient. The modelling process adopted a progressive simplification strategy: the four primary glass elements forming the chute were first reduced to a geometric "skeleton" to establish load paths and symmetry. Symmetry conditions were exploited where applicable to reduce meshing effort, and particular emphasis was placed on regions surrounding mechanical connections, where stiffness interaction was expected to be most influential. Minor geometric features, such as rounded edges at the base of the vertical fins, were simplified where they were not expected to materially affect global behavior.

The final global finite-element model comprised approximately 170,000 nodes, with an average runtime of 4.5 h per load case. As a conservative modelling assumption, glass thicknesses were taken as the minimum values permitted by ASTM E1300, while differences in self-weight relative to the final specification were accounted for through adjusted material density or supplementary loads. This ensured that stresses and deformations were not underestimated at system level.

The modelling of structural sealant represents a distinct case and is addressed separately in the subsequent section.

6.3 *Treatment of the Chute Floor and Thermal Considerations*

During design development, preliminary computational fluid dynamics (CFD) analyses performed by a third-party consultant identified elevated surface temperatures at the chute floor under peak summer conditions, raising concerns regarding operational usability. An insulated glass unit (IGU) configuration was therefore initially proposed to mitigate heat transfer. For structural analysis purposes, the IGU was not explicitly modelled in the global finite element model, as only the inner glass pane was intended to participate in load bearing. The external pane was considered nonstructural, and its contribution was represented conservatively through the addition of equivalent permanent load acting on the primary structural glass. Subsequent

CFD analyses conducted at the construction stage demonstrated that the IGU was no longer required to meet thermal performance objectives, and the configuration was therefore removed from the final design without compromising structural safety.

Modeling of Structural Sealant

Similar to the approach adopted for the glass panels, the structural silicone sealant was initially assessed using simplified load distribution methods and joint force calculations derived from available standards. While these checks provided an initial bounding assessment, it became evident that the safety factors prescribed by guidance documents such as ETAG 002 [7] and ASTM C1401 are primarily intended for generic façade applications and prescriptive verification and were not directly compatible with the specific load-sharing mechanisms, redundancy strategy, and performance objectives of the glass chute system.

A more refined analytical approach was therefore required to accurately characterize the stress state within the sealant joints and to rationalize the required joint dimensions while maintaining an appropriate level of safety [8]. In particular, the intent was to avoid excessive conservatism that would unnecessarily increase sealant widths and joint lengths, while still ensuring robust performance under serviceability, ultimate, and accidental design scenarios. On this basis, the global finite-element model was identified as the most suitable framework for assessing sealant behavior within the context of the overall structural system.

Direct three-dimensional modelling of the sealant using solid elements was initially considered. However, this approach proved impractical due to both computational limitations and supplier requirements for sealant discretization, which typically mandate element sizes not exceeding approximately 10 mm to resolve stress gradients accurately. Applying such refinement consistently across the full chute geometry would have led to prohibitive model sizes and runtimes, compromising the iterative design process.

An additional complexity arises from the hyper-elastic mechanical behavior of structural silicone sealants, which exhibit large, non-linear deformations and time-dependent response. These characteristics are not readily represented within conventional linear elastic frameworks and require the use of specialized constitutive models. Sealant suppliers typically provide material parameters for established hyper-elastic formulations such as Neo-Hookean or Mooney–Rivlin models; however, these parameters are product-specific and often limited in scope. For the purposes of numerical representation within the global model, a sealant product with the most comprehensive and transparent material data availability was selected as a reference material, enabling calibration of stiffness behavior.

Following parametric trials using detailed solid-element sub-models, the structural sealant joints were ultimately represented within the global finite-element model using equivalent spring elements. This abstraction allowed the essential axial and shear stiffness characteristics of the sealant to be captured while avoiding the computational burden associated with explicit volumetric modelling. The approach ensured compatibility with the overall modelling strategy, enabled efficient evaluation of load redistribution and relative movement between components, and provided a rational basis for assessing sealant stresses within the performance-based design framework.

In the global finite-element model, the structural silicone sealant was represented by spring elements defined through axial and shear stiffness components. These stiffness values are not intrinsic material constants but effective properties that depend on the joint geometry, specifically the sealant width, thickness, and effective bonded area. To derive representative stiffness parameters for the major sealant joints, a calibration procedure based on detailed solid-element modelling was adopted.

Representative sealant joints were first modelled using three-dimensional solid elements to capture the non-linear deformation behavior under controlled loading. Eight-node brick elements with a refined mesh of approximately $1 \times 1 \times 1$ mm were employed to ensure accurate resolution of strain gradients within the joint. The sealant material was modelled using a hyper-elastic constitutive formulation, adopting material parameters corresponding to the selected reference silicone. In the present case, a Mooney–Rivlin material model was implemented using coefficients derived from manufacturer documentation, with parameters selected to represent the non-linear stress–strain response of the cured sealant under large deformations.

Incremental axial and shear loads were applied to the solid-element sub-models, and the resulting relative displacements between bonded faces were extracted from non-linear analysis results. These displacement responses formed the basis for determining equivalent joint stiffness values, expressed as force–displacement relationships rather than material-level stresses.

From the solid-model results, effective axial and shear stiffnesses were calculated using energy-equivalent relationships:

- Axial stiffness: $k_{\mathrm{n}} = E \cdot A/t$
- Shear stiffness: $k_{\mathrm{v}} = G \cdot A/t$.

where E and G represent the effective Young's and shear moduli of the joint, A is the bonded area associated with the finite-element discretization, and t is the sealant thickness. Depending on the analysis requirements, stiffness behavior may be represented either by non-linear force–displacement curves derived directly from the solid-model response or by linearized stiffness values obtained through averaging over the relevant deformation range.

Because effective stiffness is dependent on the tributary area of each spring element, stiffness values vary with mesh resolution. To maintain consistency and avoid unnecessary proliferation of parameters, only two discrete meshing sizes were adopted in the global model: a coarser mesh for the central regions of the chute and a finer mesh in the vicinity of supports and connection interfaces. Correspondingly, two calibrated sets of axial and shear stiffness values were applied within the global model. This approach preserved geometric and mechanical consistency between the solid sub-models and the global representation while keeping computational complexity within practical limits.

Interlayer

The glass interlayer (stiff ionoplast) is a well-known viscoelastic polymer, which mechanical properties are highly dependent on temperature of application and load duration (time). Typically, the longer the duration and the higher the temperature, the lower (worse) are the mechanical properties. While the load duration is clearly defined by the load type (e.g. self-weight means very long duration, wind load means 3-s duration) the temperature of application required a more detailed study [9].

At the initial stage of the design, only generic climatic data for Dubai was available. To obtain more representative input for the analysis, an in-house experimental study was conducted to evaluate the variation in surface temperature of different glass types under summer exposure conditions. The test samples comprised clear coated glass from two different suppliers and two fritted glass panes with distinct patterns (see Fig. 7). All specimens were installed in a horizontal orientation and continuously exposed on the rooftop of the company's office. Surface temperature measurements were recorded daily at 9:00, 13:00, and 17:00, over the period from 12 June to 29 October of the same year, corresponding to the peak summer conditions.

Chart below shows the measurements for all the samples at 1 PM (highest daily values). As we can see, most of the values lie in the 50–60 °C range, with some peaks above 70 °C. As no other data was available it was decided to carry out the design with the highest temperature for which the interlayer supplier provides data: 80 °C. This could give a sufficiently high level of confidence and an adequate margin on the FEA results to account for possible variations that could not be noticed otherwise during this stage of the design. It is worth noting that FEA was also conducted for a

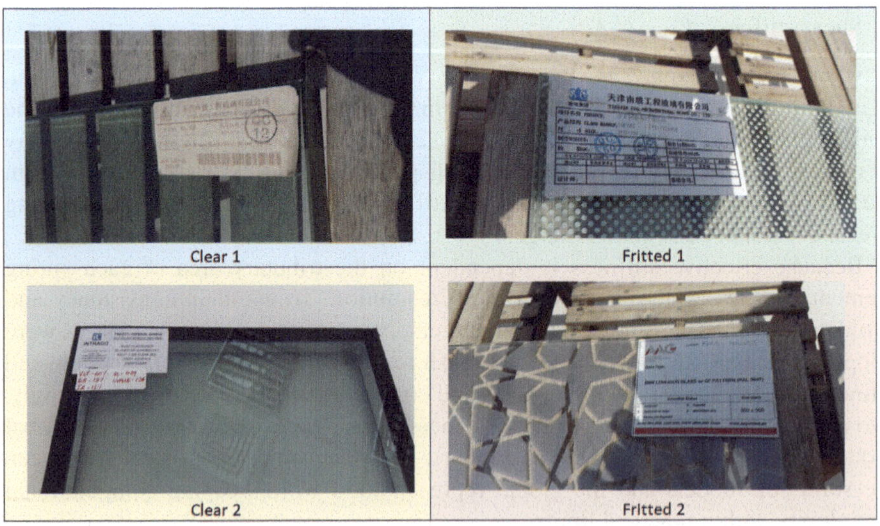

Fig. 7 Test samples for surface temperature monitoring, exposed and measured over summer months

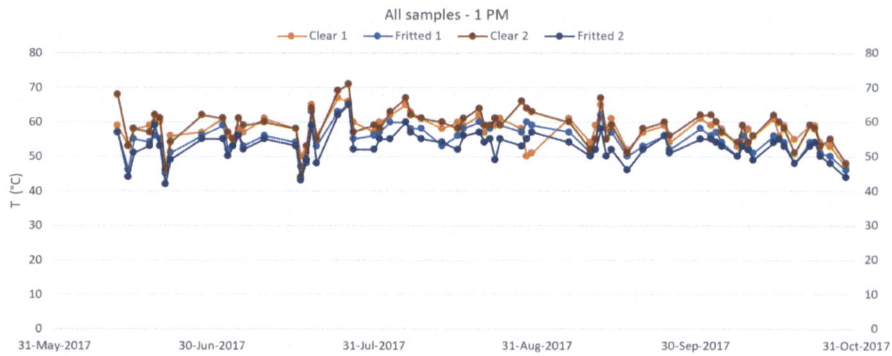

Fig. 8 Surface temperature readings

low temperature scenario (10 °C) to understand the behavior of the glass panels and its attachments during the winter season.

Further simulation using CFD showed localized peaks of surface temperature in the excess of 70 °C, confirming the overall reliability of the results obtained from the Koltay Facades in-house test (Fig. 8).

Load Cases

The basis of design for the Load Cases was taken from the ASCE 07-10. However, since for each Load Combinations, different load types and durations are acting together, the property of the interlayer and the strength of the glass had to be considered separately. After a rigorous consultation of the available standards, it was decided to adopt a conservative approach, by which each load type corresponded to a single model that included the exact material properties. The combination was then achieved as sum (with the proper load factors) of the utilization ratios of each simulation, and for each node (see below).

$$\sum_j \varphi_j \cdot \frac{\sigma_{i,j}}{\sigma_{Ri,j}} \leq 1$$

where:

φ_j is the load factor for the load j;

$\sigma_{i,j}$ is the resulting stress at node i, as extracted from the results of the model loaded using the load j;

$\sigma_{Ri,j}$ is the resisting strength (with appropriate FoS) at node i and considering the material properties for a load j.

Similar approach was used to check the deflection.

Further to the basic load combinations, the risk assessment highlighted some possible scenarios of components failure that required a proper analysis. List below summarizes the principal combinations. For each combination, the loads were applied based on the basic combinations, but the FoS was reduced to a minimum value.

- 1 glass lite broken on the floor panel
- 1 glass lite broken on the vertical panel (external)
- 1 glass lite broken on the vertical panel (internal on the lower shelf)
- Complete failure of the interlayer on the vertical panel
- Complete failure of the interlayer on the floor panel
- Complete failure of the roof-to-side sealant connection.

Should any of the failure cases described above occur, the load path would be re-directed from the primary elements to the redundancy item, which are provided in the form of stainless steel fittings along the chute.

6.4 Results

Figure 9 present representative results from the global finite element analysis, illustrating stress distributions and deflection patterns in the glass panels forming the chute assembly. The results are shown for governing load combinations and are used to assess both serviceability and structural performance. Stress contours are evaluated to identify critical regions influenced by geometry, support conditions, and interaction with bonded and mechanical connections, while deflection results are reviewed to verify deformation compatibility between adjacent panels and compliance with defined performance limits. Together, these outputs provide confirmation that the glass components act in accordance with the intended load path strategy and that localized effects remain within acceptable bounds under the analyzed scenarios.

Evaluation of the structural sealant performance was carried out through post processing of the global finite element results, focusing on the response of the spring elements representing the bonded joints. Axial and combined shear forces acting on each spring were extracted from the analysis and normalized by the corresponding effective tributary area to obtain equivalent stress levels within the sealant. This approach enabled a joint-by-joint assessment of sealant demand consistent with the calibrated stiffness representation adopted in the model.

For study purposes, the allowable capacity of the sealant was assessed using a reduced factor of safety compared to the prescriptive values typically adopted in simplified design checks. In agreement with the sealant supplier, a factor of safety of 4.0 was adopted in lieu of the conventional value of 6.0. This reduction was justified by the increased resolution of the numerical analysis, which explicitly captures joint deformation, relative movements, and load redistribution effects that are not represented in simplified hand calculations. The refined modelling framework therefore provides a more accurate representation of sealant demand, allowing rational optimization of safety margins without compromising reliability.

Long term loading effects associated with the self weight of the chute were intentionally excluded from the sealant stress evaluation. This exclusion reflects

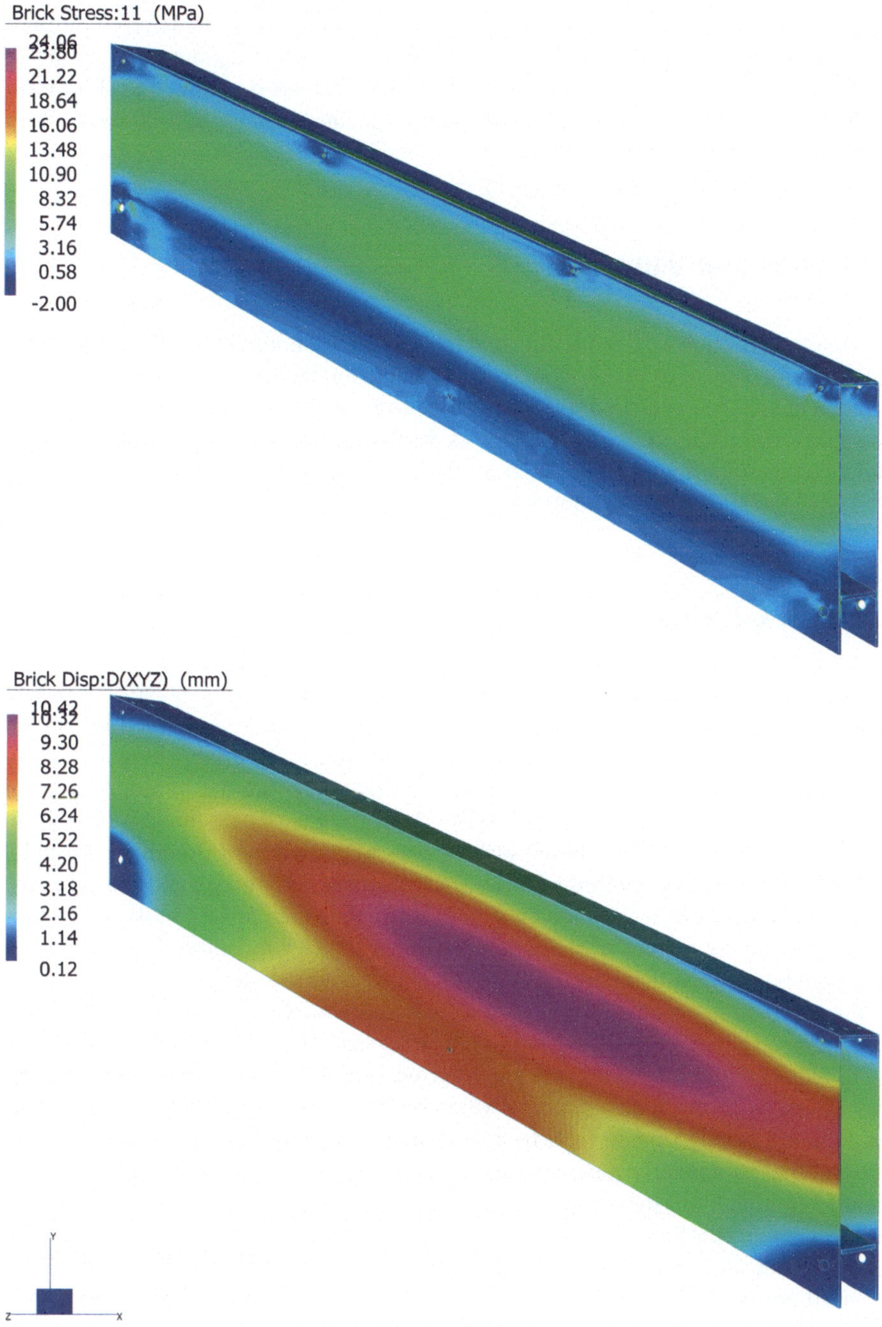

Fig. 9 General results (stress and deflection) on the glass panels composing the chute

the adopted detailing strategy, whereby the permanent loads are transferred through mechanical support mechanisms rather than through the bonded joints. By preventing sustained stress in the sealant under dead load, the analysis is confined to short term and transient loading scenarios for which the sealant behavior and derived safety margins are directly relevant.

7 Testing and Validation

To complement analytical modelling and address uncertainties associated with material behavior and user interaction, a targeted customized testing programme was implemented. This included full-scale mock-up trials to validate geometry and overall performance, user experience, as well as component-level tests-covering routel connections and glass floor panels under out-of-plane and in-plane loading-to verify critical structural assumptions.

7.1 Mock-Up Test

Initial analytical calculations were undertaken during the concept stage to estimate sliding duration, user velocity, and run off distance based on simplified kinematic assumptions. While these calculations were adequate for establishing an initial geometric envelope, their validity was recognized to be limited by uncertainty in key parameters governing sliding behavior. In particular, the coefficient of friction between the user and the sliding surface was identified as a dominant variable, strongly influenced by clothing fabric, user posture, body mass, and the effective contact area between the user and the chute. Given the inherent variability and unpredictability of these factors, it was concluded that user behavior and slide performance could not be reliably characterized through analytical modelling alone.

Physical testing was therefore introduced as a necessary validation tool to calibrate geometric assumptions and assess qualitative aspects of user experience and safety.

To reduce variability in friction and improve repeatability, an operational control was agreed with the future operator whereby all users would slide on mats. This decision effectively transformed friction from an uncontrolled behavioral variable into a design parameter. A range of mat materials was subsequently tested in order to identify combinations that produced controlled speeds, predictable run off distances, and acceptable comfort levels. Along with a mat, a protective adhesive film was also applied to the slide floor. This is intended to provide a wearable surface that can be replaced periodically by the operator, and protect from accidental scratches while allowing control over the friction coefficient of the surface.

A full scale trial mock up was developed during the concept and schematic design stages to evaluate alternative geometric configurations. Inclinations of 24, 26, and 28° were selected to represent the practical range compatible with space constraints, user

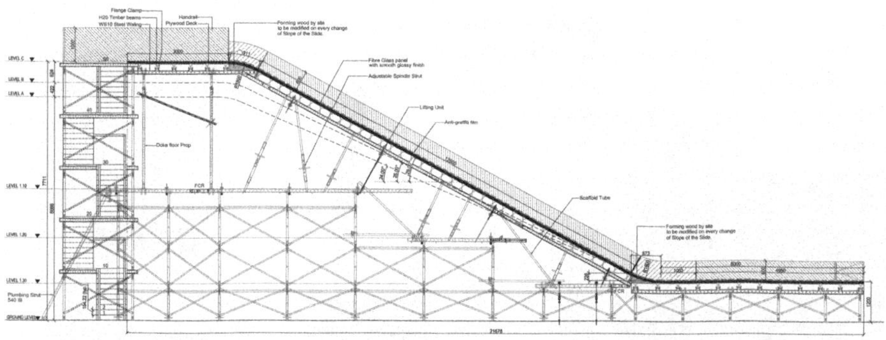

Fig. 10 Diagram of mock up with adjustable scaffolding

comfort, and anticipated speed development. The primary purpose of the mock up was not to replicate construction detailing, but to assess the combined performance of transition geometry, chute inclination, and run off configuration. The results directly informed the selection of the final slide geometry adopted at detailed design and tender stages.

The mock up utilized a fiberglass slide mounted on adjustable scaffolding, selected for its ability to approximate the target geometry while allowing rapid modification of geometry and smooth surface. Due to the adhesive film application, the surface friction matched well with project conditions. Although the materials differed from the final glass construction, the mock up provided a sufficiently representative behavioral benchmark to identify potential safety issues, validate run off distances, and support performance based decision making. In addition, the testing phase provided an opportunity for the developer and operator to evaluate operational considerations and confirm that the selected geometry met both safety expectations and experiential objectives (Fig. 10).

7.2 Routel

Testing of the routel connections was undertaken to validate their structural performance under combined tensile, shear, and bending actions representative of service and ultimate load conditions within the glass chute assembly. The objective of the test programme was to confirm the load-bearing capacity, deformation characteristics, and failure modes of the point-fixing system, particularly in relation to stress transfer between glass panels and supporting steel elements. Test specimens were configured to replicate the actual installation geometry, including glass thickness, hole tolerances, and edge distances, as well as interaction with associated fittings and washers. The testing was carried out using internal testing protocols that fully replicate the actual project conditions, under controlled laboratory environment. Load was applied incrementally until reaching the design load and further until the ultimate

Fig. 11 Component testing at laboratory to verify the design

load (design load × FoS), with continuous monitoring of displacement. Particular attention was given to potential glass edge stresses and the behavior of the connection under eccentric loading. The results were used to verify design assumptions, confirm safety margins, and ensure compatibility between routel stiffness and the overall load path strategy. Once the test was completed under nominal conditions, the same was carried out for post-breakage behavior and up until critical conditions where all the glass panels are broken to understand the residual capacity in such extreme scenarios even if such cases are nonrealistic (Fig. 11).

7.3 *Glass Floor Out-of-Plane Loads (High Temperature Test)*

Out-of-plane load testing of the glass floor panels was conducted to assess their structural performance under elevated temperature conditions representative of peak environmental exposure. Given the sensitivity of laminated glass behavior to temperature, particularly due to the reduction in interlayer stiffness, the test aimed to validate the load-bearing capacity and deflection response under worst-case thermal scenarios. Test specimens were fabricated to match the final build-up of the floor panels, including glass thickness, interlayer type, and support conditions. The testing was performed at a controlled temperature of 80 °C, achieved using an insulated chamber underneath the glass unit. Uniformly distributed loads (e.g. sandbags) were applied to simulate user loading conditions, with measurements taken for deflection.

Fig. 12 Glass floor testing at laboratory to verify the design (out-of-plane loads) with heating mat and sand bag weights

The test sequence included sustained loading to capture time-dependent effects associated with interlayer creep. The outcomes were used to validate the conservative assumptions adopted in the numerical model and to confirm that serviceability and strength criteria were satisfied under elevated temperature conditions. Similarly to the routel, once the test was completed under nominal conditions, the same was done for post-breakage behavior (Fig. 12).

7.4 Glass Floor In-Plane Loads

In-plane load testing of the glass floor panels was performed to evaluate the capacity of the glass and its brackets to resist shear forces. Although glass floor panels are typically governed by out-of-plane loading, the specific configuration of the Sky Slide required verification of in-plane performance due to the specific load path. Test specimens were prepared to replicate the actual panel geometry, support conditions, and connection detailing. In-plane loads were applied progressively along one axis on the two bearing surface as per design, with instrumentation used to measure the displacement. Particular attention was given to edge conditions and potential stress concentrations around fixings or bonded zones. The results provided

Fig. 13 Glass floor testing at laboratory to verify the design (in-plane loads)

confirmation that the panels could accommodate anticipated in-plane actions without compromising structural integrity or inducing premature cracking (Fig. 13).

8 Technology Limitations

8.1 General Limitations

The design and analysis of the Sky Slide highlight several inherent limitations in current analytical tools and material modelling approaches when applied to unconventional, publicly exposed glass structures. While global finite element analysis provided a rational framework for evaluating system-level behavior, the complexity of the assembly—combining brittle glass elements, viscoelastic interlayers, and hyperelastic sealants—necessitated several simplifying assumptions. In particular, the representation of structural silicone through equivalent spring elements and the treatment of interlayer properties using temperature- and duration-dependent parameters introduce a level of abstraction that limits the accuracy of localized stress predictions, despite providing reliable global response.

A further limitation arises from the absence of directly applicable design standards for this type of structure—due to its rarity. Existing codes address isolated aspects of glass design, such as panel strength or façade loading, but do not capture the combined effects of dynamic human interaction, bonded structural behavior, and post-breakage performance. As a result, the design relies on selective interpretation of multiple standards and the development of project-specific criteria, reducing the extent to which outcomes can be generalized without additional validation.

Material behavior introduces additional uncertainty. The mechanical response of interlayers and sealants is highly sensitive to environmental conditions, particularly temperature, as well as load duration. Although conservative assumptions and experimental calibration were adopted, the variability inherent to these materials limits predictive precision and necessitates a performance-based verification approach. Collectively, these constraints underline the current technological boundaries of structural glass engineering in high-risk, non-standard applications.

8.2 Evaluation of Alternative Bonding Technologies

During early design stages, alternative adhesive technologies were investigated to assess their suitability for the application [10].

Hot-melt adhesives were considered due to their promising appearance and novelty use in structural applications. However, this option was rejected based on insufficient certification under established ETAG and ASTM frameworks, the absence of a long-term performance record in comparable load-bearing glass structures, and practical challenges associated with application and quality control. Moreover, the relatively high stiffness of hot-melt systems compared to silicone adhesives was identified as a critical limitation, as it would reduce deformation capacity and increase the risk of stress concentration and overstressing at glass interfaces.

A two-part grey structural silicone system with established certification and widespread façade application was also evaluated. Despite its proven material properties, feedback from fabricators and installers highlighted difficulties in achieving consistent mixing ratios during application, leading to variability in cured strength and stiffness. Given the sensitivity of the chute system to bonded joint performance and the need for uniform behavior across multiple interfaces, this variability was considered an unacceptable risk (Figs. 14 and 15).

9 Conclusion

9.1 Lessons Learnt

The development and realization of the Sky Slide highlighted several key lessons relevant to the design of non-standard, human-occupied glass structures operating beyond the limits of codes. First, the project confirmed that performance-based design is a process requiring continuous calibration between testing, numerical modelling,

Fig. 14 The completed Sky Slide

Fig. 15 Standard operation, a rider sliding using a mat

and engineering judgment. Neither finite element analysis nor simplified calculations alone were sufficient to capture the full complexity of system behavior; reliable outcomes were achieved only through iterative validation against physical testing and targeted assumptions.

A second critical lesson concerns the importance of early definition and preservation of load paths. The success of the glass chute as a coherent structural system depended on maintaining clarity in how forces were transferred through the bonded glass assembly and into the supporting structure. Attempts to introduce additional restraints or overly rigid connections would have resulted in unintended stress concentrations and compromised performance. This reinforces the principle that, in structural glass systems, simplicity and determinacy in load transfer are more robust than over-constrained configurations.

The project also demonstrated that redundancy must be treated as a primary design driver rather than a secondary safety measure. Given the brittle nature of glass and the absence of ductility, post-breakage behavior must be explicitly engineered at system level. The integration of steel elements as secondary load paths, while maintaining architectural transparency, proved essential in ensuring robustness without compromising design intent. This approach establishes a clear hierarchy between primary (glass) and secondary (steel) load-carrying mechanisms.

Another key insight relates to the limitations of material modelling, particularly for structural silicone and interlayers. Accurate representation of these materials required simplification strategies, such as equivalent stiffness modelling, combined with calibration through detailed sub-models. This highlights that in complex hybrid systems, modelling precision must be balanced with computational feasibility, and that over-refinement does not necessarily lead to more reliable results.

From a user-centric perspective, the project underscored the need to treat human interaction as a governing design parameter. Variables such as friction, posture, and behavioral unpredictability introduced significant uncertainty that could not be resolved analytically. Full-scale mock-up testing was therefore indispensable in defining geometry, validating safety margins, and aligning engineering performance with user experience.

Finally, the absence of directly applicable standards required selective interpretation of existing codes combined with project-specific criteria. This reinforces the broader lesson that innovation in façade engineering increasingly depends on the ability to synthesize partial guidance, experimental data, and risk-based reasoning into a coherent and defensible design methodology.

9.2 Design Transfer and Validation Across High-Rise Structural Glass Applications

Looking ahead, the design and delivery of the Sky Slide provide a transferable framework for future transparent, human occupied glass installations in high-rise and complex architectural contexts. Similar applications may include glass bridges, inclined walkways, vertical descent elements, or façade mounted viewing capsules integrated into observation decks, hotels or mixed use developments. The project demonstrates how performance based structural design, differentiated safety factors, and explicit consideration of human interaction can be successfully combined to address the limitations of prescriptive codes. As demand grows for immersive architectural experiences that blur the boundary between structure, enclosure, and user engagement, the Sky Slide establishes a scalable design methodology that can be used as model for structures of varying geometries, heights, and operational requirements while maintaining robust safety, durability, and public confidence.

To give examples, the same performance based design approach developed for the Sky Slide was subsequently applied to two major projects with comparable technical characteristics. The first is a rotating glazed viewing platform at the "Top of the Rock," located on the 70th floor of Rockefeller Center in New York, where human occupancy, dynamic movement, and continuous operation governed the structural glass design. The second comprises a series of cantilevered glass viewing platforms projecting from the top floor of "The View at the Palm Jumeirah" in Dubai, requiring similar consideration of glass behavior under public loading, interface detailing, and risk based allocation of safety factors. Just as for the Sky Slide, custom developed component testing and overall mock up load test were implemented for design finalization and verification. In both projects, the methodology developed for the Sky Slide was applied through a performance based structural glass design, differentiated safety factors focused on user critical components, and careful interface detailing to manage movements, tolerances and loads. Although the geometries and operational functions differed, the same risk driven design framework governed the assessment of glass behavior, post breakage performance, and overall user safety.

Acknowledgements The authors would like to thank for the contribution and input—in no particular order—of Michael Ludvik Engineers, Dow Corning, and the team of Koltay Facades during design; Lindner Steel and Glass, LSL Laboratory and Sedak in construction phase.

References

1. OUE Skyspace (2016–2020) Los Angeles, glass engineering. Michael Ludvik Engineering, New York
2. Patterson M (2011) Structural glass facades and enclosures. Wiley
3. Dow Corning® 993. Advance analysis and design. www.dowcorning.com
4. Outukumpu (2015) Atmospheric corrosion resistance of stainless steel: results of a field exposure program in the Middle-East
5. Strand7 (2017) Reference manual
6. Strand7 (2017) Theoretical manual contents
7. ETAG 002 (2012) Guideline for European technical approval for structural sealant glazing kits. European Organisation for Technical Approvals
8. Descamps P, Hayez V, Chabih M (2017) Next generation calculation method for structural silicone joint dimensioning. Glass Struct Eng

9. Kuraray. Technical data elastic properties
10. Dias V, Odenbreit C, Hechler O, Scholzen F, Zineb TB (2014) Development of a constitutive hyperelastic material law for numerical simulations of adhesive steel-glass connections using structural silicone. Int J Adhes Adhes 48

Extreme Performance, Elegant Solutions: Frameless Systems, Thermally Broken Profiles, and Curved Glazing in Hurricane Zones

Jalal Farooq, Angela Mejorin, Jason T. Quimby, Jason Eastwood, Vanessa Guerrero, Vaughn Schauss, and Sophia Liu

Abstract Over the past decade, façade engineering within hurricane-prone regions has undergone a profound transformation driven by evolving codes, advances in testing standards, and rapid material innovation. This chapter examines the convergence of Miami-Dade County requirements for high-velocity hurricane zones, ASTM standards (notably E1886/E1996), and the emerging ISO framework for impact resistance and pressure performance, with particular focus on how these regulatory frameworks are translated from compliance requirements into drivers of façade innovation. When codes and test protocols are interpreted as performance targets rather than prescriptive constraints, they can catalyze new architectural and technological solutions—enabling increased façade transparency, improved thermal efficiency, and complex geometries without compromising resilience in extreme wind environments. This discussion is structured around a selection of case studies either recently completed or currently under construction that exemplify a new generation of hurricane-resistant façade technologies. The featured buildings, located within the coastal United States, illustrate how stringent wind-borne-debris and cyclic pressure requirements can be reconciled with evolving architectural ambitions. Examples include frameless retail envelopes designed to maximize visual transparency, high-rise systems engineered for extreme lateral loads and energy performance, and

J. Farooq (✉) · S. Liu
Al-Farooq Corporation, Miami, FL, USA
e-mail: jfarooq@afceng.com

A. Mejorin
Performance-Based Façade Design, Venice, Italy

J. T. Quimby
SLIMPACT By Faour Glass Technologies, Tampa, FL, USA

J. Eastwood
REFLECTION WINDOW + WALL, Chicago, IL, USA

V. Guerrero
ES Windows LLC, Miami, FL, USA

V. Schauss
Kuraray America, Houston, TX, USA

A. Mejorin et al. (eds.), *Performance-Based Façade Design*, Springer Tracts in Civil Engineering, https://doi.org/10.1007/978-3-032-28835-6_16

façades incorporating large-radius curved glazing under high design pressures. Each case study is analyzed in terms of testing approach, materials, connection detailing, and the balance between performance, aesthetics, and constructability. Through this comparative lens, the chapter identifies key patterns in façade design evolution—toward greater transparency, improved energy efficiency, and enhanced resilience—demonstrating how the dialogue between codes and design teams is shaping a new generation of hurricane-resistant building envelopes. The findings highlight a broader paradigm shift from prescriptive compliance toward performance-based façade design and demonstrate how rigorous testing frameworks can serve as catalysts for technological and architectural innovation in hurricane-resistant building envelopes.

Keywords Impact-resistant curved façade modules · Façade engineering · Hurricane glazing · Envelope performance · Impact-resistant frameless façades · Curtain wall systems · Thermally broken hurricane glazing · High-rise buildings

1 Codes and Standards Development and History of Impact Resistance

On 25 December 1974, Tropical Cyclone Tracy struck Darwin, in the Northern Territory of Australia, causing significant damage and fatalities with recorded gusts of 217 km/h [1]. The city of Darwin had previously been hit by cyclones in January 1897 and March 1937, and in these disaster events significant damage had already been recorded. The Australian Department of Housing commissioned a report [2], which highlighted the loss of roof cladding as significantly contributing to the extensive damage to housing. Damage to cladding was also weakening the structural performance of the building, causing in many cases the collapse of the whole building. Breach of building envelopes was also generating a significant amount of debris, which had a vital role in the further destruction of structures and set off a domino effect. It was clear that damage to windows caused by wind-borne debris represented a serious threat to building occupants' safety and increased the post-event recovery costs [3]. It was decided to enhance the performance and integrity of structures in cyclonic areas [4].

After Cyclone Tracy, laboratories had to follow the Australian Standard requirements or the National Association of Testing Authorities. Therefore, Australia came up with an early and novel method of shielding windows from wind-borne debris in tropical cyclone-prone regions. The Darwin Area Building Manual [5] introduced requirements for protection of doors, windows, and cladding from wind-borne debris in cyclonic regions. Building envelope technologies capable of resisting the impact of a 4-kg lumber mass with a 100×50 mm impacting cross-section, at a velocity of 20 m/s have been introduced in tropical cyclone-prone areas [5]. Technical observations demonstrated that this impact speed, combined with the use of the 4-kg lumber projectile, was able to produce a disruptive effect similar to the survey observations. Following these first wind-borne debris-resistance performance requirements, the original so-called "cyclone-resistant glazing" was developed in 1975. After the impact of the 100 × 50 mm lumber missile, this glass type had to withstand the entire design wind pressure of its cyclonic zone. If the glass edges were correctly secured to the frame using an adhesive glazing solution, the glass would have been able to withstand the pressure without any air escape [3].

Guidelines for the Testing and Evaluation of Products for Cyclone-Prone Areas were issued as Technical Record (TR) 440 by the Experimental Building Station, Department of Construction, in 1978 [6]. Using the 100 × 50 mm lumber projectile, TR 440 stated that the debris protection performance should decrease from an impact at 20 m/s (of the Darwin Manual [5]) to 15 m/s [6]. The impact velocity requirement was based on qualitative observations from impact testing campaigns conducted in the Australian testing laboratories [3]. In TR 440, it was also agreed that metal roof cladding in cyclone-prone areas must endure the impacts of dynamic wind loading, but there was no mention of the need for pressure cycling for windows, doors, or curtain walling. Although TR 440 was not a legislative document, it was adopted as a reference standard for cyclone-prone Australian locations shortly after it was published. The events that followed Cyclone Tracy led to progressive revisions of Australian wind actions provisions for cyclone-prone regions, including explicit consideration of wind-borne debris impact resistance for building envelopes in later code editions.

Hurricane Andrew hit Florida, the Bahamas, and Louisiana in August 1992, 18 years after the Darwin event in Australia. The strongest wind gusts reached 270 km/h on August 23–24, 1992, in Miami-Dade County, Florida. Hurricane Andrew was a Category Four hurricane (Saffir–Simpson Scale) and 44 people were killed in Florida alone. It caused losses exceeding $27 billion [7]. After Andrew, 250,000 people were left homeless, the infrastructure for communication and transportation was severely damaged, and 1.4 million people were left without electricity for up to six months [8]. The psychological impact was severe, and instead of fixing their houses and businesses, many individuals chose to relocate to other towns and states. When individuals chose to rebuild their structures, the process took years to complete.

Post-disaster event evaluations identified a number of problems associated with severe wind events [9, 10]. Similar to Australia, the Florida Building Code added curtain-wall regulations in the years following Hurricane Andrew, and introduced provisions to add building resilience against wind-borne debris impacts [11]. For impact-resistant façade systems, the Miami-Dade rules and any updates since then continue to be among the most demanding in the U.S. [12].

The Florida impact test procedures for building enclosures impacted by wind-borne debris were an improvement over Australia's, as reported in TR 440 [6]. The procedure improved upon earlier developed Australian standards by requiring the test specimen to withstand cycling of positive and negative pressure after the impact test. Debris projectile and impact speeds were aligned with the Australian approach, while façade performance criteria were set in the Testing Application Standards TAS

201-94, TAS 202-94, and TAS 203-94 as adopted within the Florida Building Code framework [11]. Furthermore, the U.S. testing procedure defined missile impact locations on the specimen. Other non-mandatory reference standards such as SSTD 12–94 "Test Standards for Determining Impact Resistance from Wind-borne Debris", published by the Southern Building Code Congress International, also served as foundational material for the requirements implemented at the national level [13], including later editions of the Florida Building Code [14]. Following TAS 201-94, TAS 202-94, and TAS 203-94, ASTM standards were developed: ASTM E1996, Standard Specification for Exterior Windows, Curtain Walls, Doors, and Impact Protective Systems Impacted by Wind-Borne Debris in Hurricanes [15], and ASTM E1886, Standard Test Method for Performance of Exterior Windows, Curtain Walls, Doors, and Impact Protective Systems Impacted by Missile(s) and Exposed to Cyclic Pressure Differentials [16].

These ASTM standards provide additional criteria, including allowable temperature ranges for conducting the tests, which were not included in the original TAS documents. Currently, according to ASCE/SEI 7-22, buildings erected in U.S. wind-borne debris regions must follow ASTM E1886 and ASTM E1996 specifications, or local standard requirements, whichever is more demanding [17]. The weights and velocity of the testing projectiles change according to the building's wind zone and level of protection (ranging from Wind Zone 1, the least wind-prone, to Wind Zone 4, the most wind-prone, and from basic to enhanced protection, per ASCE/SEI 7-22 [17]).

The design wind pressure (inward and outward) from the building code is used to determine the pressure cycling to be performed on the façade once the impact test is passed. It is necessary to do 4500 cycles of positive and negative pressure, with each cycle lasting 1–3 s [16]. Through a process of consensus-building, ASTM E1886 was created with the participation of manufacturers, consultants, building code authorities, and other specialists. The permitted tolerances for the testing criteria for both debris projectile impacts and the pressure cycle program have been developed, and the testing loading sequence and conditions have been defined [16].

In Australia, the 2002 revision of AS/NZS 1170.2 "Structural design actions: Wind actions" led to the first unified Australian and New Zealand design wind code [18]. A number of substantial modifications were made to AS/NZS 1170.2 in the 2011 revision [19], including the introduction of criteria for impact loading of wind-borne debris. The latest studies on wind profiles in tropical storms and hurricanes developed in the United States served as the basis for Amendment 4 of AS/NZS 1170.2:2011 (2016). These studies found a significant correlation between horizontal missile velocity and distance traveled. The study "Trajectories of Wind-Borne Debris in Horizontal Winds and Applications to Impact Testing" provides a key foundation for the specification of missile speeds for windows and façades, based on experimental studies [20]. In terms of wind gust speeds, the research proposes ratios for missile speed based on the wind zone of the building to be designed and tested in its impact performance for wind-borne debris. Plots by Lin et al. [20] show that the terminal velocity of the projectiles is represented by $0.4 \times$ VR, with VR being the 10-min mean regional design wind velocity. VR is, therefore, used to calculate the missile impact velocity in the Design Guidelines for Queensland Public Cyclone Shelters. These studies consider wind-borne debris flight in uniform flow and, since the regional velocity can reach 36 m/s for the Australian Severe Cyclonic Region D, the impact velocity of the 4-kg projectile reaches impact speeds of 36 m/s. This impact velocity is more than double the previous one adopted in Australia, which was equal to 15 m/s. The fenestration assembly, according to AS/NZS 1170.2:2011, should be tested for wind-borne debris-impact resistance for a height of 25 m above the ground level [19]. No provisions are in place for cyclic pressure testing post-debris impact for glass façades and debris screens, unlike in ASTM E1886 [16].

The small projectile, a 2-g steel ball, was chosen since it is representative of roof gravel. AS/NZS 1170.2:2011 does not specify a single mandatory testing method or acceptance criterion for debris impact, indicating instead that test methods and acceptance criteria "may vary according to the purpose of the test", and referring to guidance developed by the Cyclone Testing Station at James Cook University (as referenced within the standard text and amendments) [19]. The steel ball impact test followed Minor's observations about high-rise building window breakage [21]. Florida requirements and the ASTM standard procedures that are widely adopted to test hurricane/cyclone-resistant windows, doors, and curtain walls include the pressure cycling test following the impact phase because it is considered a good representation of the natural phenomenon [16]. Pressure cycling testing is required in Australian roofing tests for cyclonic areas, but it is not in place for wind-borne debris resistance for fenestration assemblies [19].

In the Asia–Pacific region, different countries are currently increasing their level of preparedness for climate change and for issues related to building envelope breakage due to wind-borne debris impact, pushed especially by insurance-driven risk mitigation strategies. In Japan, the Disaster Prevention Research Institute (DPRI) at Kyoto University investigated the behavior of wind-borne objects in typhoon winds and several impact tests have been conducted on building envelope glazing panels [3, 22].

Currently, there are no requirements for wind-borne debris resistance in Europe because the continent has experienced relatively few extratropical cyclones and tornadoes, but in recent years there has been an increased number of such events as noted earlier. This trend led to the term "medicanes", which are "hurricanes" occurring in the Mediterranean Sea. The necessity to use this new terminology occurred in the last two decades to localize these phenomena to areas where extreme wind events did not happen in the past. Other international product standards related to wind-borne debris resistance have been developed, such as ISO 16932, which outlines specifications for destructive-windstorm-resistant security glazing [23], and ISO 16316, on "Windows, doors and curtain walling—Impacted by wind-borne debris in windstorms—Test method and classification" [24]. ISO 16932 currently does not test the window, door, or curtain walling assembly, but the glazing installed in a standardized frame, therefore not considering the actual assembly of the glass in the window/ curtain walling, nor the product geometry. In Japan, JIS R 3109 references ISO 16932 [25]. Both these standards differ from the ASTM and AS/NZS approaches discussed above by focusing on the glazing specimen rather than the full installed system. By contrast, ISO 16316 is developed to test the whole assembly, aligning more closely with the system-level philosophy of ASTM standards [15, 16, 24]. Moreover, ISO 16316 provides the designer with a design flow chart for alternative impact test requirements that are based on engineering assumptions [24].

The main regional codes and standards addressing wind-borne debris impact and façade system performance are summarized in Table 1, illustrating differences in testing philosophy, system validation approaches, and areas where further development is expected.

Despite the progress represented by these regional regulatory frameworks, several performance dimensions relevant to contemporary façade engineering remain underdeveloped internationally. In many jurisdictions, wind-borne debris standards primarily address structural resistance and occupant safety, but they do not yet fully integrate additional performance objectives that increasingly define modern building envelopes. These include the simultaneous consideration of impact resistance, post-fracture behavior, water penetration resistance, thermal and energy performance, and architectural requirements such as minimal sightlines and large-format glazing. As a result, regulatory frameworks often evaluate façade systems primarily through safety-driven criteria rather than through integrated performance-based approaches capable of accommodating multiple design objectives within a single façade system.

The High-Velocity Hurricane Zone (HVHZ) regulatory environment in South Florida remains among the most rigorous frameworks for impact resistance of building enclosures. Florida HVHZ testing procedures are generally more prescriptive in terms of test sequencing and verification protocols than comparable standards in other regions. In addition to the laboratory testing procedures defined in TAS 201, TAS 202, and TAS 203, products seeking approval must undergo a detailed technical review process administered through the Miami-Dade Notice of Acceptance (NOA) or Florida Product Approval (FPA) program, in which the governing jurisdiction evaluates test data, engineering calculations, and system configurations prior to approval for use in the HVHZ.

HVHZ-based impact testing also incorporates enhanced performance verification features beyond those commonly required in other regional frameworks. These include higher water-penetration resistance requirements for systems subjected to elevated commercial design pressures, longer load durations for static pressure testing, and additional impact verification applied to framing members and system components. Collectively, these requirements create a testing environment in which façade systems must demonstrate not only impact resistance but also robust environmental performance under extreme wind loading conditions.

Within this context, the case studies presented in this chapter illustrate how performance-driven façade design can reconcile stringent impact-resistance requirements with architectural ambitions such as increased transparency, thermally efficient envelopes, and complex geometries. By examining projects developed under the HVHZ regulatory framework, the chapter highlights how rigorous testing protocols can function as performance targets that enable technological innovation rather than acting solely as prescriptive constraints. These examples therefore contribute

Table 1 Comparison of major regional standards for wind-borne debris impact and façade system performance

Region/ regulatory framework	Main standards/ codes	Key features	Limitations/gaps	Future development potential
United States (Florida/ HVHZ)	Florida building code (FBC), miami-dade TAS 201/ 202/203, ASTM E1886, ASTM E1996, ASCE/SEI 7	Full-system testing of windows, doors, and curtain wall assemblies; large- and small-missile impact tests; cyclic pressure testing following impact; defined missile locations and pressure cycling protocols	Testing largely focused on specific system configurations approved through product approval pathways; limited flexibility for unconventional geometries unless additional testing is conducted	Continued integration with performance-based façade engineering methods and broader international adoption through ASTM-based procedures
Australia/ New Zealand	AS/NZS 1170.2 Wind Actions; cyclone testing guidelines from Cyclone Testing Station	Wind-borne debris impact criteria based on projectile velocity derived from wind speed correlations; regional wind zone classification	No standardized cyclic pressure testing for glazing systems after debris impact; testing methods and acceptance criteria can vary depending on the project context	Potential harmonization with system-level testing methodologies similar to ASTM standards
International (ISO framework)	ISO 16316 ISO 16932	ISO 16316 introduces system-level testing for windows, doors, and curtain wall assemblies impacted by wind-borne debris; classification approach intended for international application	ISO 16932 evaluates glazing only rather than full assemblies; ISO 16316 introduces minimum test specimen sizes that may require additional testing relative to ASTM or TAS procedures	Potential to act as a global reference framework aligning regional standards and enabling broader performance-based façade validation
Japan/ Asia–Pacific	JIS R 3109	Experimental studies of wind-borne debris trajectories and glazing impact performance; adaptation of ISO-based glazing impact procedures	Limited regulatory requirements mandating façade system-level impact resistance testing; reliance on engineering judgement and research findings	Increasing interest in envelope resilience driven by typhoon risk and insurance-driven mitigation strategies

to advancing the broader understanding of performance-based façade engineering in extreme wind environments.

This chapter provides a comparative analysis of major international regulatory approaches to wind-borne debris impact resistance and façade system testing, highlighting differences in testing philosophy and performance verification across regions. It documents how the High-Velocity Hurricane Zone (HVHZ) regulatory framework has enabled the development of façade systems capable of simultaneously addressing extreme wind loading, impact resistance, and architectural performance objectives such as transparency, energy efficiency, and geometric complexity. Through detailed case studies of recently completed or ongoing projects, the chapter illustrates how performance-driven design approaches can translate rigorous testing protocols into practical engineering solutions for contemporary building envelopes.

2 From Standards to Façade Innovation: Frameless Systems, Thermally Broken Window Walls, and Curved Glazing Solutions

The evolution of testing standards and regulatory frameworks has progressively shifted façade engineering in hurricane-prone regions toward a performance-based design paradigm. In contrast to prescriptive approaches that define specific construction assemblies, performance-based façade design evaluates building envelope systems against measurable performance objectives, including resistance to wind-borne debris impact, post-fracture structural behavior, cyclic pressure durability, water penetration control, and long-term serviceability. Laboratory testing protocols such as ASTM E1886/E1996 [15, 16] and the Miami-Dade TAS procedures [14] therefore serve not only as compliance requirements but also as experimental validation frameworks enabling the development of innovative façade technologies. Within this context, performance-based façade design provides a pathway through which architectural ambitions, such as increased transparency, thermally efficient envelopes, and complex geometries, can be reconciled with the stringent resilience requirements of high-velocity hurricane zones. The case studies presented in the following sections illustrate how this approach has enabled new generations of façade systems to be realized while maintaining rigorous performance verification.

This chapter proposes that façade innovation advances within hurricane-prone regions despite regulatory and climatic constraints. Through specific design constraints, new performance-based frameworks are actively being developed to solve a variety of novel architectural needs.

Building on this regulatory and technical context, the chapter develops a comparative analysis through three representative case studies from Florida's hurricane market that span distinct façade typologies and performance drivers. The first case study, Kith Miami Design District, examines the engineering constraints and detailing strategies required to realize frameless, high-transparency storefront glazing in an

HVHZ environment, with emphasis on load path definition through top/bottom channels, geometric discontinuities (trapezoids), installation tolerances, and system limitations governed by tested configurations. The second case study, South Flagler House, documents the scale-up of a thermally broken, captured-glazed unitized window wall to resist extreme design pressures and to accommodate door-scale operable vents qualified for large- and small-missile impact, highlighting receptor redesign, glazing retention as a structural mechanism under cyclic loading, and operability control in high-count multi-point locking hardware. The third case study, 830 Brickell, focuses on curved curtain wall modules subjected to very high design pressures, detailing the interaction between curvature tolerances, fabrication and assembly sequencing, and full-scale laboratory qualification under Miami-Dade/TAS requirements. These case studies were chosen because they combine performance goals often considered to be in conflict with impact resistance: large format unobstructed glazing modules, thermal efficiency, and complex geometries. Together, these projects frame the chapter's central argument: when codes and test protocols are treated as performance targets rather than prescriptive constraints, they can catalyze innovation in façade transparency, thermal efficiency, and geometric complexity, without compromising resilience in extreme wind environments.

The selection criteria for these case studies considered projects located within or adjacent to the HVHZ, representing one of the strictest regulatory environments and some of the most extreme climatic conditions worldwide. The case studies are relevant for specific performance criteria in hurricane-prone regions. These performance attributes include: transparency, thermal and structural performance, complex geometries, constructability, impact resistance, and the ability to withstand post-fracture cyclic loading. Additionally, each case study considers project requirements where limited or no standard solutions were available. For each case study, the products described were either at the extreme edge of the 'state of the art' or were required to exceed it, in order to accomplish their objectives at the time of their development. Within this context, best practices were to meet minimum standard requirements, but in a context where project requirements exceeded 'standard application' in terms of either size, energy efficiency, or geometry. Because these were committed commercial projects contingent on their successful implementation, careful coordination was required throughout the product design process to ensure successful and timely testing, regulatory approval, and realization of architectural objectives.

2.1 Assumptions and Limitations of the Case Study Approach

The case studies presented in this chapter are based on façade systems developed for real commercial projects and validated through High-Velocity Hurricane Zone (HVHZ) certification testing. Due to the destructive nature of impact testing under TAS and ASTM procedures, testing outcomes are inherently binary: systems either satisfy the acceptance criteria ("pass") or do not ("fail"). As a consequence, façade assemblies are typically designed and tested to achieve certification thresholds rather than to determine their ultimate structural or impact capacity. Because the projects discussed in this chapter were developed within defined construction schedules and budgets, testing programs were generally limited to configurations expected to satisfy project-specific performance requirements. Therefore, the certified performance ratings of the systems discussed should not be interpreted as representing their absolute performance limits, but rather the verified performance levels required for regulatory approval and project implementation.

In addition, system configurations and test specimens are developed through a combination of analytical calculations, laboratory testing protocols, and the practical experience of façade engineers, manufacturers, and system designers. While structural calculations and engineering analysis provide a baseline for predicting system behavior, façade performance under extreme loading conditions—particularly during impact events and subsequent cyclic pressure loading—can exhibit complex behaviors that are not always fully captured through analytical models alone. For this reason, experience-based design adjustments are frequently incorporated during the development process to account for factors such as material behavior, connection detailing, fabrication tolerances, and post-fracture glass performance. The systems described in this chapter therefore reflect an integration of calculated design, empirical testing validation, and professional engineering judgment.

3 Kith Miami

3.1 Frameless Glass Façades in South Florida

Frameless glass façades in hurricane zones are an increasingly common choice for commercial and residential projects. While frameless façades have long been used in non-hurricane markets because they provide unobstructed views and minimal visual interruption, comparable solutions have become broadly available in hurricane jurisdictions only within the past decade. SLIMPACT by Faour Glass Technologies is a manufacturer that has focused on this segment. Early adopters included a number of high-end car dealerships, who introduced SLIMPACT to luxury retail applications. By 2020, frameless hurricane-rated window walls gained wider acceptance, with SLIMPACT used in over 300 projects. One such project is the Miami Design District retail store for Kith of New York City, a lifestyle brand founded by Ronnie Fieg in 2011, with façade design by Porto Architecture (New York City). This case study documents constraints that are specific to frameless hurricane fenestration.

The Kith Miami Design District store project was developed based on an existing building envelope and a new architectural intent for greater transparency. AND Studio (Miami, Florida) served as the Architect of Record and was responsible for translating the design vision established by Porto Architecture. The site, 69 NE 41st Street, is a highly trafficked location within the district and features a bold two-story building

Fig. 1 Original façade of Stardust West in the Miami Design District, showing the pre-existing curtain wall subdivided by vertical and horizontal mullions within double-height trapezoidal openings; this baseline condition illustrates the limited transparency that the Kith redesign sought to overcome in a High-Velocity Hurricane Zone (HVHZ) context *(Courtesy of Kith)*

with a three-dimensional fluted façade. Two large trapezoidal, double-height openings define the primary façade, with an exposed intermediate slab dividing the two levels (Fig. 1).

The original building by Dacra, known as Stardust West, incorporated a traditional curtain wall assembly spanning the trapezoidal openings. The trapezoids were approximately 22 ft wide and 26 ft high, subdivided into four vertical stacks with both vertical and horizontal black-painted aluminum mullions. From an exterior viewpoint, the mullion grid substantially reduced visual openness. This was considered suboptimal for Kith, particularly in comparison with larger façades used by the brand in non-hurricane markets. As noted by Kith brand architect Lesley Loo, "the façade plays a huge role in symbolizing the unity and strong presence for Kith within the Design District."

3.2 *Performance Drivers and Challenges*

Prior to 1992 and Hurricane Andrew's impact in Miami-Dade County, frameless glass façades were common in non-hurricane markets. Many such façades incorporated 1/4 in to 3/8 in annealed glass within window systems, and frameless assemblies often used 9/16 in annealed-laminated glass at heights of approximately 10 ft. After Hurricane Andrew, Miami-Dade County tightened construction requirements for hurricane zones, and these requirements were incorporated into the Florida Building Code. While these changes accelerated impact-related qualification of fenestration, the limited availability of impact-rated frameless systems reduced their use relative to more conventional framed solutions.

Frameless hurricane systems introduced to the Florida market in 2010 were not immediately adopted at scale. The frameless nature of these systems required engineering and assembly approaches relying on materials and solutions from suppliers such as Kuraray and DuPont, resulting in higher costs than conventional framed aluminum systems. Installation also required a high degree of attention to detail, including reliance on heavy equipment capable of handling glass panel weights of up to 5000 lb. Together, these factors placed frameless systems at a significantly higher price point than alternatives, limiting early deployment to a relatively small number of projects.

Following recovery from the 2008 financial crisis, Florida hurricane jurisdictions experienced increasing demand from designers and clients seeking façade concepts with reduced vertical and horizontal framing. This coincided with significant development activity, including projects within the Miami Design District envisioned by Craig Robins and the Dacra Group. The district is characterized by a dense concentration of luxury retail and art, much of it in outdoor settings; consequently, façade solutions frequently require hurricane-certified products.

From the outset, Porto Architecture proposed a façade concept ("Façade View—Option #2") intended to increase openness and transparency, with more minimal vertical and horizontal mullions spanning both floors (Fig. 2). AND Studio advanced this direction and contacted SLIMPACT, with whom it had recently completed another Miami Design District façade. Dakota Hendon (AND Studio) described the core constraint as the need to reconcile high transparency and large, uninterrupted openings with the structural and anchorage demands typical of High-Velocity Hurricane Zones (HVHZ).

Following iterative design development, the permit set captured the revised façade intent and included the following hurricane-zone-related elements (Fig. 3):

A. Large frameless SLIMPACT XL window wall modules in 6′W × 12′H
B. Trapezoidal frameless shapes in SLIMPACT XL window wall
C. Large curved SLIMPACT XL-curve (non-impact) window wall
D. Single large SLIMPACT Automatic Door in 4′W × 10′H size.

Fig. 2 Porto Architecture rendering of façade "Option #2", proposing larger glazed openings with reduced mullion density to increase visual openness *(Courtesy of Kith)*

3.3 Engineering Solutions

In 2010, Faour Glass launched SLIMPACT as a frameless window wall solution with Florida Product Approval and HVHZ certification. The first patented SLIMPACT module was designed as a 5 ft × 10 ft triple-laminated glass unit supported by top and bottom channels. The channels were intended to be as minimal as possible and concealed by floor and finish conditions to achieve the appearance of a floor-to-ceiling frameless glass wall. By 2025, SLIMPACT had expanded its offerings under Florida Product Approvals, including larger flat modules, curved modules, pivot doors, and postless/capless glass railings.

In the Kith Miami Design District store, the SLIMPACT frameless pivot door was an important component of the overall façade intent. The ability to combine large frameless glazed modules with a minimalist entrance system and curved glazing zones supported the architectural objectives for the storefront (Fig. 4).

Unlike typical window wall applications, SLIMPACT is a window wall system designed to transfer imposed loads through top and bottom channels (Fig. 5). Existing impact testing criteria do not clearly define special considerations when evaluating and testing systems with this type of loading pattern; thus, a performance-based design approach must be taken. These channels are configured to be minimal at both sill and head. The head channel is deeper than the sill channel to accommodate installation tolerances, providing the space required to slip-fit glass panels into position. Head conditions are typically more readily concealed using interior drywall and exterior stucco finishes. In several applications, these components are clad using architectural metals consistent with project-specific visual intent; at Kith Miami

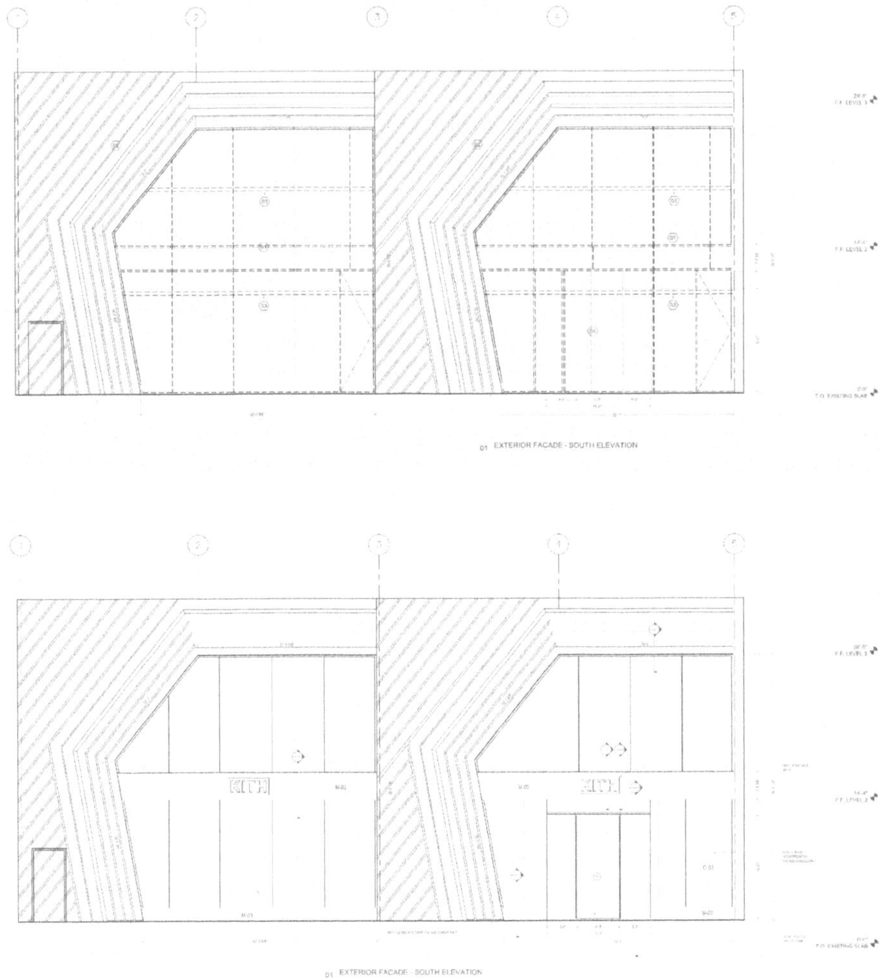

Fig. 3 Studio elevations of the demolition plan and new SLIMPACT façade composition, including large frameless modules, trapezoidal glazing, and an automatic entrance door

Design District, black-painted metal was used. NV5, acting as the third-party engineer for the project-specific application, paid particular attention to load transfer at the trapezoidal openings and to verifying adequate glass bite in areas requiring trapezoid-shaped glass.

SLIMPACT automatic doors were also incorporated. Ideally, the single door opening would have aligned with the 12 ft window wall module height; however, a design adjustment was required to accommodate the 10 ft tested and approved door height. This drove the use of the SLIMPACT "transom frame" (Fig. 6), consisting of a

Fig. 4 Completed Kith Miami Design District storefront, showing the realized frameless hurricane-rated façade with large-format glazing and minimal visual obstruction; the built result demonstrates how impact-resistant fenestration can support a highly transparent retail envelope in an HVHZ location *(Courtesy of Kith)*

slim 3/4 in × 6 in steel component running along the jambs from floor to ceiling, with a horizontal cross member that provides door load transfer and supports a smaller transom lite.

Overall, the project illustrates how frameless hurricane-rated fenestration can be integrated into a storefront façade where transparency and minimal sightlines are primary architectural drivers, while still accommodating system-specific constraints related to module geometry, installation tolerances, and tested component configurations. In this case, key project decisions were shaped by the structural and detailing implications of trapezoidal openings, the load-transfer strategy associated with top/bottom channel systems, and height limitations associated with tested door configurations and the resulting need for a transom condition.

3.4 Laminated Glass Behavior and Façade Performance Considerations

After the destructive Hurricane Andrew impacted South Florida in 1992, the Florida building code—particularly in Miami-Dade County—began to implement new code provisions and test standards aimed at reducing envelope breach and associated

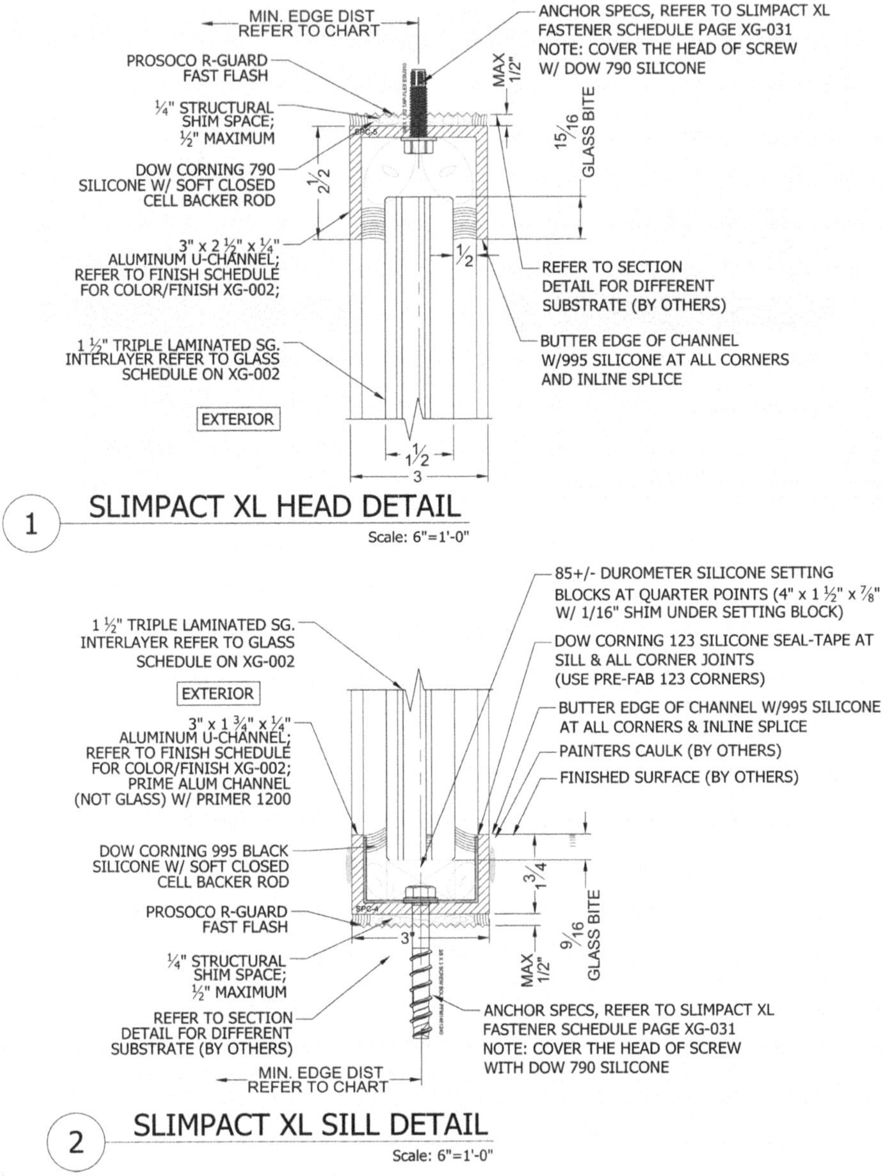

Fig. 5 Head (1) and sill (2) details of the SLIMPACT frameless system, showing the top- and bottom-channel load-transfer strategy and the differential channel depths used to accommodate structural demands and installation tolerances; these details are critical to the performance of frameless glazing under hurricane loading

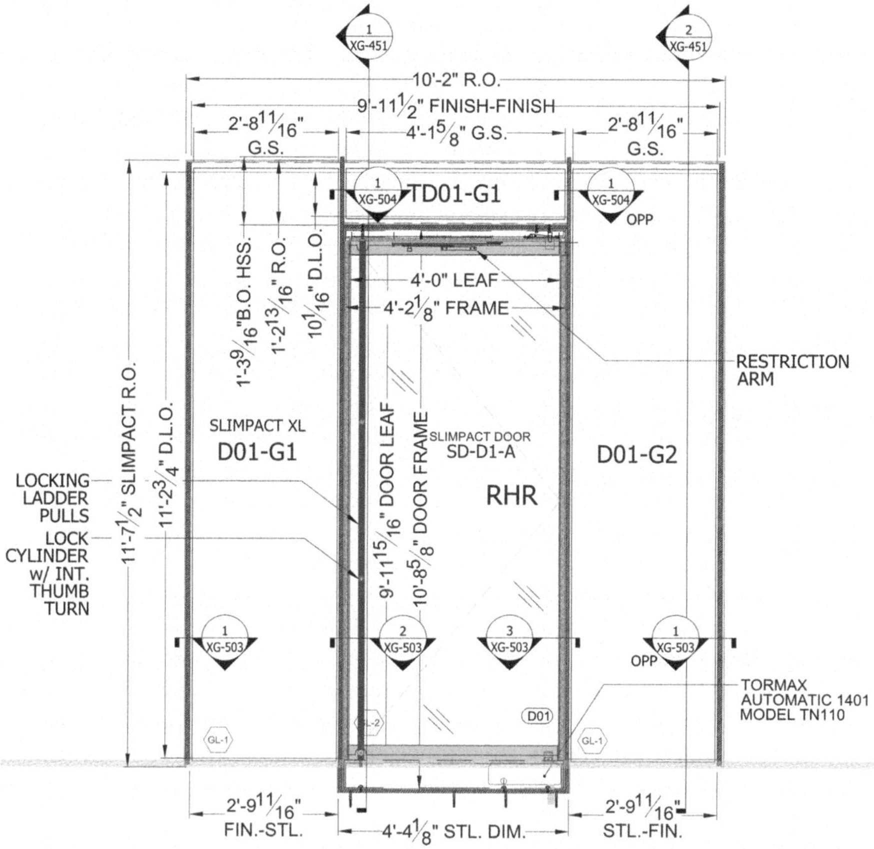

Fig. 6 SLIMPACT door with transom frame, showing the slim steel jamb-and-transom assembly introduced to accommodate the maximum tested hurricane-rated door height; the detail illustrates how dimensional limits of certified components can shape façade composition and load transfer in frameless systems

damage during future events. A principal early mitigation strategy for glazing failure was the adoption of laminated glass. If glazing could be retained in the opening—even when cracked—buildings could reduce internal pressurization and limit water intrusion.

However, the viscoelastic response of conventional polyvinyl butyral (PVB) interlayers—while beneficial for post-fracture energy dissipation in safety glazing—has been shown to be sub-optimal when subjected to large-missile impact followed by cyclic pressure loading, as required for glazing installed within 9 m (30 ft) of grade in hurricane-prone regions. Under this protocol, which involves impact by a nominal 4.1 kg (9 lb) 50 × 100 mm timber missile and subsequent pressure cycling to the full design wind load, PVB-laminated assemblies may exhibit excessive interlayer deformation, crack opening, and progressive loss of composite action, particularly at elevated temperatures. These behaviors can compromise load redistribution across the fractured glass plies and reduce the ability of the glazing to sustain the required cyclic pressure regime without excessive deflection or edge distress.

The need for an interlayer with enhanced post-fracture stiffness and tear resistance led to the early adoption of ionoplast interlayers, most notably SentryGlas, in hurricane-resistant glazing systems. Compared to PVB, ionoplast interlayers exhibit significantly higher shear modulus, improved creep resistance, and superior resistance to crack propagation under sustained and cyclic loading. These mechanical characteristics enable laminated glass assemblies to retain a higher degree of composite behavior after fracture, facilitating the development of impact-resistant systems with increased panel spans, higher design pressure capacities, and reduced dependence on auxiliary framing or mechanical restraints. Furthermore, the enhanced edge stability provided by ionoplast interlayers is critical in minimizing interlayer pull-out and delamination at the glass-to-frame interface, particularly in applications where edge restraint governs post-impact performance. This property is especially relevant for geometrically complex glazing configurations—such as curved or doubly curved glass panels—where non-uniform stress distributions and torsional effects amplify the demands on interlayer stiffness and edge integrity.

Introduced by DuPont in 1999 as SentryGlas Plus (SGP), the product later became known as SentryGlas (or SG) interlayer. In 2014, Kuraray acquired DuPont's Glass Laminating Division, adding SentryGlas to its Trosifol PVB portfolio. Over the past 25 years, the adoption of ionoplast interlayers has significantly influenced laminated-glass applications in hurricane regions, supporting the qualification and deployment of windows and doors, storefronts, and curtain wall systems intended to provide continuous protection against extreme wind events. This interlayer has been extensively used in Kith Miami.

3.5 Transferable Lessons

The Kith Miami case study highlights several transferable lessons relevant to the harmonization of performance-based façade design within the context of frameless hurricane-rated fenestration in retail and other high-transparency applications. First, architectural requirements for minimal sightlines and large, uninterrupted openings translate directly into system-level constraints, including limits associated with tested and approved product configurations, installation tolerances, and handling logistics for heavy glazing units. Second, non-rectilinear geometry, such as trapezoidal openings, introduces localized detailing and verification needs, including project-specific assessment of load transfer and adequate glass bite at atypical edge conditions. Third, load-path definition becomes particularly explicit in top- and bottom-supported frameless systems, where head and sill channel design must simultaneously address structural demands and installation allowances. Finally, the integration of doors within frameless compositions may require dimensional compromises relative to adjacent glazed modules, and transitional elements should be treated as performance-critical components.

3.6 Technology Readiness Level

The Technology Readiness Level of the SLIMPACT frameless glass façades and door system by Faour Glass Technologies is TRL-9. The SLIMPACT Frameless Glass Façades are commercially available and have been deployed throughout North America and the Caribbean. The products are certified for use with Florida Approval by the Florida Building Commission. Authors Farooq and Quimby were involved in the development and/or deployment of these products.

The Technology Readiness Level of SentryGlas by Kuraray is TRL-9. SentryGlas is commercially available and has been deployed worldwide. The product is certified for use with Notice of Acceptance by Miami-Dade County. Authors Farooq and Schauss were involved in the development and/or deployment of this product.

4 South Flagler House

4.1 Thermally Broken, High-Performance Façades for High-Rise Buildings

South Flagler House is a high-rise residential development in West Palm Beach, Florida, developed by Related Ross. Following the completion of multiple tower projects across the United States, Related Ross engaged Reflection Window + Wall (RWW) to execute the façade vision for this project in Florida. The architectural design by Robert A.M. Stern Architects consists of two 28-story residential towers rising from a shared base. The design intent was to integrate contemporary architectural expression with high-performance façade systems capable of meeting the environmental and regulatory demands of a coastal hurricane-prone location [26].

The façade concept balances stone-clad precast concrete panels with large-scale, captured, unitized window wall systems supplied by RWW (Fig. 7). The project thus required façade solutions that simultaneously addressed architectural articulation, structural performance under extreme wind loading, and the operational requirements of a high-rise residential building.

Fig. 7 Rendering of South Flagler House, showing the architectural integration of stone-clad precast elements with large captured-glazed unitized window wall systems; the façade concept illustrates the project's dual requirement for architectural expression and high-performance envelope behavior in a coastal hurricane-prone environment

4.2 Performance Drivers and Challenges

A primary performance driver for the façade design emerged from the project-specific wind tunnel study. Peak design pressures of + 110 psf and − 162 psf (ASD) were reported for unitized window wall assemblies measuring up to 176–1/4 in in height and 84 in in width. These pressures exceeded the capacity of RWW's standard 7000 series captured, unitized window wall system, necessitating the development of a higher-capacity variant.

The 7000 series system is a captured-glazed, thermally broken window wall assembly incorporating multiple reveals for architectural expression. The chassis consists of 6063-T6 aluminum profiles separated by polyamide thermal struts. While the base system had been successfully deployed in other high-rise applications, the combination of extreme design pressures and large module sizes at South Flagler House (Fig. 8) required re-evaluation of both structural capacity and load-transfer mechanisms. Thus, in order to satisfy the competing needs of both very-high design pressures coupled with thermal efficiency requirements, the need for a performance-based design modification approach was recognized at an early stage within the project's life cycle.

Fig. 8 South Flagler House in-progress construction as of 17 February 2025

A further critical challenge involved the outswing casement vents incorporated into the façade. These vents were sized at a scale comparable to doors and represented the largest vent assemblies, at the highest design pressures, known to be qualified under any Florida Product Approval at the time of the project. Individual vents reached areas of up to 39 ft2, with heights up to 126 in and widths up to 64 in (Fig. 9). Large-missile-impact (LMI) vents were qualified to ± 110 psf, while small-missile-impact (SMI) vents were qualified to + 110 psf and − 162 psf. In addition to impact resistance, the vents were required to resist water infiltration at a pressure differential of 16.5 psf.

Fig. 9 Large-scale outswing casement vent assemblies developed for South Flagler House, showing vent sizes approaching door-scale dimensions; these assemblies are significant because they combine operability, impact resistance, and high design-pressure capacity within a thermally broken residential façade system

4.3 *Engineering Solutions*

To accommodate the increased spans and wind loads while maintaining the architectural proportions of the system, RWW increased the nominal mullion sightline to 3 in, thereby increasing the moment of inertia of the primary framing members (Fig. 10). This adjustment allowed the system depth to remain compatible with the architectural intent while providing the required structural capacity.

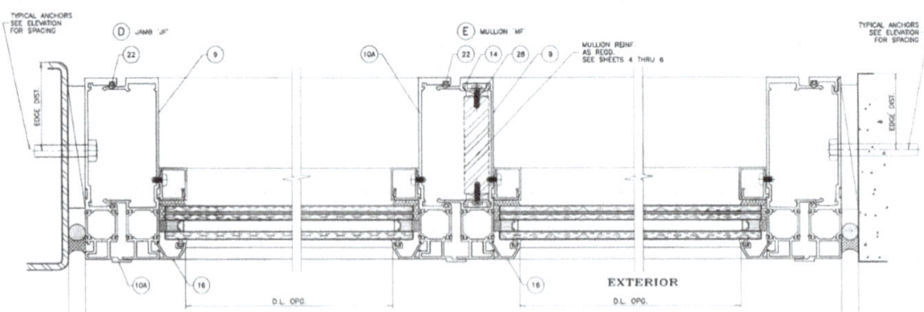

Fig. 10 Typical mullion cross-sections of the modified 7000-series system, showing increased member size and section properties adopted to resist extreme design pressures while maintaining the architectural proportions of the façade; the figure illustrates how structural enhancement was used to extend the capacity of a thermally broken window wall platform

Beyond increasing member stiffness, the project necessitated a comprehensive redesign of the receptors that transfer loads from the unitized window wall panels to the building structure. Structural calculations and finite element analysis (FEA) were employed to develop a receptor configuration that could be extruded, manufactured, and installed efficiently while transferring extreme wind loads and accommodating vertical and horizontal inter-story movements.

The glazing retention strategy of the 7000 series system was also revised to perform under the cyclic loading requirements of TAS 203 and ASTM E1886/E1996. The South Flagler House façade incorporates multiple infill types and glazing configurations, including laminated insulated glazing units, solid aluminum panels, and extruded nesting aluminum panel assemblies. The system was required to allow efficient shop glazing and field reglazing from either the interior or exterior, depending on access conditions.

In its base configuration, the 7000 series utilizes snap-in glazing stops with a heel bead of silicone serving as the primary air- and water-seal line. To retain infill panels under missile impact and cyclic pressure loading, the glazing stops were modified to include mechanical fastening, and the weather sealant was replaced with a structurally glazed retention method (Fig. 11). This adjustment ensured that infill retention was governed by a defined structural load path rather than by sealant performance alone.

Fig. 11 Isometric section views of vision- and spandrel-zone assemblies, showing revised infill-retention strategies including mechanically secured glazing stops and structurally engaged retention methods; these details are important because post-impact cyclic performance depends on a defined structural load path rather than sealant-only retention

4.4 Vent Hardware and Operability Considerations

The outsized vents introduced additional challenges related to hardware selection and operability. Conventional concealed four-bar hinges were determined to be insufficient to carry the dead load of the impact-rated infill at the required dimensions. RWW therefore adopted exposed barrel hinges mounted within Euro-Groove-style channels on both sash and frame components (Fig. 12).

These Euro-Groove channels were incorporated as applied elements on the master panel frame to maintain consistency with the base 7000 series system and to avoid additional machining when integrating fixed transom lites within the same panel. Depending on vent size, each vent assembly incorporated 4 to 5 barrel hinges to support dead load and resist wind-induced forces.

Each vent was also equipped with a multi-point locking system engaging all four sides of the sash. The locking assemblies combined compression-adjustable keepers and burglar-resistant keepers actuated by a single handle through transmission bars and corner drives. Depending on vent size and loading, between 20 and 32 locking points were required to resist design wind pressures, representing nearly four times the number typically used for comparable vents in lower-pressure regions.

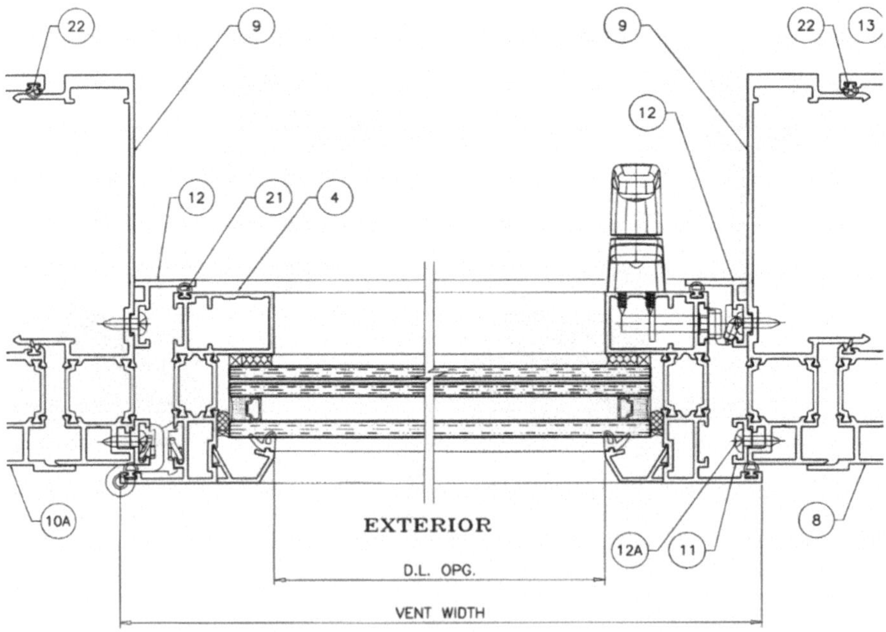

Fig. 12 Vent assembly section detail showing the hinge, sash, frame, and hardware interface for large outswing casement vents; the section illustrates how the vent design addresses dead-load support, multi-point locking, water resistance, and operability under high hurricane design pressures

The high number of locking points necessitated careful control of tolerances and engagement sequencing. Factory-set jigs were used to establish clearances between lock pawls and keepers, ensuring consistent engagement while accounting for cumulative manufacturing tolerances in the panel frame, sash frame, transmission bars, and associated components. This approach maintained manageable operating forces despite the large number of engagement points.

4.5 Transferable Lessons

The South Flagler House project presented façade engineering challenges exceeding those typically encountered in U.S. high-rise construction, including within hurricane-prone regions. Addressing extreme design pressures, door-scale operable vents, and cyclic impact-loading requirements required targeted adaptation of an existing system rather than reliance on prescriptive solutions, thus demonstrating how code-driven requirements can spur adaptation in areas like material and hardware selection to achieve certain performance-based design objectives.

Key transferable lessons from the project include the importance of scaling structural capacity in proportion to vent size and pressure demand, explicitly defining glazing retention as part of the structural load path under cyclic loading, and managing operability through careful tolerance control in multi-point locking systems. The project demonstrates how performance-based design enables large-scale, operable, and architecturally expressive façade systems to be realized in high-rise residential buildings subjected to extreme wind environments.

The 7000-series window wall and casement windows by Reflection Window + Wall were developed to meet unusually high design-pressure and impact performance requirements within this project's context, notably for thermally-broken systems. To the authors' knowledge, the approved vent dimensions of the 7000-series casement are among the largest reported for this application type.

4.6 Technology Readiness Level

The Technology Readiness Level of the 7000-series window wall and casement systems by Reflection Window + Wall is TRL-9. The 7000-series window wall and casement systems are commercially available and have been deployed throughout North America. The products are certified for use with Florida Approval by the Florida Building Commission. Authors Farooq and Eastwood were involved in the development and/or deployment of these products.

5 Brickell

5.1 Curved Glazing at Extreme Design Pressures

High-rise office buildings in the United States require façade systems capable of balancing architectural intent, energy performance, and structural resistance under demanding environmental conditions. Curtain wall systems in this context must comply with evolving building codes, accommodate extreme wind events, and integrate high-performance glazing to limit thermal loads while maintaining transparency and visual quality.

830 Brickell is a Class A office tower located at 830 Brickell Plaza in Miami, Florida, developed by OKO Group in partnership with Cain International. The project comprises a 55-story high-rise building characterized by a contemporary architectural language and stringent façade performance requirements. ES Windows' scope of work commenced at Level 14 and included the design, fabrication, and supply of a high-performance curtain wall system based on the GW-7000 platform.

A defining feature of the façade design is the use of curved glazing at the northeast corner of the tower (Fig. 13), extending over a significant vertical extent. The integration of curved curtain wall modules at high elevation and under extreme wind loading conditions presented engineering, manufacturing, and testing challenges relative to conventional flat façade systems.

Fig. 13 Upper levels of the 830 Brickell tower, showing the façade zone where curved glazing modules were developed to satisfy high wind-pressure demands in a hurricane-prone high-rise context; the image establishes the project setting for the integration of curved curtain wall assemblies within the High-Velocity Hurricane Zone (HVHZ)

5.2 Performance Drivers and Challenges

One of the principal performance drivers for the 830 Brickell façade was the requirement to implement curved curtain wall modules at the building's northeast corner, extending from Level 14 through Level 56. Across more than forty floors, eight curved units per level were incorporated, resulting in a vertically continuous curved façade zone (Fig. 14). This represented a first application of curved modules at this scale and geometry for the GW-7000 curtain wall system.

The size, weight, and curvature of the units introduced challenges related to handling, transportation, fabrication tolerances, and on-site installation logistics. In addition to geometric complexity, the curtain wall assemblies were required to satisfy stringent structural performance criteria. Test specimens were evaluated in accordance with Miami-Dade County testing protocols (TAS 201, TAS 202, and TAS 203) and qualified to a design pressure of 180 psf.

Curved curtain wall systems impose additional demands relative to planar systems due to non-uniform load distribution, sensitivity to dimensional tolerances, and the interaction between curved framing members and glazing units. Addressing these factors required close coordination among façade engineering, external metal-curving specialists, and in-house manufacturing and assembly operations.

5.3 Engineering Solutions

The façade solution implemented for the project was based on the GW-7000 curtain wall system (Figs. 15 and 16). This system was selected to meet requirements for structural resistance, environmental performance, and architectural expression while accommodating the project-specific curved geometry.

Saint-Gobain KS 138 glass units were incorporated into the façade, including curved insulated glass units that contribute to the tower's architectural identity. To realize the required curvature and dimensions, the system was customized for this project through the development of new dies specific to the intended geometry.

Aluminum profiles were externally curved by Techniform Metal Curving (Texas) and subsequently assembled at ES Windows' manufacturing facility. This process sequence allowed for control of curvature tolerances and dimensional accuracy prior to unit assembly, supporting compliance with local code requirements and alignment with architectural intent.

Engineering analysis for the façade and building envelope addressed structural behavior under wind loading, envelope performance considerations, and integration with the primary building structure. These analyses supported validation of the curtain wall design and ensured compatibility between the façade system and the tower's structural framework.

Fig. 14 Construction progress of 830 Brickell in Miami, Florida, photographed on 5 April 2024; the image documents the upper portion of the high-rise during enclosure works and provides project context for the implementation of curved glazing solutions designed to withstand high wind-pressure demands in a hurricane-prone environment. Photograph by Phillip Pessar. *Source* Wikimedia Commons, from Flickr, licensed under CC BY 2.0

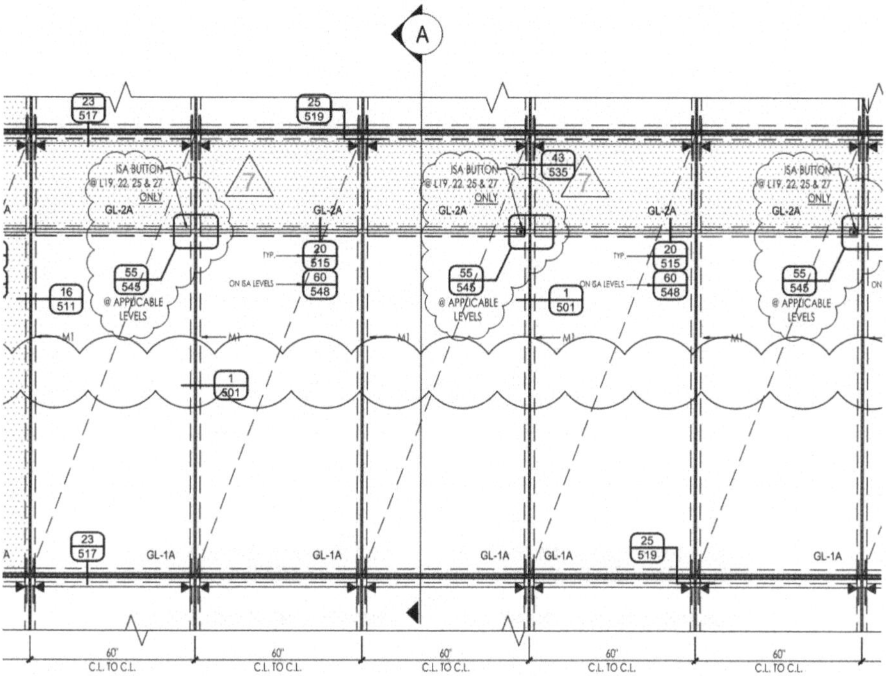

Fig. 15 Typical window wall elevation of the 830 Brickell façade system, illustrating the module layout and proportions

The curtain wall system underwent comprehensive performance testing at Blackwater Lab, including evaluation of air infiltration, water penetration, structural wind load resistance, and impact performance. The testing program verified compliance with applicable performance criteria related to environmental resistance, structural integrity, and occupant safety.

5.4 Transferable Lessons

The 830 Brickell project illustrates the technical and logistical complexities associated with implementing curved curtain wall systems in high-rise office buildings subject to hurricane winds. This case study highlights the importance of early coordination among design, engineering, manufacturing, and testing teams when curved geometries are introduced at scale (Fig. 17).

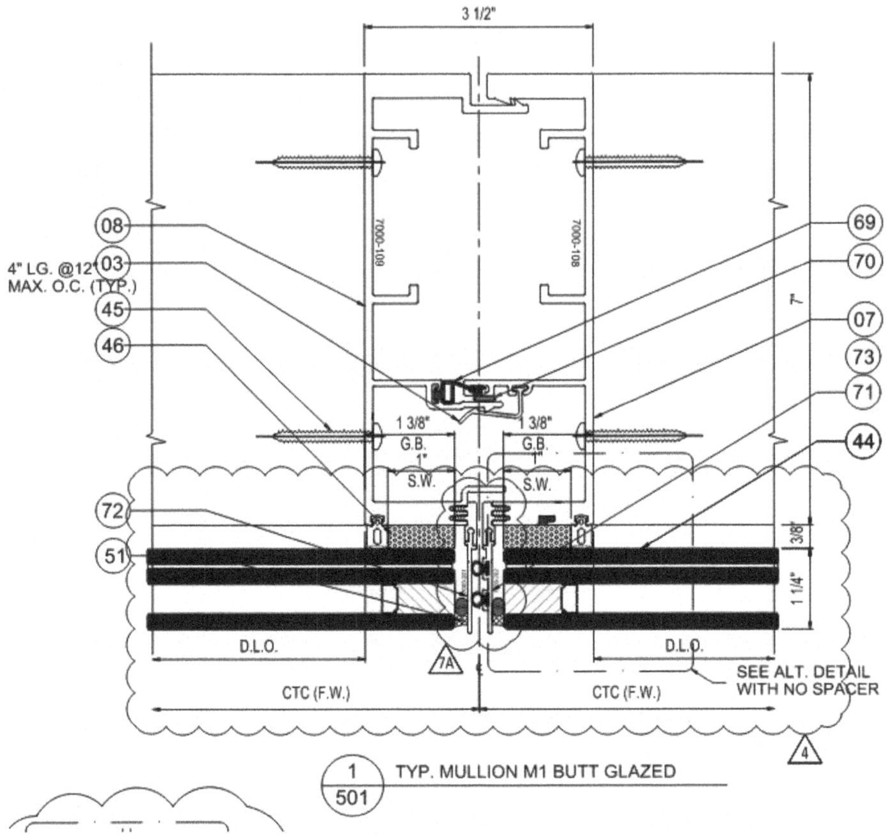

Fig. 16 Typical mullion detail of the GW-7000 curtain wall system, showing the framing configuration used to support glazed units and transfer wind loads to the primary structure

Key transferable lessons include the need for precise dimensional control throughout the curving and assembly process, explicit validation of curved assemblies through full-scale testing, and integration of façade engineering considerations early in the architectural design process. The project demonstrates how performance-based façade design and rigorous validation can enable the successful realization of architecturally expressive curved glazing systems while still meeting the extreme design pressure requirements of hurricane-prone regions (Fig. 18).

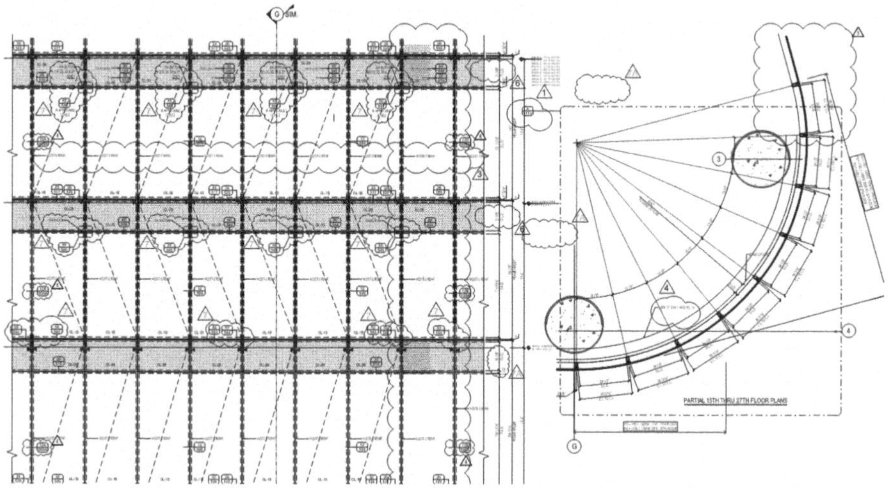

Fig. 17 Radius mullion detail and corresponding façade elevation of the curved curtain wall system at 830 Brickell, showing the geometric control of curved aluminum profiles and glazing interfaces; the detail is significant for understanding fabrication tolerances, alignment requirements, and structural continuity necessary to achieve impact resistance and high-pressure performance in curved façade applications

This project represents a rare, documented example of a commercially deployed impact-rated curved curtain wall within the HVHZ.

5.5 Technology Readiness Level

The Technology Readiness Level of the GW-7000 series curtain wall system by ES Windows is TRL-9. The GW-7000 series curtain wall system is commercially available and has been deployed throughout the Americas. Author Guerrero was involved in the development and deployment of this product.

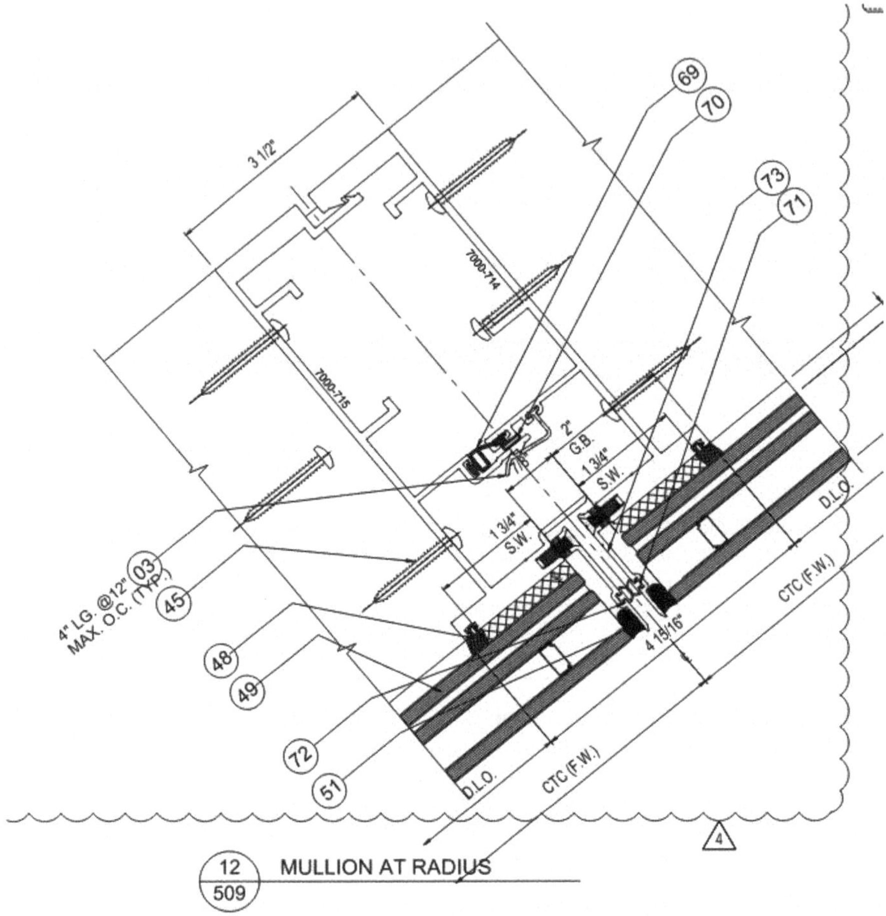

Fig. 18 Section through the curved curtain wall assembly at 830 Brickell, illustrating the mullion geometry at radius and the framing relationship between curved glazing and supporting profiles; the detail is significant because dimensional control and geometric continuity are essential to fabrication accuracy, structural performance, and qualification of curved façade systems under high design pressures

6 Conclusion

Across this chapter, the three case studies presented have demonstrated how innovation in façade technology within hurricane-prone regions has advanced through a performance-based design approach despite the unique climatic constraints of such regions. In this framework, testing protocols, material science, and constructability constraints directly shape architectural outcomes. The historical trajectory from early debris-impact provisions to today's HVHZ framework—anchored in TAS procedures and the ASTM E1886/E1996 system-level methodology, and increasingly complemented by the ISO 16316 approach—has established a robust verification pathway for preventing envelope breach, limiting internal pressurization, and maintaining post-impact integrity under cyclic pressure differentials. Within this context, the three case studies examined in this chapter illustrate a clear technical direction: frameless storefront systems used in projects similar to those at Kith Miami can be executed in the HVHZ when load paths, tolerances, and bonding/retention details are explicitly engineered while preserving architectural transparency in large-format sizes for retail applications; thermally broken unitized systems used in projects similar to those at the South Flagler House can be scaled to extreme design pressures and vent sizes while still incorporating thermal efficiency through targeted increases in section properties, receptor design, and the strategic use of hardware, with glazing retention treated as a structural mechanism rather than a sealing function; and curved curtain wall modules similar to those used at 830 Brickell can be qualified at very high pressures when curvature tolerances, fabrication sequencing, and full-scale validation to Miami-Dade County's rigorous testing protocols are integrated early in design and procurement.

Generally, the products referenced within the chapter are tailored to the climatic requirements of hurricane-prone regions of the Americas and Caribbean with particular focus on South Florida, except for SentryGlas by Kuraray, which is available for use worldwide. These products are generally transferable to other hurricane-prone regions as standard, with the caveat that differences in testing standards between TAS 201, TAS 202, and TAS 203 for HVHZ (which is the testing criteria used by many of the products within the case studies), ASTM E1886/E1996 and ISO 16316 may result in some degree of additional re-testing requirements before approval where these products are intended for use in regions requiring those standards. For example, while ISO 16316 was written with the intent of making TAS-based testing transferable, this ISO standard introduces a testing requirement for minimum module size in addition to the maximum module size restrictions that are a part of both TAS and ASTM E1886/E1996-based testing. This additional limitation was introduced in ISO 16316 (to which authors Farooq, Mejorin and Schauss contributed) based on observations that very small glazing units tend to have less predictable post-impact characteristics due to reduced availability for energy dissipation.

Collectively, these three projects highlight several key principles involved in façade design innovation within hurricane regions. The core enabling factor is the alignment of stakeholders—owner, design team, façade consultants, EOR, manufacturers, third-party engineering, laboratories, and specialized installers—around measurable performance objectives and testable details. The resulting façade solutions demonstrate that "extreme performance" and "elegant solutions" are not opposing goals. Transparency, energy performance, and complex geometries can be achieved in hurricane regions when system behavior is understood as a coupled problem of impact resistance, post-fracture load redistribution, cyclic pressure durability, and edge stability. As standards continue to converge internationally, these project-based lessons provide transferable strategies for future hurricane-resilient envelopes, supporting a broader shift toward performance-based design as the organizing framework for innovation.

Acknowledgements The authors of this chapter would like to extend special thanks to the various project teams and other stakeholders for each respective case study. Their collective efforts and contributions were vital in delivering the resulting project outcomes. A special thanks to:

Kith Miami

Landlord: Dacra.
Tenant: Kith
Design Architect: Porto Architecture
Architect of Record: AND Studio
Structural Engineer: YHCE Consulting
Fenestration Engineer: NV5

South Flagler House

Developer: Related Ross
Design Architect: Robert A.M. Stern Architects
Façade Consultant: Paramount Consulting & Engineering
Structural Engineer: DeSimone Consulting Engineering

830 Brickell

Architect: Revuelta
Design Architect: Adrian Smith + Gordon Gill Architect
Developer: OKO Group and Cain International
Structural Engineer: Façade + Envelope
Façade Consultant: Deerfield Technical.

References

1. Walker GR (2010) A review of the impact of cyclone tracy on building regulations and insurance. Aust Meteorol Ocean Jour-Nal 60:199–206
2. Walker GR, Reardon GF (1987) Technical Report No. 29—a discussion of criteria for the structural design of buildings to resist tropical cyclones. Cyclone Testing Station, Townsville
3. Mejorin A, Trabucco D, Stelzer I (2019) Cyclone-resistant façades. best practices in Australia, Hong Kong, Japan, and the Philippines. CTBUH Research Report, 160. Images Publishing
4. Mason M, Haynes K (2010) Historical case studies of extreme events—adaptation. Lessons from Cyclone Tracy. Gold Coast: National Climate Change Adaptation Research Facility
5. Darwin Reconstruction Commission (1975) Darwin area building manual. Darwin Reconstruction Commission, Darwin
6. Experimental Building Station (EBS), Department of Construction (1978) Technical Record (TR) 440: Guidelines for the Testing and Evaluation of Products for Cyclone-Prone Areas. North Ryde: EBS
7. FEMA (1992) Building performance: hurricane Andrew in Florida. Observations, recommendations, and technical guidance. Federal Emergency Management Agency, Washington, DC

8. Cochran L, Levitan M (1994) Lessons from hurricane andrew. Archit Sci Rev 37(3):115–121
9. FEMA (1993) Building performances: hurricane Andrew in Florida. Federal Emergency Management Agency, Washington, DC
10. Powell MD, Huston SH (1996) 'Hurricane Andrew's Landfall in South Florida. Part II: Surface Wind Fields and Potential Real-Time Applications', Weather and Forecasting 11(3)
11. Florida Building Code (FBC) (2001) Florida building code: building. International Code Council (ICC)
12. Marshall TP, Gilvary K, Kestner J (2012) 'Hurricane Andrew 20 years later: what have we learned?. In: 30th conference on hurricanes and tropical meteorology, Jacksonville, 16–20 April 2012
13. SBCCI (1997) SSTD 12–94 Test standards for determining impact resistance from wind-borne debris. Southern Building Code Congress International
14. Florida Building Code (FBC) (2020) Florida Building Code: Building. International Code Council (ICC)
15. ASTM (2023) ASTM E1996-23: standard specification for performance of exterior windows, curtain walls, doors and storm shutters impacted by Wind-borne Debris in Hurricanes. ASTM International
16. ASTM (2019) ASTM E1886-19: standard test method for performance of exterior windows, curtain walls, doors and storm shutters impacted by missile(s) and exposed to cyclic pressure differentials. ASTM Int
17. ASCE/SEI (2022) ASCE/SEI 7-22: Minimum design loads and associated criteria for buildings and other structures. American Society of Civil Engineers, Reston, VA
18. Standards Australia and Standards New Zealand (2002) AS/NZS 1170.2: structural design actions, Part 2: Wind Actions
19. Standards Australia and Standards New Zealand (2011) AS/NZS 1170.2: structural Design Actions, Part 2: Wind Actions
20. Lin N, Holmes JD, Letchford CW (2007) Trajectories of wind-borne debris in horizontal winds and applications to impact testing. J Struct Eng 133:274–282
21. Minor JE (1974) Analysis of the window damaging mechanism in windstorms. PhD thesis. Texas Tech University
22. Mejorin A (2022) Wind-borne debris resistance of façades: identification of alternative impact test requirements. PhD thesis. Western University and Università Iuav di Venezia
23. ISO (2007) ISO 16932: Glass in Building—Destructive-Windstorm-Resistant Security Glazing
24. ISO (2024) ISO 16316: windows, doors and curtain walling—impacted by wind-borne debris in windstorms—test method and classification
25. Japanese Industrial Standards (2018) JIS R 3109: glass in building—destructive-windstorm-resistant security glazing—test method
26. South Flagler House (nd) Homepage. Available at: https://www.southflaglerhouse.com (Accessed: 10 April 2026)

The manufacturer's authorised representative in the EU is Springer Nature Customer Service Centre GmbH, Europaplatz 3, 69115 Heidelberg, Germany. If you have any concerns regarding our products, please contact ProductSafety@springernature.com

Printed and bound by CPI Group (UK) Ltd, Croydon, CR0 4YY
17/09/2026
02218458-0011